S0-AHE-036

**Proteomics
in Drug Research**

*Edited by
M. Hamacher, K. Marcus,
K. Stühler, A. van Hall,
B. Warscheid, H. E. Meyer*

QP
551
,P75669
2006

65201691

Methods and Principles in Medicinal Chemistry

Edited by R. Mannhold, H. Kubinyi, G. Folkers

Editorial Board
H.-D. Höltje, H. Timmerman, J. Vacca, H. van de Waterbeemd, T. Wieland

Previous Volumes of this Series:

H. van de Waterbeemd, H. Lennernäs,
P. Artursson (Eds.)

Drug Bioavailability

Vol. 18

2003, ISBN 3-527-30438-X

H.-J. Böhm, G. Schneider (Eds.)

Protein-Ligand Interactions

Vol. 19

2003, ISBN 3-527-30521-1

R. E. Babine, S. S. Abdel-Meguid (Eds.)

**Protein Crystallography
in Drug Discovery**

Vol. 20

2004, ISBN 3-527-30678-1

Th. Dingermann, D. Steinhilber, G. Folkers (Eds.)

**Molecular Biology
in Medicinal Chemistry**

Vol. 21

2004, ISBN 3-527-30431-2

H. Kubinyi, G. Müller (Eds.)

Chemogenomics in Drug Discovery

Vol. 22

2004, ISBN 3-527-30987-X

T. I. Oprea (Ed.)

**Chemoinformatics
in Drug Discovery**

Vol. 23

2005, ISBN 3-527-3075-2

R. Seifert, T. Wieland (Eds.)

**G-Protein Coupled Receptors
as Drug Targets**

Vol. 24

2005, ISBN 3-527-30819-9

O. Kappe, A. Stadler

**Microwaves in Organic
and Medicinal Chemistry**

Vol. 25

2005, ISBN 3-527-31210-2

W. Bannwarth, B. Hinzen (Eds.)

Combinatorial Chemistry

Vol. 26

2005, ISBN 3-527-30693-5

G. Cruciani (Ed.)

Molecular Interaction Fields

Vol. 27

2005, ISBN 3-527-31087-8

Proteomics in Drug Research

Edited by
Michael Hamacher, Katrin Marcus, Kai Stühler,
André van Hall, Bettina Warscheid, Helmut E. Meyer

CABRINI COLLEGE LIBRARY
610 KING OF PRUSSIA ROAD
RADNOR, PA 19087

WILEY-VCH

WILEY-VCH Verlag GmbH & Co. KGaA

 # 6520 1691

QP
551
. P75669
2006

Series Editors:

Prof. Dr. Raimund Mannhold
Biomedical Research Center
Molecular Drug Research Group
Heinrich-Heine-Universität
Universitätsstrasse 1
40225 Düsseldorf / Germany
raimund.mannhold@uni-duesseldorf.de

Prof. Dr. Hugo Kubinyi
Donnersbergstrasse 9
67256 Weisenheim am Sand / Germany
kubinyi@t-online.de

Prof. Dr. Gerd Folkers
Department of Applied Biosciences
ETH Zürich
Winterthurerstrasse 19
8057 Zürich / Switzerland
folkers@pharma.anbi.ethz.ch

Volume Editors:

Dr. Michael Hamacher
Michael.Hamacher@ruhr-uni-bochum.de

Jun. Prof. Dr. Katrin Marcus
Katrin.Marcus@ruhr-uni-bochum.de

Dr. Kai Stühler
Kai.Stuehler@ruhr-uni-bochum.de

Dipl.-Oec. André van Hall
Andre.vanHall@ruhr-uni-bochum.de

Jun. Prof. Dr. Bettina Warscheid
Bettina.Warscheid@ruhr-uni-bochum.de

Prof. Dr. Helmut E. Meyer
Helmut.E.Meyer@ruhr-uni-bochum.de

Medical Proteom-Center
Ruhr-University Bochum
Universitätsstr. 150
44780 Bochum
Germany

All books published by Wiley-VCH are carefully produced. Nevertheless, authors, editors, and publisher do not warrant the information contained in these books, including this book, to be free of errors. Readers are advised to keep in mind that statements, data, illustrations, procedural details or other items may inadvertently be inaccurate.

Library of Congress Card No.: Applied for

British Library Cataloging-in-Publication Data
A catalogue record for this book is available from the British Library.

Bibliographic information published by Die Deutsche Bibliothek
Die Deutsche Bibliothek lists this publication in the Deutsche Nationalbibliografie; detailed bibliographic data is available in the internet at http://dnb.ddb.de.

© 2006 Wiley-VCH Verlag GmbH & Co. KGaA, Weinheim

All rights reserved (including those of translation in other languages). No part of this book may be reproduced in any form – nor transmitted or translated into a machine language without written permission from the publishers. Registered names, trademarks, etc. used in this book, even when not specifically marked as such, are not to be considered unprotected by law.

Printed in the Federal Republic of Germany.
Printed on acid-free paper.

Composition   Manuela Treindl, Laaber
Printing   Strauss GmbH, Mörlenbach
Bookbinding   Litges & Dopf Buchbinderei GmbH, Heppenheim

ISBN-13   978-3-527-31226-9
ISBN-10   3-527-31226-9

Contents

Proteomics in Drug Research
Edited by M. Hamacher, K. Marcus, K. Stühler, A. van Hall, B. Warscheid, H. E. Meyer
Copyright © 2006 Wiley-VCH Verlag GmbH & Co. KGaA, Weinheim
ISBN: 3-527-31226-9

A Personal Foreword

Ten years ago Marc Wilkins, at that time an Australian student, coined the word proteome. Since then this research field has evolved ever more rapidly and continues to do so. Worldwide, many conferences and research programmes cover this topic and a series of new journals are publishing more and more proteomics articles.

However, to best profit from the results of this relatively new field it is important to consider the right strategies and experimental setups before the laboratory work starts. This is even more important when performing drug research. Proteomics workflows are relatively new – very promising, but time- and cost-intensive. The identification of suitable drug targets as well as preclinical development of drugs easily consumes US$ 450–800 million today, committing the pharmaceutical industry to launching successful products on the market.

Unfortunately in these times of high-throughput analysis, very often the meaning of scientific work and statistically validated data is not understood and consequently results are published describing one-off experiments. In order to achieve highly reliable data in proteomics studies, a lot of laboratory work as well as time to evaluate the data of single experiments are necessary. Five to ten independent repetitions of the experiments are a must to elucidate the deviation of a single data point, including the biological variance. In this way the number of candidate proteins that are reproducibly up- or down-regulated drops drastically and a lower number of candidates have to be followed up thereafter. It should be borne in mind that only statistically significant proteomics data are likely to be validated by follow-up experiments like Western blot analysis, immunohistology, RNAi knock-down, or other techniques.

In this book, three different aspects of this new research field are surveyed: administration, technology and application. All three are equally important in achieving these valuable results that furnish new knowledge for further developments and thereby justify the extraordinary expense.

The first chapter describes why consortia, networking, and standardization are of increasing impact in the complex world of proteome research, and why the optimization of administrative structures, particularly in academia, will be essential for successful and rapid research.

The need for standardization, most notably in bioinformatics, e.g., data formats and exchange, is stressed in the chapter "Proteomic Data Standardization,

Proteomics in Drug Research
Edited by M. Hamacher, K. Marcus, K. Stühler, A. van Hall, B. Warscheid, H. E. Meyer
Copyright © 2006 Wiley-VCH Verlag GmbH & Co. KGaA, Weinheim
ISBN: 3-527-31226-9

Deposition and Exchange" where basic problems and approaches are addressed. The following articles deal with proteomics technologies, explaining the most common methods and techniques, including 2-dimensional gel electrophoresis (2-DE) and 2-D fluorescence difference gel electrophoresis (DIGE), high performance liquid chromatography (HPLC) and its widespread approaches, as well as an in-depth view on qualitative as well as quantitative mass spectrometry (MS). In addition, insights into toponomics and the identification of protein patterns in cells, the use of protein biochips, the validation of protein interactions by in vitro characterization via, e.g., BIAcore, and into the emerging field of peptidomics are offered.

The "Applications" part focuses on the adaptation of proteomics technologies to the special demands of drug research, with a particular regard to the different diseases of interest, namely rheumatoid arthritis (including a general strategy from target to lead synthesis), human platelets and inflammatory processes (with a special focus on the phosphoproteome analysis), Alzheimer's disease and other forms of dementia (analyzing brain tissue as well as human body fluids), renal cell carcinoma (covering several approaches used), breast cancer (with special focus on drug resistance), and cardiovascular diseases.

The book is rounded off by an article presenting the considerations and strategies for handling innovation processes that should generally be considered when aiming at introducing new products into the market.

We hope that you will both enjoy the articles presented in this book and profit from them in your daily work.

Michael Hamacher
Katrin Marcus
André van Hall
Bettina Warscheid
Kai Stühler
Helmut E. Meyer

Preface

When the expression Proteom had been coined more than a decade ago many scientists looked at the new field very critically and didn't consider it as a "real" science. And it turned out that some of the criticism was justified, since some early papers in the field overdid the methods and ended in mere speculation rather than in scientific insight.

This attitude however is seen in every new emerging methodology - remember the colorful front page illustrations in the early eras of force filed microscopy or computer-aided molecular design- but it helps in its somewhat naïve enthusiasm to establish a new field of research.

Proteomics has grown up and has very well established itself among the essential tools in life science research. But several pre-condition had to be fulfilled in order to give the methodology trust and efficiency. The present volume reflects exactly this, in showing the unprecedented interdisciplinarity of the young field of research.

It is not only doing the experiment. It is as always, doing it at the right time, at the right place, by the right means. Having living cells, tissues or organisms serving as the objects of study, does not facilitate the task. To retrieve interpretable, reliable and sustainable data a variety of experts is needed. As it is nicely depicted in the pearl necklet of papers this volume comprises kind of a handbook, all the necessities to perform good proteomics.

It starts with administrative optimization and ends up with the question for the innovation process itself. In between, methodology, development and its application is addressed. Among that, important future challenges are raised. Proteomics might be the appropriate tool for a much deeper insight into systems, be it on cellular or organismic level. Systems biology is one of the most exciting fields to be established and proteomics one of the most important methods to study biological systems at molecular level. Still, there is a long way to go. Time resolution remains to be a major challenge and the data management of and the simulation of cellular networks is still in its infancy.

It is without doubt that, beyond basic research, the whole field is driven by a vision of fundamental amelioration in the finding and creating of new therapeutics. Individualized treatment at manageable costs will be the future challenge of health care. Everything that makes us understand better individual response to drugs, xenobiotics and different kinds of inputs from the context, will improve both, personal health and cost structures of the health care systems.

Proteomics in Drug Research
Edited by M. Hamacher, K. Marcus, K. Stühler, A. van Hall, B. Warscheid, H. E. Meyer
Copyright © 2006 Wiley-VCH Verlag GmbH & Co. KGaA, Weinheim
ISBN: 3-527-31226-9

Authors come from a broad variety of affiliations, be it big industries, small and half-way grown up start-ups, clinics and academia. All of them integrate and focus on the question how proteomics methodology can be improved and serves best in their field of expertise.

The series editors are indebted to the authors and the editors who made this comprehensive issue possible. We are convinced that the book represents an important contribution to the body of knowledge in the field of proteomics.

We also want to express our gratitude to Renate Doetzer and Frank Weinreich from Wiley-VCH for their invaluable support in this project.

Raimund Mannhold, Düsseldorf
Hugo Kubinyi, Weisenheim am Sand
Gerd Folkers, Zürich

List of Contributors

Rolf Apweiler
EMBL Outstation European
Bioinformatics Institute
Welcome Trust Genome Campus
Hinxton
Cambridge
CB10 1SD
UK

Marcus Bode
MelTec GmbH & Co. KG
ZENIT-Building
Leipziger Str. 44
39120 Magdeburg
Germany

Petra Budde
BioVisioN AG
Feodor-Lynen-Str. 5
30625 Hannover
Germany

Elke Butt-Doerje
Institut für Klinische Biochemie und
Pathobiochemie
Universitätsklinik Würzburg
Josef-Schneider-Str. 2
97080 Würzburg
Germany

Dolores Cahill
Centre for Human Proteomics
Royal College of Surgeons in Ireland
123 St. Stephen's Green
Dublin 2
Ireland

Reinhard Ditz
Merck KGaA
Life Science & Analytics
New Business
Frankfurter Str. 250
64293 Darmstadt
Germany

Michael J. Dunn
Proteome Research Centre
Conway Institute of Biomolecular
and Biomedical Research
University College Dublin
Belfield
Dublin 4
Ireland

Erich Eigenbrodt
University of Giessen
Institute for Biochemistry and
Endocrinology
Frankfurter Str. 100
35392 Giessen
German

Hugo Fasold
University of Frankfurt/M.
Biozentrum
Institute for Biochemistry
Marie-Curie-Str. 9
60439 Frankfurt/M.
German

Catherine C. Fenselau
Greenebaum Cancer Center
Biochemistry Building, Room 1504
University of Maryland
College Park
MD 20742
USA

Proteomics in Drug Research
Edited by M. Hamacher, K. Marcus, K. Stühler, A. van Hall, B. Warscheid, H. E. Meyer
Copyright © 2006 Wiley-VCH Verlag GmbH & Co. KGaA, Weinheim
ISBN: 3-527-31226-9

Michael Fountoulakis
Foundation for Biomedical Research
of the Academy of Athens
Soranou Ephessiou 4
11527 Athens
Greece

Manuela Friedenberger
Medizinische Fakultät
Universitätsklinikum
Otto-von-Guericke-Universität
Magdeburg
Leipziger Str. 44
39120 Magdeburg
Germany

Frank Gesellchen
Institut für Biochemie
Universität Kassel
Heinrich-Plett-Str. 40
34109 Kassel
Germany

Michael Hamacher
Medical Proteom-Center
Ruhr-University Bochum
Bldg. ZKF E. 143
Universitätsstr. 150
44780 Bochum
Germany

Friedrich W. Herberg
Institut für Biochemie
Universität Kassel
Heinrich-Plett-Str. 40
34109 Kassel
Germany

Henning Hermjakob
EMBL Outstation European
Bioinformatics Institute
Welcome Trust Genome Campus
Hinxton
Cambridge
CB10 1SD
UK

Klaus Jung
Department of Statistics
University of Dortmund
Vogelpothsweg 87
44227 Dortmund
Germany

Roland Kellner
Target Research & Biotechnology
Pharma R&D
Merck KgaA
Frankfurter Str. 250
64293 Darmstadt
Germany

Sophia Kossida
Foundation for Biomedical Research
of the Academy of Athens
Soranou Ephessiou 4
11527 Athens
Greece

Norbert Lamping
BioVisioN AG
Feodor-Lynen-Str. 5
30625 Hannover
Germany

Piotr Lewczuk
Labor für molekulare Neurobiologie
Klinik mit Poliklinik für Psychiatrie
und Psychotherapie
Universität Erlangen–Nürnberg
Schwabachanlage 6 und 10
91054 Erlangen
Germany

Gert Lubec
Medical University of Vienna
Department of Pediatrics
Währinger Gürtel 18
1090 Vienna
Austria

Angelika Lücking
Protagen AG
Otto-Hahn-Str. 15
44227 Dortmund
Germany

Egidijus Machtejevas
AK Unger, FA19
Institute of Analytical Chemistry
Johannes Gutenberg University
Duesbergweg 10–14
55099 Mainz
Germany

Katrin Marcus
Medical Proteom-Center
Ruhr-University Bochum
Bldg. ZKF E. 043
Universitätsstr. 150
44780 Bochum
Germany

Sybille Mazurek
University of Giessen
Institute for Biochemistry and
Endocrinology
Frankfurter Str. 100
35392 Giessen
Germany

Emma McGregor
Institute of Psychiatry
16 De Crespigny Park
London
SE5 8AF
UK

Helmut E. Meyer
Medical Proteome Center
Ruhr-University Bochum
Universitaetsstr. 150
44801 Bochum
Germany

Daniela Moll
Institut für Biochemie
Universität Kassel
Heinrich-Plett-Str. 40
34109 Kassel
Germany

Stefan Müllner
Protagen AG
Otto-Hahn-Str. 15
44227 Dortmund
Germany

Päivi Niskanen
University of Frankfurt/M.
Biozentrum, Institute for
Pharmaceutical Chemistry
Marie-Curie-Str. 9
60439 Frankfurt/M.
Germany

Sandra Orchard
EMBL Outstation European
Bioinformatics Institute
Welcome Trust Genome Campus,
Hinxton
Cambridge
CB10 1SD
UK

Manuela Pruess
EMBL Outstation European
Bioinformatics Institute
Welcome Trust Genome Campus
Hinxton
Cambridge
CB10 1SD
UK

Horst Rose
BioVisioN AG
Feodor-Lynen-Str. 5
30625 Hannover
Germany

Sven Rüger
Protagen AG
Otto-Hahn-Str. 15
44227 Dortmund
Germany

Burghardt Scheibe
GE Healthcare Bio-Sciences
Discovery Systems
Munzingerstr. 9
79111 Freiburg
Germany

Michael Schrader
Fachhochschule Weihenstephan
Fachbereich Biotechnologie
85350 Freising
Germany

Alexander Schramm
Universitätsklinikum Essen
Kinderklinik/Pädiatrische Onkologie
Hufelandstr. 55
45122 Essen
Germany

Walter Schubert
Medizinische Fakultät
Universitätsklinikum
Otto-von-Guericke-Universität
Magdeburg
Leipziger Str. 44
39120 Magdeburg
Germany

Peter Schulz-Knappe
BioVisioN AG
Feodor-Lynen-Str. 5
30625 Hannover
Germany

Barbara Seliger
Institute of Medical Immunology
Martin Luther-University
Magdeburger Str. 2
06112 Halle
Germany

Barbara Sitek
Medical Proteom-Center
Ruhr-University Bochum
MA2/47
Universitätsstr. 150
44780 Bochum
Germany

Holger Stark
University of Frankfurt/M.
Biozentrum, Institute for
Pharmaceutical Chemistry
Marie-Curie-Str. 9
60439 Frankfurt/M.
Germany

Kai Stühler
Medical Proteom-Center
Ruhr-University Bochum
MA01/251
Universitätsstr. 150
44780 Bochum
Germany

Klaus K. Unger
AK Unger, FA19
Institute of Analytical Chemistry
Johannes Gutenberg University
Duesbergweg 10–14
55099 Mainz
Germany

André van Hall
Medical Proteom-Center
Ruhr-University Bochum
Bldg. ZKF E. 143
Universitätsstr. 150
44780 Bochum
Germany

Bettina Warscheid
Medical Proteom-Center
Ruhr-University Bochum
Bldg. ZKF E. 042
Universitätsstr. 150
44780 Bochum
Germany

Jens Wiltfang
Labor für molekulare Neurobiologie
Klinik mit Poliklinik für Psychiatrie
und Psychotherapie
Universität Erlangen–Nürnberg
Schwabachanlage 6 und 10
91054 Erlangen
Germany

Bastian Zimmermann
Biaffin GmbH & Co. KG
Heinrich-Plett-Str. 40
34132 Kassel
Germany

Zongming Fu
Professor of Chemistry and
Biochemistry
Greenebaum Cancer Center
Biochemistry Building, Room 1504
University of Maryland
College Park
Md 20742
USA

Hans-Dieter Zucht
BioVisioN AG
Feodor-Lynen-Str. 5
30625 Hannover
Germany

I
Introduction

Proteomics in Drug Research
Edited by M. Hamacher, K. Marcus, K. Stühler, A. van Hall, B. Warscheid, H. E. Meyer
Copyright © 2006 Wiley-VCH Verlag GmbH & Co. KGaA, Weinheim
ISBN: 3-527-31226-9

1
Administrative Optimization of Proteomics Networks for Drug Development

André van Hall and Michael Hamacher

Abstract

Administrative structures are gaining more and more importance in the complex world of modern science. This article will define the terms administration and networking, describing the aims and tasks of project management. The analysis of neurodegenerative diseases with proteomics technologies will be looked at from the administrative point of view with a focus on the different phases of strategy development, human resources, project control and networking. The realization of these tasks is illustrated by short presentations of a national funded network, the German Human Brain Proteome Project (HBPP) within the National Genome Research Network (NGFN), as well as of the international Brain Proteome Project of the Human Proteome Organisation (HUPO BPP).

1.1
Introduction

In modern science, the importance of administration has increased steadily over the last few decades. Nevertheless, administrative work and its influence on the success of projects as well as on financial aspects (e.g., refunding) are still undervalued in the academic field. While industry recognized the importance of organizational aspects long ago, positions responsible for administrative tasks within scientific research groups (excluding administrative departments of the universities themselves) are rare. The number of operative relative to administrative personnel is still much higher in academia than in companies (at least in Europe). As a consequence, these tasks are often done by the coordinator of a given project or one of his coworkers, who are often overloaded with work, sometimes unmotivated and mostly untrained in this field. A picture of the typical administrative research scientist as being exhausted by research, teaching and organization is emerging. In addition, staff turnover in these positions is often high, resulting in loss of knowledge, lack of continuity, and commonly, a lack of

Proteomics in Drug Research
Edited by M. Hamacher, K. Marcus, K. Stühler, A. van Hall, B. Warscheid, H. E. Meyer
Copyright © 2006 Wiley-VCH Verlag GmbH & Co. KGaA, Weinheim
ISBN: 3-527-31226-9

perception as to where responsibility lies. At the same time, such positions could be extremely important for the overall success of the group, e.g., in the crosslinking of basic research and commercialization.

Owing to the increasing complexity of modern science, e.g., international networking and large consortia, and the urgent need to present scientific research to the public and to the governmental project management/advisory board, a department-spanning administration should be implemented. Most research efforts in the health sciences are extremely complex and are difficult to explain to nonscientists, which often leads to misunderstanding, antipathy or even hostility from the public (e.g., see stem cell discussion, gene technologies, etc.). As the last 20 years have clearly shown, the support of a common administrative staff leads to the scientific personnel being relieved of additional work to which they are not suited, to an optimization of the scientific output (increasing added value) and to a broader acceptance in society. The need for management expansion has also been recognized by the European Union and its advisory councils, as expressed by Ernst-Ludwig Winnacker, the president of the European Heads of Research Council, in an interview with *The Scientist*: "The networks of excellence are big enterprises that require a great deal of management, and these have not been appreciated by scientists as much as the smaller, short-term programs that are less complex to manage and that facilitate work with smaller partners." (*The Scientist* online, 25 August 2004: http://www.biomedcentral.com/). The reasons for this development will be shown in the next paragraphs.

1.2
Tasks and Aims of Administration

The following chapter will present a short overview about modern scientific administration, mainly focusing on the academic side of research. To set a common starting point of what "administration" is about, the following definition is used:

- The act or process of administering, especially the management of a government or large institution.
- The group of people who manage or direct an institution.

Simultaneously with the increasing complexity of life science, the tasks and aims of the administration have steadily grown and evolved to a much more active management role. Originally mainly involving finances and human resources, these tasks have been joined by numerous other duties and responsibilities. Many projects demand large groups or consortia resulting in network systems (see below), thus making the organization and the feedback of teamwork as well as facilitation of the flow of information within a network an essential part of work. Additionally, interactions between the network and other national as well as international research projects, research institutions and private enterprises have to be handled. This includes so-called lobby work, the discussion with and convincing of policy makers e.g., within the European Union, to support the kind of research one advocates as the most promising approach.

Further tasks required of an administration are the composition of progress reports/business plans and final reports on schedule, the organization and calling of coordination meetings, the coordination and active participation in public relations (conferences, seminars, TV, radio, journals, etc.). This includes the planning and realization of training courses concerning technologies and topics provided by the consortium members, and the publication of the subproject results obtained at the respective time points. Moreover, (existing) homepages should be improved and optimized steadily, so that they serve not only as an information platform, but also as an interchange and communication portal.

Taken together, the administration has to
- build up a network offering fast and efficient information flow;
- elaborate business plans, evaluate the progress of subprojects and co-ordinate efforts;
- implement infrastructures (see evaluation, Section 1.4);
- serve as a central contact and administration point (added value);
- increase public knowledge and acceptance of proteomics;
- implement a bioinformatics infrastructure that will serve as a basis for further data base projects.

The aims of the administration – particularly in universities – are obviously to optimize processes and workflows within the respective department or network. Though implementation of controlling and monitoring could be hard to adopt in academia (in regard to the strong group autonomy), both processes are inevitably mandatory, especially in times of decreasing budgets and funding, as a consequence of which some US universities have started to gather discarded or not-required high-tech equipment from local departments and offer it to all other groups for free, avoiding unnecessary investments and expenses.

There are several other domains that have to be carefully considered when aiming at successful projects, most notably in human resources, where the generation of job specifications and the consequent identification of adequate coworkers should not be underestimated. Qualified and motivated employees who fit into the group structure are the basic requirement for planning, performing and finishing work packages in a defined schedule. These have to be generated carefully and in regard to several questions, e.g., medical need, potential return of investment, proof of concept and commercialization.

Commercialization was more or less been ignored in academia until the 1980s, when more and more scientists came to the opinion that research and marketing do not necessarily exclude each other. Several processes around the world now show the increased importance of marketing. No application within the EU can be submitted to obtain grants without presenting utilization strategies. Scotland started a Proof of Concept Fund in 1999 to advance promising ideas from university to readiness for marketing (www.scottish-enterprise.com/proofofconceptfund).

More than 140 projects have already been funded with €36 million, resulting in six existing and ten planned spin-offs. In Germany, universities and research organizations, e.g., the Fraunhofer Gesellschaft have implemented utilization departments specializing in regard to patents, licensing, consortia contracts, etc. The Ruhr University in Bochum, Germany, for example founded the Research and Collecting Society "RUBITEC – Society for Innovation and Technology" in March 1998 (http://www.ruhr-uni-bochum.de/rubitec/start.htm), consulting the numerous groups at the campus. The National Institutes of Health has elaborated a complex organization structure including the Office of Technology Transfer (http://ott.od.nih.gov) dealing with 341 invention disclosures and US$ 53.7 million in royalties in 2004. These centers offer competent help in realizing products and patents, but leave the initial efforts to the research groups. Scientists have to inform themselves about possible strategies and have to evaluate the putative success. An administrative coworker assuming this time-consuming job will function as a bridge between the groups and the central transfer departments. Thus, taken together, the optimal realization of these tasks will lead to the relief of the operative coworker, enabling the researcher to concentrate on the actual scientific work, to shorten the time from idea to output, and to commercialize his output successfully.

As already mentioned above, research efforts are more and more bundled in consortia and networks. Owing to the importance of this circumstance, it is necessary to discuss some theoretical aspects of networking and the consequences resulting from its nature.

1.3
Networking

Networks are an organizational structure with at least two independent entities being in a repetitive, long-lasting exchange/interaction status (see also Burt, 1980). Owing to the independency of the entities the network is more or less bound together by social relations, according to one or more motivations:

- Necessity: interaction is initiated by law or regulatory prescription.
- Asymmetry: to gain influence and control over the partner/its resources.
- Reciprocity: to achieve bilateral aims and interests.
- Efficiency: to gain higher input/output-ratio by utilize synergistic effects.
- Stability: to reduce/absorb/predict uncertainties.
- Legitimization: to gain or improve reputation, image or prestige.

The process of composing and inspiring a network can be divided into seven phases:

- Self-analysis: what is the goal?
- Specification: which resources are missing?
- Preselection: who offers the lacking resources?

- Partner analysis: does the new partner fit in the overall concept?
- Definition of goals: what do the partners expect from each other?
- Process modeling: how can the goals be reached?
- Realization.

Industry is again on the cutting edge in establishing strategic alliances or regional clusters. In Switzerland more than 80% of all biotechnology companies are concentrated in the four regions Basel (Biovalley), Zurich (MedNet), Lake Geneva (bioalps) and Tessin (biopolo) (Veraguth, 2004), profiting from the "big pharma" industry that offers potential financiers, manpower and licensees.

In academia, the factors asymmetry, reciprocity, legitimization and efficiency probably have to be considered as the main motivation for building up networks. Nevertheless, most cooperative enterprises follow from personal relationships or historically derived projects that have been performed in the group several years ago. The need for combining synergistic resources is often unseen, sometimes hampered by ignorance of which potential partners are working in the same field or could offer complementary techniques. The identification of key players and potent partners is therefore an essential task in organizing a powerful network.

In addition to this selection mode, the management has to deal with regulation between the partners as well as between the consortium and external entities, with allocation concerning the access of given resources and with evaluation in regard to the output (profit, innovation, proof of concept). Problems within networks often evolve from the opportunistic behavior of one or more partners or due to different strategic targets, thus demanding complex agreements and interaction/communication right from the beginning to generate confidence between the partners. Throughout the whole project, several quality control steps concerning work packages, finances, etc., have to be performed.

1.4
Evaluation of Biomarkers

In general, the struggle for understanding and fighting e.g., neurodegenerative diseases, is intended to find either drug targets involved in the pathological processes or diagnostic markers that allow sensitive identification of disease stages (Zolg and Langen, 2004). Diagnostic markers can be subdivided into:

- Screening markers: allow indication of the transit from health to disease [e.g., maternal serum invasive trophoblast antigen for Down syndrome during the second trimester (Palomaki et al., 2004)].
- Prognostic markers: allow prediction of the disease process [e.g., survivin expression in pancreatic cancer patients (Kami et al., 2004)].
- Stratification markers: allow prediction of the response to a medication strategy [NQO1 genotype in adenocarcinoma of upper gastrointestinal tract (Sarbia et al., 2003)].

- Efficacy markers: allow monitoring of the efficacy of a given drug treatment [serum CYFRA 21-1 (cytokeratin-19 fragments) in breast cancer (Nakata et al., 2004)].

Before starting research, several questions have to be answered in a detailed business plan when aiming at a successful utilization concept in industry (Zolg and Langen, 2004), e.g.,

- Do competitive markers already exist on the market?
- Will the marker be easily accepted in the market?
- Will the marker cover/exceed the research costs?

As academic research usually is much more philanthropic than industrial, these considerations are normally secondary for scientists in universities. Nevertheless, it is highly advisable to elaborate a business-plan-like approach concept dealing with pros and cons, work packages and possible contingency plans to increase efficacy and output.

The interconnection between the basic research and commercialization is structured most efficiently in an innovation process organized with clear stage gate decisions (see Chapter 17). An estimated 50% of all life science companies are using this structure. Here, product ideas originating from the research will be judged by a decision board in regard to economically relevant features (e.g., market need, competition, etc.). People and know-how will be transferred in several stages to the commercialization branch. This milestone-oriented process will be reviewed constantly by a board. After passing all criteria including concept, market attractiveness, competitive market position, competitive technology position, reward, and risk, the project will go into the next stage of the innovation process with clear planning for milestones and budgets. This phased project planning was developed by NASA in the early period of crewed spaceflights and propagated by product development experts such as R. G. Cooper (Cooper, 2001). Work is divided into sequential phases avoiding overlapping activities, but as every gate has to be carefully evaluated, it is inherent to the process that there will be a relatively long time from the idea to the market.

To bypass this, the so called bounding box approach (management by exceptions) can be implemented: prior to the beginning of a project, all internal and external factors are surveyed (budget, profit margin, schedule, etc.) and boundaries are fixed in which the project is regarded as on-track. If these boundaries are crossed, a decision board has to reevaluate the work. As the team is free to move within the boundaries, time-consuming evaluation processes are minimized, as is time to market. Alternatively, the well known project risk management can be chosen. Risk management is a process of thinking systematically about all possible undesirable outcomes before they happen and setting up procedures that will avoid them, minimize their impact, or cope with their impact. Thus, risk assessment and risk control are two important concepts that have to be kept in mind.

1.5
A Network for Proteomics in Drug Development

The concepts described so far, *administration, networking* and *bio markers* represent fundamental cornerstones for considering how to establish a scientific program for drug development within the field of proteome analysis.

The identification of bio markers by proteomics and proteomics-associated technologies is the key approach for drug development on the protein level. It is obvious that a higher number of identified proteins will increase the chances of finding relevant markers regarding a specific pathogenic question. After validation these marker proteins can then be used as starting-point for a drug development process.

To meet this challenge, it is necessary to combine a wide range of technologies including "classical" proteomics, e.g., 2D-PAGE and mass spectrometry, new proteomics approaches like multidimensional chromatography, and technologies for transcriptional analysis. As very few institutes provide all these applicable approaches, the reasonable procedure is to combine groups with outstanding expertise in the different fields to form a network of excellence.

In Figure 1.1, a possible structure for such a consortium is shown. First, general considerations lead to a hypothesis which comprises an approach for understanding pathogenic mechanisms of a specific disease. Based on this hypothesis, the appropriate tissue as well as the suitable model organism must be defined, and providers of the relevant samples found. For networking reasons, integration

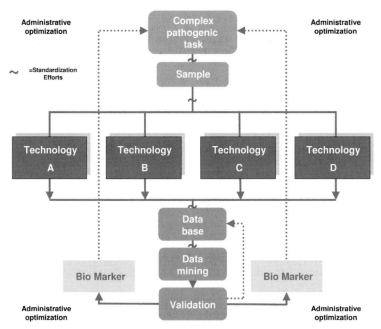

Figure 1.1 Structured workflow within a consortium for disease-oriented proteome analysis.

of the tissue providers into the consortium is recommended. The standardized samples will then be distributed to the single-technology partners within the network. As already mentioned, the range of this established technology portfolio is crucial for the possible impact of the entire consortium on the pathogenic relevance. Thereafter, the generated data will be incorporated into the project data base and reanalyzed by data mining experts (see Chapter 2). Using this process, proteins will be identified as bio markers to provide potential drug targets. Another crucial step is to validate the candidate proteins. Here, technologies for analyzing protein functions and protein–protein interactions are the instrument of choice. The comparison with the initial hypothesis will then hopefully lead to a feasible clinical approach for drug development.

Within the entire workflow described above the standardization of sample preparation [implementation of standard operating procedures (SOPs)], analyzing procedures and data handling will assure the comparability of results within the network as well as with results outside the consortium. Thus, the standardization is essential for efficient networking.

In establishing a scientific network, it is indispensable to bring together experts in the required fields. Recapitulated, partners for the following tasks have to be identified: tissue provision, technology-based analysis, pathology, data management, data reanalysis, and validation. Owing to the complexity of such an accumulation of heterogeneous partners that are also locally separated, implementation of a goal-oriented coordination is necessary. In the following, the realization of the described network structure will be illustrated by presenting German initiatives in proteomics networking in both the national and international environments.

1.6
Realization of Administrative Networking: the Brain Proteome Projects

The need for large international collaborations is obvious when analyzing the human proteome. The reasons are manifold, e.g., the low abundance of the majority of most cellular proteins (10% of all genes probably encode for 90% of all prevalent proteins), the absence of suitable high-throughput techniques for increasing sensitivity [polymerase chain reaction (PCR) equivalent for proteins] as well as the enormous number of protein species as the consequence of differential splicing, posttranslational modifications, etc. (Humphery-Smith, 2004). In addition, most diseases might not be monogenetic, but may be caused by multiple genes, modifier genes, the genetic background, etc. As a consequence, the most promising and synergistic approach is the analysis of the protein complements via transcriptome, proteome and toponome profiling.

One of the most striking tasks to start with is standardization (Meyer et al., 2003; see Chapter 2). Although it may not be feasible to elaborate fixed SOPs for all imaginable setups and questions, the key parameters of each experiment have to be annotated at least, so that possible differences can be traced back to variable steps in the chain of work (Hamacher and Meyer, 2005a). Some elements of standardization cannot be realized employing human material. Each of us is supplied with a diverse set of genes (polymorphisms) and has undergone a different history within his lifespan, entailing varying proteomes. This might be solved via studying numerous human samples and statistical methods. In general, single groups or technical approaches are not sufficient to overcome the complexity of this challenge or to describe a given (disease) status properly. Instead, the simultaneous efforts of numerous, but standardized working groups are essential for this huge challenge and to develop a knowledge base of the normal human proteome (Hanash, 2004a,b). Two mainly academic examples will demonstrate the attempt to understand and to ease/cure the malfunction of the diseased brain, namely a national funded consortium as well as an international, voluntarily driven project (Klose et al., 2004).

1.6.1
National Genome Research Network: the Human Brain Proteome Project

In 2001, the German Federal Ministry of Education and Research (BMBF) initiated the National Genome Research Network (NGFN) as a nation-wide multidisciplinary platform network aiming at the analysis of common human diseases, as well as aging. Within the NGFN the so-called Human Brain Proteome Project (HBPP) focuses on the analysis of the human brain in health and disease. The concept is based on three consecutive steps:

- Elaborating and establishing the necessary technology platforms: HBPP1 (2001–2004).
- Proteome analysis of Alzheimer's and Parkinson's diseases: HBPP2 (2004–2007).
- Validation of target proteins and analysis of disease mechanisms: HBPP3 (planned for 2007–2010).

The HBPP1 has been funded for a period of three years with approximately. €10.5 million (2001–2004). In this project 12 partners formed a strategic consortium, consisting of nine academic groups and two companies (Marcus et al., 2003, 2004) The main focus was on the improvement of proteomics-related technologies on the basis of brain analysis. One aim of the consortium was the characterization of the human and mouse brain proteomes in regard to the identification of proteins, generating mRNA profiles, studying protein/protein interactions and validating possible targets. Data gained was used to compare mouse models and relevant human tissues for neurodegenerative diseases. To achieve these aims, the essential technological methods had to be improved and new technologies identified.

The interest of the consortium in developing and testing new tools for proteome analysis was directed to solutions for particular technical problems concerning sample preparation, the 2D PAGE system, protein quantification, and the development of UniClone sets (nonredundant cDNA expression library) from the adult human brain to be used for creating clinically relevant biochips. These techniques were intended to be combined to develop a fully integrated Proteomic Workstation in which samples are prepared and processed automatically, e.g., by establishing 2D/3D biochips on which the samples are immobilized for further analyses. To overcome the large number of data sets that were generated by the different groups, the bioinformatics activities were expanded, e.g., a build-up of the project data base in which all data files provided by the project partners will be stored. Owing to the annotated information, for instance the link variation in protein expression to particular genes, hopefully it will be possible to elucidate the regulatory network acting between the genome and the proteome. This will create new insights for drug development concerning neurodegenerative diseases like chorea Huntington's, Parkinson's and Alzheimer's diseases or multiple sclerosis, also resulting in marketable technology products in these fields.

In the second funding period of NGFN the Systematic Methodical Platform (SMP) Human Brain Proteome Project 2 continues the work in a new formation of nine academic partners and one company. The aim of HBPP2 is to optimize developed technology and gain knowledge that, once applied, enables the development of new strategies for the diagnosis and treatment of neurodegenerative diseases. To achieve this goal, HBPP2 has gathered a critical mass of interdisciplinary German research groups with extensive experience, an unprecedented research infrastructure, a global science network within the Human Proteome Organisation (HUPO) and a solid record in clinical and preclinical work encompassing human genetics, cell biology, animal models, molecular biology, and biochemistry, thus encompassing the integration of large-scale functional genomic and proteomics approaches.

In this second funding period of the HBPP the three main goals are:
- The advancement of already established technologies: based on advances in technology achieved within NGFN1, HBPP2 will further advance its technology platform for the planned scientific program. Proteomics technologies (large 2D-PAGE, multidimensional chromatography, mass spectrometry), toponomics and functional assays such as cellular overexpression, pharmacological inhibition, RNAi and optical methods such as green fluorescent protein (GFP)-labeling, immunofluorescence and fluorescence resonance energy transfer FRET/bioluminescence resonance energy transfer (BRET) will be employed to analyze the functional implications of gene mutations selected in collaborations with the clinical partners.
- Investigating neurodegenerative diseases: HBPP2 will emphasize on applications of genomic and proteomics technology. A focus will be systematic analysis of proteins in human/mouse brain and nervous-system-related proteins in bodily fluids under normal and pathological conditions. Alzheimer's and Parkinson's diseases will be studied on the basis of human material and selected mouse models. The technology portfolio provided by the HBPP2 offers a conceptually novel opportunity to understand disease mechanisms in that it attempts to progress from current reductionist approaches to an integrated understanding of biological systems.

- Networking within NGFN-2: tight collaborations with clinical groups will allow the performance of clinically relevant proteomics studies or protein analyses that offer the most advanced proteomics technologies to the NGFN. In addition, HBPP2 is open to cooperation with other systematic-methodical platforms, namely bioinformatics, RNAi, and mammalian models. Data obtained in the project will be collected in a new type of database. Standards, SOPs and software for data management and integration will be developed. Together, these tools will form the basis for an efficient analysis and the generation of knowledge on the fundamental biological processes in normal and disease-affected brain.

Taking the current phase 2 of the HBPP as an example, the workflow within the consortium is shown in Figure 1.2. Starting with considerations about pathogenesis of neurodegenerative diseases the consortium will be provided with mouse, ape and human brain tissue. The data sets derived from the different available technologies will be incorporated into the project data base which will be presumably linked to the Data Collection Center (DCC) of the HUPO Brain Proteome Project (BPP).

After reanalyzing the data using customized software tools the identified proteins will be validated by partners within the network.

The coordination structure already established in the first funding period deals with administrative issues on different levels: Firstly, the activities within the HBPP are managed by the coordination team. Furthermore, the crossbridging to the

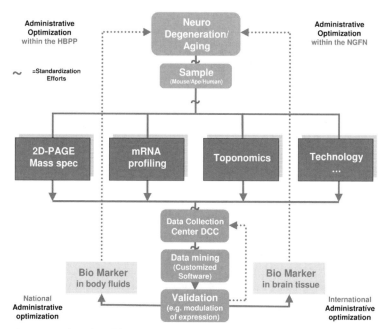

Figure 1.2 Adapted workflow of the Human Brain Proteome Project (HBPP) consortium within the second funding period (compare with Figure 1.1).

NGFN is also part of the administrative task force. In addition, most of the connections to national and international partners are coordinated centrally.

In the next step of the Human Brain Proteome Project it is planned that identified disease-associated proteins will be validated using several different techniques. The pathways they play a role in will be analyzed. This will lead to help in understanding the analyzed diseases and to develop diagnostic and/or therapeutic approaches.

1.6.2
Human Proteome Organisation: the Brain Proteome Project

At about the same time as the German HBPP was founded in 2001, the international HUPO was established as a part of the Human Genome Organisation (HUGO) (Hanash, 2004a; www.hupo.org). HUPO is a nonprofit organization promoting proteomics research and proteome analysis of human tissues. Several initiatives have been established under the roof of HUPO that analyze the proteome of a distinct human organ, e.g., the Plasma Proteome Project (PPP), the Liver Proteome Project (LPP) (Hanash, 2004b), and the Brain Proteome Project (BPP) (Meyer et al., 2003; www.hbpp.org). The HUPO Proteomics Standards Initiative (HUPO PSI) aims to establish definitive bioinformatics standards (Hermjakob et al., 2004a) and is therefore an overlapping project chaired by Rolf Apweiler from the European Bioinformatics Institute (EBI, Hinxton, UK). Standards include mass spectrometry (mzData, mzIdent), protein–protein interaction (IntAct) (Hermjakob et al., 2004b) and General Proteomics Standards (GPS), e.g., minimum information about a proteomics experiment (MIAPE) (Orchard et al., 2004). More information about this modular system is available at http://psidev.sourceforge.net.

The HUPO initiative concentrating on the brain is the HUPO BPP. After the 1st HUPO World Congress in Versailles, it was started by Helmut E. Meyer, Bochum, and Joachim Klose, Berlin, both in Germany in 2003. At a kick-off meeting in Frankfurt, Germany, in late April 2003, the first interested colleagues from around the world met to discuss the shape of the project. Since then, numerous meetings and discussions have taken place, often in close collaboration with the HUPO PSI and the EBI (e.g., Stephan et al., 2005; Hamacher et al., 2004). At the 2nd HUPO BPP workshop at the ESPCI in Paris, April 2004, attendees expressed the HUPO BPP vision as "Towards an understanding of the pathological processes of the brain proteome in neurodegenerative diseases and aging". The postulated vision of the HUPO BPP is the understanding of the pathological processes of the brain proteome in neurodegenerative diseases and aging. This will be achieved by deciphering the normal brain proteome, by correlating the expression pattern of brain proteins and mRNA and by the identification of disease-related proteins involved in neurodegenerative diseases. A pilot phase began in 2004 that addresses a quantitative proteome analysis of mouse brain of three different ages (all samples obtained and prepared by one source) and a differential quantitative proteome analysis of biopsy and autopsy human brain samples (Hamacher et al., 2004).

1.6.2.1 **The Pilot Phase**

Several conditions have to be met before the main project can be commenced, e.g., a broad community and reliable infrastructure. Without question, a detailed phenotyping of mouse models/patients, a complete characterization of tissue samples before proteome analysis and a high degree of standardization are extremely important in obtaining reliable results. Thus, in the HUPO BPP two pilot studies were initiated, limited to December 2004 (practical work) and March 2005 (data submission), respectively (Stephan et al., 2005; Hamacher and Meyer, 2005b): In order to collect the heterogeneous data of the HUPO BPP pilot studies in one database, the right database concept had to be chosen. The software ProteinScape (Bruker Daltonics Bremen & Protagen AG Dortmund, both Germany; free licenses by Bruker Daltonics) has been chosen for handling the heterogeneous data, as it is a feasible system for importing all the different data of a proteomics study, e.g., 1D gel electrophoresis, 2D liquid chromatography, 2-D difference gel electrophoresis (DIGE), etc. To learn the handling of this software, several ProteinScape training courses took place at periodic intervals and more will be held in the future.

The DCC is installed and a modified version of ProteinScape is running at 12 laboratories taking part in the pilot phase of the HUPO BPP. At the DCC all data has been imported with user-specific IDs. Dozens of gels and more than one million MS spectra were generated and transferred into the DCC. Data are being reprocessed according to a stringency set (Reprocessing Guideline, http://www.hbpp.org) and will be interpreted by invited analysts. Subsequent to the analysis phase all collected data will be exported by a newly designed exchange tool based on a mzData format into the database PRIDE hosted by the European Bioinformatics Institute (EBI) for worldwide access.

After the reprocessing phase the analysis phase will start with different task forces and different goals. The major analysis aspects are, among others, to match mRNA array data and protein data as well as peptidomics data, to analyze identified regulated proteins by interpretation of submitted protein lists (by participating groups) and gel images, to perform data mining and an overall analysis (summary, comparison pilot studies HUPO BPP and HUPO PPM, matching the results with literature).

The next steps prior to the master phase will be the completion of the pilot studies by presenting the results at the 4th HUPO World Congress in Munich (27 August–1 September 2005), by finalizing the interpretation at a bioinformatics jamboree and by preparing a joint publication similar to the HUPO PPP (Omenn, 2004a,b).

The activities of the HUPO BPP have been reported in several publications (Stephan et al., 2005; Hamacher and Meyer, 2005b; Marcus et al., 2004; Bluggel et al., 2004; Habeck, 2003; Service; 2003), newspapers, and other media. One of the most important interfaces with the scientific community is the HUPO BPP homepage, htpp://www.hbpp.org, as well as the discussion forum http://forum.hbpp.org, that offer an overview, news and the contact address of the project.

The results and considerations of the pilot phase will be used as the basis for the activities in the main phase. The network and the bioinformatics infrastructure will allow the performance of standardized differential analysis of neurodegenerative diseases.

In order to choose suitable and freely available (mouse) models for this next phase, the "HUPO BPP Symposium on Mouse Models" took place during the 4th Dutch Endo-Neuro-Psycho Meeting in Doorwerth/Arnhem, The Netherlands on 1 June 2005. Here, the most promising mouse models for Alzheimer's and Parkinson's diseases were presented and discussed, revealing the advantages as well as pitfalls of the different strains. Currently (at the time this review was written) the selection of the models to be analyzed is in progress, but will be finished by the 5th HUPO BPP Workshop that is planed for Dublin in February 2006.

The DCC and the bioinformatics tools, the network of the consortium and the developing structure of HUPO itself will definitely facilitate the reliable and reproducible analysis of neurodegenerative diseases by proteomics means. Nevertheless, HUPO BPP has several inherent peculiarities that are typical for large consortia projects, especially in regard to how willing the participating groups are to volunteer. First of all, active key players had and have to be identified throughout the scientific world by prominent intercessors, using existing email address lists and a publicity domain (announcements, articles, and contact with scientific journalists). Addressed researchers from both academia and industry had to be convinced that is essential to work together in this brain project though direct funding is not available. The motivations of the partners can be classified as follows:

- the conviction that these tasks can not be managed by single groups;
- the need for standardization and comparable results;
- contact with colleagues and the possibility for collaborations and discussions;
- increased publicity, less lobby work and national/EU funding applications.

Major problems mostly result from missing funding, e.g., most participating groups have to finance their HUPO BPP efforts from other sources, while other laboratories could not take part for this reason. As a consequence sometimes suboptimal analysis and unclear responsibilities are still prominent. This can only be overcome by the constant help and requests of the administrative partners and/or by long-term funding, e.g., by consolidation of HUPO, governmental support or industrial sponsoring.

Acknowledgements

The HUPO BPP is supported by the German Federal Ministry of Education and Research (BMBF) with grant 0313318B, the German HBPP is founded by the BMBF with grant 01GR0440.

References

BLUGGEL, M., BAILEY, S., KORTING, G., STEPHAN, C., REIDEGELD, K. A., THIELE, H., APWEILER, R., HAMACHER, M., MEYER, H. E. (2004). Towards data management of the HUPO Human Brain Proteome Project pilot phase. *Proteomics* 4, 2361–2362.

BURT, R. S. (1980). Models of network structure. *Annu. Rev. Sociology* 6, 79–141.

COOPER, R. G. (2001). *Winning at New Products: Accelerating the Process from Idea to Launch*, 3rd ed. Perseus Publishing, Cambridge.

HABECK, M. (2003). Brain proteome project launched. *Nature Medicine* 9, 631.

HAMACHER, M., KLOSE, J., ROSSIER, J., MARCUS, K., MEYER, H. E. (2004). Does understanding the brain need proteomics and does understanding proteomics need brains? Second HUPO HBPP Workshop hosted in Paris. *Proteomics* 4, 1932–1934.

HAMACHER, M., MEYER, H. E. (2005a). HUPO Brain Proteome Project: aims and needs in proteomics. *Exp. Rev. Proteomics* 1, 1–3.

HAMACHER, M., MEYER, H. E. (2005b). Great mood in proteomics: Beijing and the HUPO Human Brain Proteome Project. *Proteomics* 5, 334–336.

HANASH, S. (2004a). Building a foundation for the human proteome: the role of the Human Proteome Organisation. *J. Proteome Res.* 3, 197–199.

HANASH, S. (2004b). HUPO initiatives relevant to clinical proteomics. *Mol. Cell Proteomics* 3, 298–301.

HERMJAKOB, H., MONTECCHI-PALAZZI, L., BADER, G., WOJCIK, J., SALWINSKI, L., CEOL, A., MOORE, S., ORCHARD, S., SARKANS, U., VON MEHRING, C., ROECHERT, B., POUX, S., JUNG, E., MERSCH, H., KERSEY, P., LAPPE, M., LI, Y., ZENG, R., RANA, D., NIKOLSKI, M., HUSI, H., BRUN, C., SHANKER, K., GRANT, S. G., SANDER, C., BORK, P., ZHU, W., PANDEY, A., BRAZMA, A., JACQ, B., VIDAL, M., SHERMAN, D., LEGRAIN, P., CESARENI, G., XENARIOS, I., EISENBERG, D., STEIPE, B., HOGUE, C., APWEILER, R. (2004a). The HUPO PSI's molecular interaction format – a community standard for the representation of protein interaction data. *Nat. Biotechnol.* 22, 177–183.

HERMJAKOB, H., MONTECCHI-PALAZZI, L., LEWINGTON, C., MUDALI, S., KERRIEN, S., ORCHARD, S., VINGRON, M., ROECHERT, B., ROEPSTORFF, P., VALENCIA, A., MARGALIT, H., ARMSTRONG, J., BAIROCH, A., CESARENI, G., SHERMAN, D., APWEILER, R. (2004b). IntAct: an open source molecular interaction database. *Nucleic Acids Res.* 32, D452–D455.

HUMPHERY-SMITH, I. (2004). A human proteome project with a beginning and an end. *Proteomics* 4, 2519–2521.

KAMI, K., DOI, R., KOIZUMI, M., TOYODA, E., MORI, T., ITO, D., FUJIMOTO, K., WADA, M., MIYATAKE, S., IMAMURA, M. (2004). Survivin expression is a prognostic marker in pancreatic cancer patients. *Surgery* 136, 443–448.

KLOSE, J., MEYER, H. E., HAMACHER, M., VAN HALL, A., MARCUS, K. (2004). Human Brain Proteome Project – towards an inventory of the brain proteins. *BioForum Eur.* 8, 28–29.

MARCUS, K., HULTSCHIG, C., FRANK, R., HERBERG, F. W., SCHUCHHARDT, J., SEITZ, H. (2003/2004). Innovative Forschungsansätze im NGFN Verbund 'The Human Brain Proteome Project HBPP'. *GenomXPress*. Dec., 5–8.

MARCUS, K., SCHMIDT, O., SCHAEFER, H., HAMACHER, M., VAN HALL., A., MEYER, H. E. (2004). Proteomics – application to the brain. *Int. Rev. Neurobiol.* 61, 285–311.

MEYER, H. E., KLOSE, J., HAMACHER, M. (2003). HBPP and the pursuit of standardisation. *Lancet Neurol.* 2, 657–658.

NAKATA, B., TAKASHIMA, T., OGAWA, Y., ISHIKAWA, T., HIRAKAWA, K. (2004). Serum CYFRA 21-1 (cytokeratin-19 fragments) is a useful tumour marker for detecting disease relapse and assessing treatment efficacy in breast cancer. *Br. J. Cancer.* 91, 873–878.

OMENN, G. S. (2004a). Advancement of biomarker discovery and validation through the HUPO plasma proteome project. *Dis. Markers* 20, 131–134.

OMENN, G. S. (**2004b**). International collaboration in clinical chemistry and laboratory medicine: the Human Proteome Organisation (HUPO) Plasma Proteome Project. *Clin. Chem. Lab. Med.* **42**, 1–2.

ORCHARD, S., HERMJAKOB, H., JULIAN, R. K. JR., RUNTE, K., SHERMAN, D., WOJCIK, J., ZHU, W., APWEILER, R. (**2004**). Common interchange standards for proteomics data: Public availability of tools and schema. *Proteomics* **4**, 490–491.

PALOMAKI, G. E., NEVEUX, L. M., KNIGHT, G. J., HADDOW, J. E., PANDIAN, R. (**2004**). Maternal serum invasive trophoblast antigen (hyper-glycosylated hCG) as a screening marker for Down syndrome during the second trimester. *Clin. Chem.* **50**, 1804–1808.

SARBIA, M., BITZER, M., SIEGEL, D., ROSS, D., SCHULZ, W. A., ZOTZ, R. B., KIEL, S., GEDDERT, H., KANDEMIR, Y., WALTER, A., WILLERS, R., GABBERT, H. E. (**2003**). Association between NAD(P)H: quinone oxidoreductase 1 (NQ01) inactivating C609T polymorphism and adeno-carcinoma of the upper gastrointestinal tract. *Int. J. Cancer* **107**, 381–386.

SERVICE, R. F. (**2003**). Proteomics. Public projects gear up to chart the protein landscape. *Science* **302**, 1316–1318.

STEPHAN, C., HAMACHER, M., MEYER, H. E. (**2005**). 3rd HUPO Brain Proteome Project Workshop promises successful pilot studies. *Proteomics* **5**, 615–616.

VERAGUTH, T. (**2004**). Zukunft Biotechnol. *BIOforum* **7–8**, 20–22.

ZOLG, J. W., LANGEN, H. (**2004**). How industry is approaching the search for new diagnostic markers and biomarkers. *Mol. Cell Proteomics* **3**, 345–354.

Useful World Wide Web links

http://www.biomedcentral.com:
The Scientist online journal.

http://www.scottish-enterprise.com/proofofconceptfund:
Scottish Proof of Concept Fund.

http://www.ruhr-uni-bochum.de/rubitec/start.htm:
RUBITEC – Society for Innovation and Technology of the Ruhr-University, Bochum, Germany.

http://ott.od.nih.gov:
Office of Technology Transfer of the National Institutes of Health.

http://www.zwm-speyer.de/:
Center for Science and Research Management Speyer.

http://prod-dev.com:
The Product Development Institute.

http://www.smp-proteomics.de:
The Human Brain Proteome Project as SMP within the NGFN.

http://www.ngfn.de:
Nationales Genomforschungsnetz, Germany.

http://www.hupo.org:
Human Proteome Organisation.

http://www.hbpp.org &
http://forum.hbpp.org:
HUPO Brain Proteome Project.

http://psidev.sourceforge.net:
HUPO Proteome Standards Initiative.

2
Proteomic Data Standardization, Deposition and Exchange

Sandra Orchard, Henning Hermjakob, Manuela Pruess, and Rolf Apweiler

Abstract

The sequencing of the human genome has provided a roadmap from which it should be possible to identify those proteins that play a key role in the initiation and progression of disease processes and, as such, are potential drug targets. The attention of the scientific community has now switched to the proteome, the protein content of specific cell types at any one time. The generation of proteomic data is becoming increasingly high-throughput and both experimental design and the technologies used to generate and analyze the data are ever more complex. Workers in this field require access to tools which enable them to identify and assign function to the protein sequences which they are identifying. A parallel need for methods by which such data can be accurately described, stored and exchanged between experimenters and for public domain data repositories has been recognized. Work by the Proteomics Standards Initiative of the Human Proteome Organisation has laid the foundation for the development of standards by which experimental design can be described and data exchange can be facilitated. Once these standards have become established, public domain databases will be created where experimental data can be deposited prior to publication in the scientific literature. From this data, it will then be possible to generate reference data sets, for example of healthy tissue, which can then be downloaded for comparison with the equivalent diseased tissue in order to identify clinical markers of disease initiation and progression.

2.1
Introduction

At the time when the generation of a map of the human proteome was first seriously discussed (Clark, 1981), scientists only possessed the technical ability to generate short lists of those proteins that are expressed in relatively high abundance in cells and tissues. Many of these proved to be previously unknown proteins and

Proteomics in Drug Research
Edited by M. Hamacher, K. Marcus, K. Stühler, A. van Hall, B. Warscheid, H. E. Meyer
Copyright © 2006 Wiley-VCH Verlag GmbH & Co. KGaA, Weinheim
ISBN: 3-527-31226-9

nothing was known of their function, expression pattern or the dynamics of their metabolism within the cell. Over the subsequent 20 years, many model organism genomes have been fully sequenced and these data and associated annotations made publicly available. Protein sequence analysis tools such as InterPro (Mulder et al., 2005) have provided a means by which function may be ascribed to many novel sequences. Protein isolation techniques have improved to the extent that proteins that are only transiently expressed at very low levels may also be detected, and information on protein–protein interactions or changes in protein metabolism in response to the cellular and, indeed, extracellular environment can be detected. High-throughput technologies have enabled researchers to investigate many tissues, or tissue states, in parallel, allowing the dynamics of protein synthesis and subsequent degradation to be more closely mapped alongside corresponding changes in subcellular localization or posttranslational modification (PTM).

To a drug discovery team searching for new drug targets, the advantages of investigating the cell proteome are apparent. Whilst microarray technology has made it possible to track changes in gene expression in response to a specific agonist or in diseased tissue in comparison to normal, it has become obvious that a change in gene expression does not necessarily correlate with a change in the levels of the corresponding protein (Mehra et al., 2003). Directly measuring fluctuations in protein levels allows the investigator to pinpoint those members of the proteome that use protein synthesis as a means of metabolic control. Related techniques, for example monitoring protein phosphorylation states, identify proteins that use PTMs to regulate their activity. When several of these proteins are known to lie within a common pathway, it may be possible to identify an appropriate target molecule of that pathway for which it is feasible to design inhibitors. Once a potential chemical lead has been identified, modifications of the proteomics methodologies have been developed by which all proteins capable of binding to a particular class of molecule can be identified, thus allowing potential off-targets of a drug to be anticipated (Daub et al., 2004). As an inhibitor moves into early stage clinical trials, the use of protein biomarkers can give early stage indications of both clinical efficacy and the ability of a drug to access different compartments within the body. Comparison of the proteome of normal and diseased tissues allows such biomarkers to be identified and verified.

In order to utilize the information generated by proteomics experiments, the researcher requires a battery of bioinformatics tools to support their work. Firstly, proteins need to be unequivocally identified and a function assigned, either by accessing existing literature information or through the use of a sequence analysis tool. Secondly, the researcher needs the ability to store the large quantities of both raw and annotated data in a manner in which it can easily be retrieved and can also be exchanged with collaborating laboratories, by a mechanism recognized by both senders and recipients irrespective of their individual data storage systems. Access to previously generated proteomics data on proteins of interest, or previous experimental work on the disease/normal tissue under investigation allows groups to benchmark their own procedures and compare their results to comparative datasets. Proteomics experiments are resource-intensive and the ability to analyze current data in the light of that previously generated by other research groups can be a valuable saving of both time and money. Finally, the research group will eventually wish to publish their data and make it available for other groups to access in a manner that is both user-friendly and acceptable to journals to which the article is to be submitted. To achieve all this, common standards are required that are recognized by all parties, to allow data exchange and deposition in databases. This need extends from the assignment of stable accession numbers to unambiguously identify the protein sequences being derived and encompassing the ability to track sequence updates, through to common vocabularies by which sample preparation and analytical techniques can be described.

2.2
Protein Analysis Tools

2.2.1
UniProt

In order to track protein identities through a large volume of experimental data, stable protein accession numbers linked to fully annotated protein sequences are required. One of the most significant developments with regard to protein sequence databases is the recent decision by the National Institutes of Health to award a grant to combine the Swiss-Prot, TrEMBL and PIR-PSD databases into a single resource, the Universal Protein Resource (UniProt) (http://www.uniprot.org) (Bairoch et al., 2005). UniProt is a comprehensive catalogue of data on protein sequence and function, maintained by a collaboration of the Swiss Institute of Bioinformatics (SIB, Geneva, Switzerland), the European Bioinformatics Institute (EBI, Cambridge, UK), and the Protein Information Resource (PIR, Georgetown, USA). UniProt is comprised of three components:

* the expertly curated Knowledgebase (UniProt KB), which continues the work of Swiss-Prot (Boeckmann et al., 2003), TrEMBL (Boeckmann et al., 2003) and PIR (Wu et al., 2003);
* the archive, UniParc, into which new and updated sequences are loaded on a daily basis;
* the nonredundant databases (UniRef NREF) which facilitate sequence merging in UniProt and allow faster and more informative sequence similarity searches.

The UniProt KB is an automatically and manually annotated protein database drawn from translation of DDBJ/EMBL-Bank/GenBank coding sequences and directly sequenced proteins. Each sequence receives a unique, stable identifier allowing unambiguous identification of any protein across datasets. The KB also provides cross-references to external data collections such as the underlying DNA sequence entries in the DDBJ/EMBL-Bank/GenBank nucleotide sequence databases, 2D PAGE and 3D protein structure databases, various protein domain and family characterization databases, PTM databases, protein–protein interactions (IntAct) (Hermjakob et al., 2004a), species-specific data collections, variant databases and disease databases. UniProt/TrEMBL contains a redundant sequence set, enriched by database cross references and automatic annotation. Manual annotation of entries within UniProt/Swiss-Prot strives to augment each entry with as much information as is available, including the function of a protein, PTMs, domains and sites of importance, secondary and quaternary structures, similarities to other proteins, diseases associated with deficiencies in a protein, in which tissues the protein is found, pathways in which the protein is involved, and sequence conflicts and polymorphic variants. Sequences are merged within UniProt/Swiss-Prot to provide a single, nonredundant entry for a unique gene product from an individual organism.

2.2.2
InterPro

InterPro (Mulder et al., 2004) is an integrated resource of protein families, domains and functional sites, each entry is defined by one or more signatures derived from the member databases: PROSITE (Falquet et al., 2002), PRINTS (Attwood et al., 2003), Pfam (Bateman et al., 2004), ProDom (Bru et al., 2005), SMART (Letunic et al., 2002), TIGRFAMs (Haft et al., 2003), PIRSF (Wu et al., 2004), and SUPERFAMILY (Madera et al., 2004). Each InterPro entry corresponds to a biologically meaningful family, domain, repeat or PTM, mapped to Gene Ontology (GO) (Harris et al., 2004) terms in cases where a term applies to all proteins matching that entry. A sequence search package, InterProScan (Zdobnov et al., 2001), combines the different protein recognition methods and scanning tools of each method into one powerful resource unifying the strength of the individual signature database methods to ensure the best prediction of protein domains for a query translation. In the absence of biochemical characterization of a protein, domain predictions can be a good guide to protein function.

2.2.3
Proteome Analysis

The Proteome Analysis Database has been developed utilizing existing resources to provide comparative analysis of the predicted protein coding sequences of the complete genomes of bacteria, archeae and eukaryotes (Pruess et al., 2003). Three main projects are used, InterPro, CluSTr and GO Slim, to give an overview on families, domains, sites, and functions of the proteins from each of the complete genomes, including those of human, mouse and rat. The information is available for download, and can also be queried through the Integr8 browser (http://www.ebi.ac.uk/integr8/) (Kersey et al., 2005), which combines data from both Proteome Analysis and Genome Reviews, a standardized view of complete genomes with updated and consistent annotation, into a single view. This allows the laboratory worker to view proteomic data in the context of the entire genome and make comparisons of human data with that generated using common laboratory species.

2.2.4
International Protein Index (IPI)

IPI provides a top level guide to the main databases that describe the human, mouse and rat proteomes, namely UniProt, RefSeq and Ensembl (Kersey et al., 2004). IPI effectively maintains a database of cross references between the primary data sources, providing minimally redundant yet maximally complete sets of human, mouse and rat proteins (one sequence per transcript) and maintaining stable identifiers (with incremental versioning) to allow the tracking of sequences in IPI between IPI releases. This allows effective management of gene predictions, which vary with each release of both Ensembl and RefSeq. IPI thus provides a complete and nonredundant dataset for human and two common laboratory species, particularly suited to supporting protein identification in proteomics experiments.

2.2.5
Reactome

Drug discovery teams need to see their proteins of interest in terms of the biological pathways or processes in which they are involved. Reactome (http://www.reactome.org) provides an online, curated, free-to-access database of human biological processes, which has recently been launched as a collaboration between the Cold Spring Harbor Laboratory, the European Bioinformatics Institute and The Gene Ontology Consortium (Joshi-Tope et al., 2005). Proteins are identified by UniProt accession numbers and the protein–protein interactions are curated in collaboration with IntAct (http://www.ebi.ac.uk/IntAct), a freely available, open source database system and analysis tool for protein interaction data (Hermjakob et al., 2004a).

2.3
Data Storage and Retrieval

Protein sequence databases, whilst capable of mapping all potential sites of PTM and giving an overall picture of protein expression patterns are not designed to describe the changing patterns of protein expression and state within a living cell. Whilst the results and conclusions drawn from proteomics-based experiments are frequently published in great detail, the underlying data is often only available as supplementary material, or is stored in author-maintained databases or on websites. These databases and websites tend only to exist for the lifespan of the underlying project or grant, are often poorly maintained and the data within is often difficult to access for downloading (Whitfield, J. 2004). Some public domain databases now exist for the storage of some aspects of this data – the field of protein–protein interactions is relatively well served, for example, by IntAct (Hermjakob et al., 2004b), BIND (Alfarano et al., 2005) and DIP (Salwinski et al., 2004) – but these databases are built using different data models with the result that downloading the data from one and comparing it to that of another is both difficult and time-consuming. Other fields, such as mass spectrometry, are much more poorly served and raw data is rarely available even following publication.

2.4
The Proteome Standards Initiative

The Human Proteome Organisation (HUPO) was formed in 2001 with the aim of consolidating national and regional proteome organizations into a single worldwide body. The Proteome Standards Initiative (PSI) was established by HUPO with the remit of standardizing data formats within the field of proteomics to the end that public domain databases can be established where all such data can be deposited, exchanged between such databases or downloaded and utilized by laboratory workers (Orchard et al., 2004). All work produced by the HUPO-PSI has been generated through public meetings, discussion groups and published on the web throughout all development stages to allow input and advice from all quarters (http://psidev.sf.net).

HUPO PSI decided to develop a single data model that would describe and encompass central aspects of a proteomics experiment. This model will contain different subdomains to enable it to handle specific data types, for example 2-D electrophoresis gels or HPLC. Common processes would be described by a number of controlled vocabularies or ontologies. Where these processes are relevant to microarray data, for example in the area of sample preparation, this could be done in collaboration with the Microarray Gene Expression Data (MGED) consortium, thus facilitating the comparison of proteomic with transcript data. Each subdomain would then support a HUPO-PSI-approved interchange format which would permit the handling of data from many different sources. In the interests of making the task more manageable, the HUPO PSI agreed to concentrate their resources on two potential subdomains, mass spectrometry and protein–protein interactions, whilst concurrently developing the global proteomics data model (GPS).

2.5
General Proteomics Standards (GPS)

The context-sensitive nature of proteomic data necessitates the capture of a larger set of metadata than is normally required for genetic sequencing, where knowledge of the organism of origin will suffice. Not only is information on sample source, handling, stimulation and eventual preparation for analysis required, but the detail of the analysis itself will also need to be recorded. For example, to compare images of 2D gels, knowledge of their mass and charge ranges are required, and this information will need to be retrieved by users wishing to perform meaningful analysis of this experiment. A standard representation of both the methods used and the data generated by proteomics, analogous to a combination of the minimum information about a microarray experiment (MIAME) guidelines for transcriptomics, and the associated microarray gene expression (MAGE) object model and XML implementation, are required to facilitate the analysis, exchange and dissemination of proteomics data.

The MIAPE guidelines are currently under development and are being designed to describe all the relevant data from any proteomics experiment, such as details of the experimenter, the sample source, the methods and equipment used and all subsequent results and analyses (Orchard et al., 2004). These could be, for example, data from a laboratory performing MALDI mass spectrometry on spots of interest from comparative 2D gel electrophoresis or from a high-throughput screening facility using multidimensional liquid chromatography fed directly into a tandem mass spectrometer. An XML format for data exchange will be derived from the GPS proteomics workflow/data object model (PSI-OM). This mark-up language (PSI-ML) is designed to become the standard format for exchanging data between researchers, and for submission to repositories or journals. The object model must be flexible enough to cope with both rapidly evolving and completely novel technologies whilst fulfilling the immediate requirements of the scientists of today.

It is clearly desirable that public domain data such as that published in peer-reviewed journals should be accompanied by a defined set of information about the experiment and that all this information and the results obtained be deposited where it is available to all users. A MIAPE-compliant repository will contain sufficient information to allow users to, in principle, recreate any of the experiments stored within it and where possible, the information will be organized in a manner reflecting the structure of the experimental procedures that generated it.

Finally, it is intended that the object model would encompass, and utilize, exchange formats being developed by other groups sponsored by the PSI. For example, the mzData model for mass spectrometry and, where possible, common ontologies and vocabularies will be shared by all these varying domains and also developed in conjunction with the MGED consortium to describe common aspects of proteomic and microarray data. Together, the MIAPE guidelines, data model, ontologies and various implementations will provide a sound basis for describing proteomic experiments in their biological context.

2.6
Mass Spectrometry

The PSI-MS mzData interchange format is being written to allow both the exchange of experimental data from proteomics experiments involving mass spectrometry and also, with cooperation from instrumentation and search engine manufacturers, to enable researchers to generate data in such a standard directly from their instrumentation (Pedrioli et al., 2004). These data written by the vendor data system should be usable directly by search engines, as well as by third-party software tools such as spectral databases and other computational tools. Whilst primarily an interchange format, it is a longer term aim of this group to develop a repository for mass spectrometry data to enable deposition of data prior to publication.

It was agreed that the objectives of this group could best be achieved by aligning with the XML-based standard for analytical information exchange currently being developed by the American Society for Tests and Measures (ASTM) since both standards will have to describe mass spectrometry experiments and results. Standards for spectrometry data such as the ASTM netCDF format [E1947-98 "Standard Specification for Analytical Data Interchange Protocol for Chromatographic Data"; E2077-00 "Standard Specification for Analytical Data Interchange Protocol for Mass Spectrometric Data", ASTM International (see www.astm.org)] and the IUPAC JCAMP format (http://www.jcamp.org) are successful because of broad vendor support and, being computer-platform-neutral, they have remained readable despite changes in computer technology. Useful as these standards are, it has proved difficult to keep them up to date due to the very rapid changes in mass spectrometry technology. XML was considered the best technology for allowing extensions to keep the standard up to date, while remaining computer-platform-neutral.

The published version of the PSI-MS XML data interchange format also gives access to tools which allow the user to both convert from mass spectrometry text formats to the PSI-MS XML format and to view and browse stored data in PSI-MS XML format (Pedrioli et al., 2004).

The PSI-MS schema is flexible enough to handle a diversity of experiments with a full range of experimental descriptors whilst still remaining compliant with the ASTM model. However, the XML format has been switched to Base64 in order to produce more compact files and work is currently in hand to broaden the specification to allow a full description of acquisition, to encompass both mass array and mass intensity. The mZData format has now been released (http://psidev.sourceforge.net/ms) and will soon be accompanied by a spectral analysis output format, supporting a common syntax for peptide/protein identification and for protein modification description (mzProtID). It has been decided that wider issues such as sample preparation and separation will be part of the GPS remit rather than dealt with separately by the mass spectrometry group.

Acceptance of the export format from the instrumentation manufacturers as a direct input to search engines has been good and several vendors, for example Matrix Science, Kratos Analytical, and Bruker Daltronics, are currently experimenting with the mzData format to establish proof-of-concept for spectral and parameter storage. It has been proposed that the existing ASTM mass spectrometry standard data dictionary be adopted and updated for use as a controlled vocabulary within this model, with eventual ownership of this dictionary potentially passing to the American Society for Mass Spectrometry (ASMS) so that it could be used to support both the HUPO-PSI and the ASTM's raw data standardization efforts. MzData is now widely regarded as an acceptable format for representing mass spectrometry data and will be valuable for both data exchange and reposition. In the long term, it is recommended that the ASTM standard be used for full raw data archiving when available (2005–2006); mzData will, though, remain the appropriate format for data exchange.

2.7
Molecular Interactions

A number of protein–protein interaction databases already exist [IntAct (Hermjakob et al., 2004), BIND (Alfarano et al., 2005), DIP (Salwinski et al., 2004a), MINT (Zanzoni et al., 2002), Hybrigencs (http://www.hybrigenics.fr), HPRD (Peri et al., 2003), MIPS (Pagel et al., 2004)] which are wholly or partially in the public domain but, as discussed above, it was previously impossible to download or exchange data from any two of these databases in a common format. All these database providers, however, are committed to making their data more easily accessible and useful to the user community and actively supported the establishment of a common interchange format.

At an initial HUPO-PSI-sponsored meeting of this group it was decided that the approach to designing such a format should be multilayered, with Level 1 designed to fulfil basic requirements and be suitable for rapid implementation. Subsequent levels will supply further complexity and flexibility, for example the ability to deal with other interactors such as nucleic acids. Wherever possible, the potential values of attributes in the data model are defined by controlled vocabularies; researchers may also ascribe confidence levels to the data at various points throughout the data entry process and add free comments in appropriate places.

The model was refined at subsequent workshops and in discussion groups and the Level 1 XML-MI exchange format was published early in 2004 (Hermjakob et al., 2004b), along with accompanying tools and the required controlled vocabularies. All the databases listed above now supply some or all of their data in this format. The model has been further developed at subsequent meetings and in discussion groups and an annual schema update schedule has been agreed. Version 2.0 will allow the description of protein–nucleic acid and protein–small molecule interactions and will also encompass the description of binding features. An editorial board maintains the controlled vocabularies and adds terms when appropriate. The structure is now in place such that journals could request or demand that deposition to a participating database be a prerequisite to publication and an appropriate accession number be supplied to describe each interaction or experiment.

2.8
Summary

Significant progress has already been made in improving the accessibility and utility of proteomic data, and to date, these efforts have been enthusiastically endorsed by the scientific community. Whilst these efforts are being coordinated by the PSI, the work is being undertaken by a large body of scientists, representing the worlds of academia, industrial research and instrumentation manufacture and it is to be hoped that they are laying the groundwork for common standards to be widely adopted throughout the entire user community. As these tools and models become more widely available it can be anticipated that they will play a major role in the direction that this important area of biology takes and the eventual utility of the data generated in increasingly high-throughput biology. There are still many domain-specific areas yet to be tackled in depth by the HUPO-PSI groups, for example 2D gel electrophoresis, and it is hoped that specialists in these areas will come forward and take the lead in developing standards and exchange formats within these areas.

HUPO is also contributing to the establishment of large-scale data sets for a number of human tissues of particular interest in the pathogenesis of disease. The aim of these initiatives is to produce comprehensive lists of proteins present in normal plasma, liver and brain, whilst identifying regional and racial variants within the population. All three project groups have acquired samples from large cohorts of donors which are currently being analyzed in laboratories across the world by a standard set of procedures. The data generated will be deposited in the public domain using UniProt/IPI identifiers and HUPO-PSI data standards and will then be available to provide a reference dataset against which corresponding disease or drug-treated tissue can be compared.

The tools and standards required for the analysis of large-scale datasets generated by proteomic scientists working in the areas of drug discovery are either already available or will be released in the near future. Large reference datasets are being established and it is hoped that these will aid in the identification of a new generation of potential drug targets and assist in providing treatments for many of the life threatening or dehabilitating diseases that are common in man today.

References

ALFARANO, C., ANDRADE, C. E., ANTHONY, K., BAHROOS, N., BAJEC, M., BANTOFT, K., BETEL, D., BOBECHKO, B., BOUTILIER, K. BURGESS, E., et al. (2005). The Biomolecular Interaction Network Database and related tools 2005 update. *Nucleic Acids Res.* **33**, 418–424.

ATTWOOD, T. K., BRADLEY, P., FLOWER, D. R., GAULTON, A., MAUDLING, N., MITCHELL, A. L., MOULTON, G., NORDLE, A., PAINE, K., TAYLOR, P., UDDIN, A., ZYGOURI, C. (2003). PRINTS and its automatic supplement, prePRINTS. *Nucleic Acids Res.* **31**, 400–402.

BAIROCH, A., APWEILER, R., WU, C. H., BARKER, W. C., BOECKMANN, B., FERRO, S., GASTEIGER, E., HUANG, H.,

LOPEZ, R., MAGRANE, M., et al. (2005). The Universal Protein Resource (UniProt). *Nucleic Acids Res.* 33, 154–159.

BATEMAN, A., COIN, L., DURBIN, R., FINN, R. D., HOLLICH, V., GRIFFITHS-JONES, S., KHANNA, A., MARSHALL, M., MOXON, S., SONNHAMMER, E. L., STUDHOLME, D. J., YEATS, C., EDDY, S. R. (2004). The Pfam protein families database. *Nucleic Acids Res.* 32, 138–141.

BOECKMANN, B., BAIROCH, A., APWEILER, R., BLATTER, M. C.,ESTREICHER, A., GASTEIGER, E., MARTIN, M. J., MICHOUD, K., O'DONOVAN, C., PHAN, I., et al. (2003). The SWISS–PROT protein knowledgebase and its supplement TrEMBL in 2003. *Nucleic Acids Res.* 31, 365–370.

BRU,C. COURCELLE, E., CARRERE, S., BEAUSSE, Y., DALMAR, S., KAHN, D. (2005). The ProDom database of protein domain families: more emphasis on 3D. *Nucleic Acids Res.* 33, 212–215.

CLARK, B. F. (1981). Towards a total human protein map. *Nature* 292, 491–491.

DAUB, H., GODL, K., BREHMER, D., KLEBL, B., MULLER, G. (2004). Evaluation of kinase inhibitor selectivity by chemical proteomics. *Assay Drug Dev. Technol.* 2, 215–224.

FALQUET, L., PAGNI, M., BUCHER, P., HULO, N., SIGRIST, C. J. A., HOFMANN, K., BAIROCH, A. (2002). The PROSITE database, its status in 2002. *Nucleic Acids Res.* 30, 235–238.

HAFT, D. H., SELENGUT, J. D., WHITE, O. (2003). The TIGRFAMs database of protein families. *Nucleic Acids Res.* 31, 371–373.

HARRIS, M. A., CLARK, J., IRELAND, A., LOMAX, J., ASHBURNER, M., FOULGER, R., EILBECK, K., LEWIS, S., MARSHALL, B., MUNGALL, C., et al. (2004). The Gene Ontology (GO) database and informatics resource. *Nucleic Acids Res.* 32, 258–261.

HERMJAKOB, H., MONTECCHI-PALAZZI, L., LEWINGTON, C., MUDALI, S., KERRIEN, S., ORCHARD, S., VINGRON, M., ROECHERT, B., ROEPSTORFF, P., VALENCIA, A., et al. (2004a). IntAct: an open source molecular interaction database. *Nucleic Acids Res.* 32, 452–455.

HERMJAKOB, H., MONTECCHI-PALAZZI, L., BADER, G., WOJCIK, J., SALWINSKI, L., CEOL, A., MOORE, S., ORCHARD, S., SARKANS, U., VON MERING, C., et al. (2004b). The HUPO PSI's molecular interaction format – a community standard for the representation of protein interaction data. *Nat. Biotechnol.* 22, 177–183.

JOSHI-TOPE, G., GILLESPIE, M., VASTRIK, I., D'EUSTACHIO, P., SCHMIDT, E., DE BONO, B., JASSAL, B., GOPINATH, G. R., WU, G. R., MATTHEWS, L., et al. (2005). Reactome: a knowledgebase of biological pathways. *Nucleic Acids Res.* 33, 428–432.

KERSEY, P. J., DUARTE, J., WILLIAMS, A., KARAVIDOPOULOU, Y., BIRNEY, E., APWEILER, R. (2004). The International Protein Index: an integrated database for proteomics experiments. *Proteomics* 4, 1985–1988.

KERSEY, P., BOWER, L., MORRIS, L., HORNE, A., PETRYSZAK, R., KANZ, C., KANAPIN, A., DAS, U., MICHOUD, K., PHAN, I., et al. (2005). R. Integr8 and Genome Reviews: integrated views of complete genomes and proteomes. *Nucleic Acids Res.* 33, 297–302.

LETUNIC, I., GOODSTADT, L., DICKENS, N. J., DOERKS, T., SCHULTZ, J., MOTT, R., CICCARELLI, F., COPLEY, R. R., PONTING, C. P., BORK, P. (2002). Recent improvements to the SMART domain-based sequence annotation resource. *Nucleic Acids Res.* 30, 242–244.

MADERA, M., VOGEL, C., KUMMERFELD, S. K., CHOTHIA, C., GOUGH, J. (2004). The SUPERFAMILY database in 2004: additions and improvements. *Nucleic Acids Res.* 32, 235–239.

MEHRA, A., LEE, K. H., HATZIMANIKATIS, V. (2003). Insights into the relation between mRNA and protein expression patterns: I. Theoretical considerations. *Biotechnol. Bioeng.* 84, 822–833.

MULDER, N. J., APWEILER, R., ATTWOOD, T. K., BAIROCH, A., BATEMAN, A., BINNS, D., BRADLEY, P., BORK, P., BUCHER, P., CERRUTI, L., et al. (2005). InterPro, progress and status in 2005. *Nucleic Acids Res.* 33, 201–205.

ORCHARD, S., HERMJAKOB, H., JULIAN, R. K. JR., RUNTE, K.,

SHERMAN, D., WOJCIK, J., ZHU, W., APWEILER, R. (2004). Common interchange standards for proteomics data: Public availability of tools and schema. *Proteomics* 4, 490–491.

PAGEL, P., KOVAC, S., OESTERHELD, M., BRAUNER, B., DUNGER-KALTENBACH, I., FRISHMAN, G., MONTRONE, C., MARK, P., STUMPFLEN, V., MEWES, H. W., et al. (2005). The MIPS mammalian protein–protein interaction database. *Bioinformatics* 21, 832–834.

PEDRIOLI, P. G., ENG, J. K., HUBLEY, R., VOGELZANG, M., DEUTSCH, E. W., RAUGHT, B., PRATT, B., NILSSON, E., ANGELETTI, R. H., APWEILER, R., et al. (2004). A common open representation of mass spectrometry data and its application to proteomics research. *Nat. Biotechnol.* 22, 1459–1466.

PERI, S., NAVARRO, J. D., KRISTIANSEN, T. Z., AMANCHY, R., SURENDRANATH, V., MUTHUSAMY, B., GANDHI, T. K., CHANDRIKA, K. N., DESHPANDE, N., SURESH, S., et al. (2003). Development of human protein reference database as an initial platform for approaching systems biology in humans. *Genome Res.* 13, 2363–2371.

PRUESS, M., FLEISCHMANN, W., KANAPIN, A., KARAVIDOPOULOU, Y., KERSEY, P., KRIVENTSEVA, E., MITTARD, V., MULDER, N., PHAN, I., SERVANT, F., APWEILER, R. (2003). The Proteome Analysis Database: a tool for the in silico analysis of whole proteomes. *Nucleic Acids Res.* 31, 414–417.

SALWINSKI, L., MILLER, C. S., SMITH, A. J., PETTIT, F. K., BOWIE, J. U., EISENBERG, D. (2004). The Database of Interacting Proteins: 2004 update. *Nucleic Acids Res.* 32, 449–451.

WHITFIELD, J. (2004). Web links leave abstracts going nowhere. *Nature* 428, 592.

WU, C. H., YEH, L. S., HUANG, H., ARMINSKI, L., CASTRO-ALVEAR, J., CHEN, Y., HU, Z., KOURTESIS, P., LEDLEY, R. S., SUZEK, B. E., et al. (2003). The Protein Information Resource. *Nucleic Acids Res.* 31, 345–347.

WU, C. H., HUANG, H., NIKOLSKAYA, A., HU, Z., BARKER, W. C. (2004). The iProClass integrated database for protein functional analysis. *Comput. Biol. Chem.* 28, 87–96.

ZANZONI, A., MONTECCHI-PALAZZI, L., QUONDAM, M., AUSIELLO, G., HELMER-CITTERICH, M. CESARENI, G. (2002). MINT: a Molecular INTeraction database. *FEBS Lett.* 513, 135–140.

ZDOBNOV, E. M. APWEILER, R. (2001). InterProScan – an integration platform for the signature – recognition methods in InterPro. *Bioinformatics* 17, 847–848.

II
Proteomic Technologies

Proteomics in Drug Research
Edited by M. Hamacher, K. Marcus, K. Stühler, A. van Hall, B. Warscheid, H. E. Meyer
Copyright © 2006 Wiley-VCH Verlag GmbH & Co. KGaA, Weinheim
ISBN: 3-527-31226-9

3
Difference Gel Electrophoresis (DIGE): the Next Generation of Two-Dimensional Gel Electrophoresis for Clinical Research

Barbara Sitek, Burghardt Scheibe, Klaus Jung, Alexander Schramm and Kai Stühler

Abstract

In the last years the application of two-dimensional electrophoresis (2-DE) has often been declared outdated and a new century of gel-free proteomics was announced. Nevertheless, 2-DE is still the method of choice when analyzing complex protein mixtures. With a separation of 10 000 proteins, 2-DE gives access to high-resolution proteome analysis. Continuous development has consolidated 2-DE application in proteomics, where the introduction of difference gel electrophoresis (DIGE) is the latest improvement. DIGE is based on fluorescently tagging all proteins in each sample with one set of matched fluorescent dyes designed to minimally interfere with protein mobility during 2-DE.

For a DIGE analysis, two different fluorescence dyes, CyDye® minimal dyes and CyDye saturation dyes, are available. CyDye minimal dyes react with an NHS-ester bond of lysine ε-amino residues and enable coelectrophoresis of up to three different samples in one approach. For special applications, e.g., samples from microdissection, the CyDye saturation dyes allow complete 2-D analysis and quantification of protein abundance changes in scarce sample amounts. The dyes react via a maleimide group with all available cysteine residues in the protein sample, giving a high labeling concentration.

Here we show the application of both CyDye types for the analysis of relevant clinical samples and an approach for evaluating DIGE data statistically. We used CyDye minimal dyes to detect kinetic proteome changes resulting from ligand activation of either tyrosine kinases TrkA or TrkB in neuroblastoma cells. TrkA and TrkB are members of the family of neurotrophin receptors, which mediate survival, differentiation, growth, and apoptosis of neurons in response to stimulation by their ligands, NGF and BDNF, respectively. In addition, we applied CyDye saturation dyes for the identification of new molecular markers of the pancreatic tumor progression. One thousand microdissected cells were analyzed from different pancreatic intraepithelial neoplasias (PanIN) grades, the precursor lesion of pancreatic ductal adenocarcinoma (PDAC). Based of the multiplexing strategy and the application of an internal standard, DIGE enables one to perform

Proteomics in Drug Research
Edited by M. Hamacher, K. Marcus, K. Stühler, A. van Hall, B. Warscheid, H. E. Meyer
Copyright © 2006 Wiley-VCH Verlag GmbH & Co. KGaA, Weinheim
ISBN: 3-527-31226-9

quantitative proteomics with high accuracy allowing statistical approaches for high-confidence data analysis.

3.1
Introduction

In the early 1990s Marc Wilkins introduced the term *proteome* as a description for all *proteins*, which are coded by a gen*ome* at specific time points and under certain conditions. It is thought that in addition to the analysis of the genome that proteome analysis, with its high dynamic and complexity, can be used to describe life in more detail. A number of different methods have therefore been established in different proteomics techniques to allow analysis of complex protein mixtures. Currently, two-dimensional electrophoresis (2-DE) is the separation method with highest resolution power for protein samples. Up to 10 000 proteins can be separated in one gel and are then accessible for quantitative analysis (Figure 3.1, Klose and Kobalz, 1995). In 1975, 2-DE was independently developed by Klose et al. (1975) and O'Farrell et al. (1975) representing the combination of two different separation techniques. In the first dimension the proteins are focused according to their isoelectric points (pIs) (isoelectric focussing (IEF)) and subsequently in the second according to their molecular size (SDS electrophoresis) (see review Görg et al., 2000 for more details). For the IEF, two different techniques, carrier ampholyte IEF (CA-IEF) (Klose and Kobalz, 1995) and immobiline pH gradient (IPG) (Bjellqvist et al., 1982; Görg et al., 1988), respectively, are available. The immobilization of the pH gradient and fixation on a plastic backing simplified the IEF and therefore allowed the widespread application of 2-DE (reviewed in more detail in Anderson and Anderson, 1996; Herbert, 1999; Rabilloud, 2000).

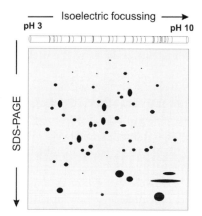

Figure 3.1 Schematic representation of 2-dimensional electrophoresis (2-DE). In the first dimension the proteins are focused according to their isoelectric point [isoelectric focusing (IEF)] and subsequently in the second dimension according to their molecular size (SDS electrophoresis or SDS-PAGE).

Meanwhile, a number of different IPG stripes, with narrow or broad as well as linear or nonlinear pH gradient, are provided by different manufacturers. In contrast, CA-IEF, a labor-intensive technique, failed to achieve widespread application but is still used in more specialized laboratories.

For quantitative analysis the proteins can be visualized using a number of staining techniques with different sensitivities and linear dynamic detection ranges (Table 3.1). In contrast to array technologies like, e.g., Affymetrix chips or protein chips where the analyte position is assignable with a high local x, y resolution, the protein (spot) position in a 2-DE gel depends on its physicochemical properties. Therefore, generating 2-DE gels for a differential analysis requires standardized operating procedures to achieve highly reproducible protein patterns (Lopez et al., 1997). However, small alterations in 2-DE processing, such as, for instance, during the gel casting process, polymerization reaction and/or temperature changes during the IEF run (21–24 h) results in reproducibility variations which have to be compensated for by appropriate image analysis software. Besides detection and normalization features, classical image analysis software like, for example, ImageMaster, PDQuest®, ProteomWeaver® or Delta2D® provide appropriate algorithms for matching of the spots between different 2-DE gels to compensate for these variations.

Table 3.1 The most commonly used protein staining methods for proteome analysis using 2-DE. Besides the low detection limit of the applied staining methods, the linear dynamic range for protein quantification is another important parameter for a global description of a proteome.

Staining method	Detection limit (ng)	Linear dynamic range	Reference
Silver	1	2 orders of magnitude	Heukeshoven and Dernick, 1988
	3–5[a]		Nesterenko et al., 1994
Zinc imidazole	10	No quantification	Fernandez-Patron et al., 1998
SyproRuby	1	3 orders of magnitude	Rabilloud et al., 2001
SyproOrange, SyproRed	30	3 orders of magnitude	Yan et al., 2000
Colloidal Coomassie	8–10	3 orders of magnitude	Neuhoff et al., 1990
DIGE minimal	0.1–0.2	3–5 orders of magnitude	Tonge et al., 2001
DIGE saturation	0.005–0.010	3–5 orders of magnitude	Shaw et al., 2003
Phosphor-imaging (^{32}P, ^{14}C and ^{35}S)	≪ 1	5 orders of magnitude	Johnston et al., 1990

a Detection limit of MS-compatible silver staining (Nesterenko et al., 1994).

3.2
Difference Gel Electrophoresis: Next Generation of Protein Detection in 2-DE

An important improvement for the application of 2-DE was the introduction of the so-called difference gel electrophoresis (DIGE) by J. S. Minden's group in 1997 (Unlü et al., 1997). DIGE circumvents some basic problems of 2-DE, for example gel-to-gel variations and limited accuracy using different fluorophores (Cy2, Cy3 and Cy5) for a multiplexed analysis. In contrast to classical detection methods (Table 3.1) this method relies on covalent derivatization of proteins in each sample with one of the set of matched CyDye that do not affect the relative mobility of proteins during electrophoresis. Thus, DIGE allows rapid identification of protein changes between two samples on the same 2-DE gel without influences of gel-to-gel variations. Additionally, DIGE covers a dynamic detection range of 3–5 orders of magnitude; while silver staining can only detect 30-fold changes (Alban et al., 2003; Knowles et al., 2003; Gharbi et al., 2002; Tonge et al., 2001; Unlü et al., 1997). For a DIGE analysis the two different fluorescence dyes CyDye minimal dyes and CyDye saturation dyes are available.

CyDye minimal dyes that react via an NHS–ester bond with ε-amino residues of lysine enable coelectrophoresis of up to three different samples in one approach. Approximately 3% of the available proteins are labeled on a single lysine per protein, whereas the rest remains unlabeled. This makes the technique robust and labeling optimization is usually unnecessary. The three different CyDye tags add approximately 450 Da to the protein mass when coupled to the protein. The resulting image patterns are comparable to poststained gels and sensitivity of the minimal dyes is similar to most sensitive silver staining. A pooled internal standard can be created by mixing aliquots of all samples to be analyzed. The use of such an internal standard on each gel that comprises equal quantities of each of the samples in the experiment allows calculation of ratios for the same protein spot within one gel as well as between gels. This largely removes experimental gel-to-gel variation leading to improved accuracy of protein quantification between samples from different gels. In a setup of two samples plus internal standard per gel, renunciation of gel repetitions and emphasis on biological repetitions brings down the total number of gels per experiment necessary for quantitative statistics (see Section 3.2.3). The bottlenecks of classical 2-DE-like methodical variation, laborious image analysis and restricted quantification are therefore minimized in the DIGE workflow (Figure 3.1).

For special applications, e.g., samples from microdissection (see Section 3.2.2), the CyDye saturation dyes enable complete 2-DE analysis and quantification of protein abundance changes in scarce sample amounts (5 μg protein/image). The dyes react via a maleimide group with all available cysteine residues in the protein sample, giving a high labeling concentration. Owing to their net zero charge, there is no charge alteration of the labeled protein. As with all DIGE experiments an internal standard sample containing an equal amount of each sample is run on each gel. Sensitivity is 20-fold higher than for silver staining. In contrast to minimal labeling with saturation dyes, a labeling optimization is necessary to

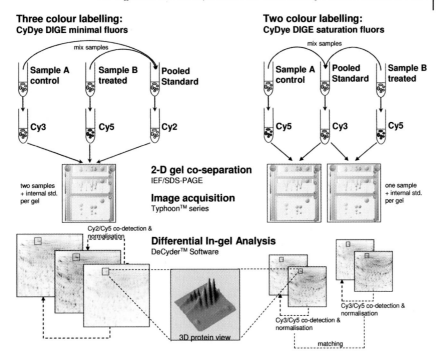

Figure 3.2 Difference gel electrophoresis (DIGE). Ettan® DIGE workflow: three-color and two-color experiments including the internal standard. For fluorescence proteins tagging, two different CyDyes techniques are available. Minimal fluors allow consideration of three different CyDyes (Cy2, Cy3 and Cy5) in a multiplexing experiment.

Applying saturation fluors multiplexing is only done for internal standardization, but allows working with low concentration protein samples. After image acquisition at CyDyes specific wavelength using a confocal fluorescence scanner (e.g., Typhoon series) protein spot pattern can be analyzed by appropriate image analysis software (DeCyder).

determine the appropriate amount of dye. With a confocal fluorescent imager the dye images are acquired at their specific wavelength without crosstalk.

Dedicated image analysis software (DeCyder) utilizes a proprietary codetection algorithm (up to triple detection) that permits automatic detection, background subtraction, quantification, normalization and intergel matching of fluorescent images. The experimental design using the internal standard effectively eliminates gel-to-gel variation, allowing detection of small differences in protein levels (Figure 3.2).

Systemic variation as well as inherent biological variation arising from patient-to-patient, culture-to-culture differences, etc., can be clearly differentiated from induced biological changes. The DIGE system allows discrimination between true biological differences on the one hand and systemic as well as interindividual differences on the other.

3.2.1
Application of CyDye DIGE Minimal Fluors
(Minimal Labeling with CyDye DIGE Minimal Fluors)

3.2.1.1 General Procedure

In the standard labeling protocol, proteins are first solubilized in a DIGE lysis buffer (30 mM Tris, 7 M urea, 2 M thiourea, 4% (w/v) CHAPS, pH 8.5). Samples produced by acidic precipitation should be lysed with DIGE lysis buffer adjusted to an appropriate pH (e.g., 9.5) resulting in a pH between 8 and 9. Lower pH during the labeling reaction affects labeling efficiency and kinetics in a negative way. The protein concentration should then be determined using a standard protein quantification method. Protein concentrations in the range 1–20 mg/mL have been successfully labeled using the standard protocol (personal communication, Amersham Bioscience). CyDye samples are then added to the protein lysate and incubated on ice in the dark for 30 min so that 50 µg of protein are labeled with 400 pmol of CyDye (prepared in a working solution containing 400 pmol CyDye μL^{-1} anhydrous dimethylformamide). Then 1 µL of 10 mM lysine is added to stop the reaction and left for 10 min on ice in the dark. Samples can now be stored for at least one month at –70 °C. Lysis buffer in classical 2-DE traditionally contains primary amines (e.g., carrier ampholytes, pharmalytes) and excess of thiols (e.g., DTT, DTE) which should be not present during the labeling reaction.

After labeling, 2×lysis buffer [8 M urea, 4% (w/v) CHAPS, 2% (v/v) carrier ampholytes, 2% (w/v) DTT] can be added to the samples in a 1+1 dilution for IEF. Combining the three samples to be separated in one IEF strip or tube gel results in a total protein concentration of 150 µg per gel (50 µg Cy3+50 µg Cy5+50 µg Cy2 labeled). From here methods are virtually identical to classical 2-DE. A peculiarity is that glass cassettes should have low intrinsic fluorescence capacity, since the gel will be scanned still assembled between the plates.

The CyDyes are matched for size and charge resulting in an overlay of identical proteins when run on the same 2-DE gel. The lysine amino acid in proteins carries a +1 charge at neutral or acidic pH. CyDye minimal dyes also carry an intrinsic +1 charge which, when coupled to the lysine, replaces the lysine's +1 charge with its own, ensuring that the pI of the protein does not significantly alter (Figure 3.3).

The experimental design is based on evidence that the experimental variation in a 2-DE experiment is mostly due to gel-to-gel variation. Running multiple samples on a single gel reduces the number of gels required to produce the same amount of data. The recommended protocol suggests that a Cy2-labeled internal standard sample should be run on all gels within an experiment. The standard sample consists of an aliquot of all the different samples in an experiment (Figure 3.2). This ensures representation of all proteins on all the gels analyzed. The standard sample will increase the reliability in matching between gels and will also allow the generation of accurate spot statistics between gels (see Section 3.2.3).

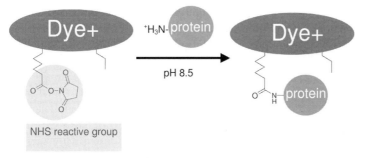

Figure 3.3 Labeling chemistry of CyDye DIGE minimal fluors.
Dye N-hydroxysuccinimid-ester (NHS-ester) reacts with a single ε-amino
residue of lysine per protein and a positively charged dye molecule replaces
the positive charged lysine on the protein so there is no net change in pI.

3.2.1.2 Example of Use: Identification of Kinetic Proteome Changes upon Ligand Activation of Trk-Receptors

DIGE exerts its full strength in settings comprising highly similar but not identical biological conditions. Potential applications include monitoring of cellular responses to all kinds of external stimuli. Those systems are mostly error-prone in classical 2-DE due to the above-mentioned interexperimental variations. We have chosen the DIGE system to detect kinetic proteome changes resulting from ligand activation of either tyrosine kinases TrkA or TrkB in neuroblastoma cells (Figure 3.4). TrkA and TrkB are members of the family of neurotrophin receptors, which mediate survival, differentiation, growth, and apoptosis of neurons in response to stimulation by their ligands, NGF and BDNF, respectively (reviewed in Miller and Kaplan, 2001). So far, little is known about the molecular mechanisms used by TrkA/TrkB to mediate this different biological behavior. Expression levels of TrkA/TrkB are important prognostic factors in a variety of embryonal tumors including neuroblastoma, the most common solid tumor of childhood (Nakagawara et al., 1993). Since TrkA/TrkB exhibit a high level of sequence similarity and use overlapping pathways for signal transduction, the existence of specific effector molecules crucial for receptor and cell-type specific response is likely (Schulte et al., 2005). To identify these effectors by analyzing biological effects of TrkA and TrkB activation in a defined model, we performed a proteome analysis using the human neuroblastoma SY5Y cell line stably transfected with the TrkA or TrkB cDNA (Sitek et al., 2005a). The resulting phenotypes of these cell lines have been extensively analyzed and are in agreement with the observed biological functions (Eggert et al., 2000, 2002). Proteomic changes were monitored in a time-course of 0, 0.5, 1, 6 and 24 h following receptor activation. These time points were chosen to monitor immediate responses to neurotrophins (NGF/BDNF) as well as late effector proteins. Considering the biological variation, we focused on identification of significantly regulated protein spots in five biologically independent experiments at each time point. At each time point a differential comparison was performed between neurotrophin-treated and untreated cells. Technically, 50 µg protein of

A

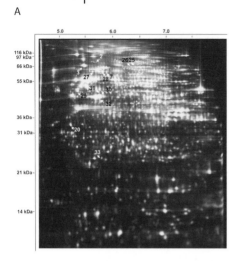

B

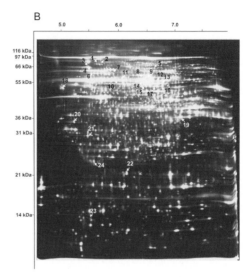

Figure 3.4 2DE-gels of SY5Y–TrkA cell lysate (A) or SY5Y–TrkB cell lysate (B). Separation of proteins was performed in the first dimension (*horizontal*) by IEF and then in the second dimension (*vertical*) by SDS-PAGE. Representative pictures of whole cell lysates of SY5Y–TrkA (A) and SY5Y–TrkB (B) are shown. Differentially expressed proteins following activation of neurotrophin receptors TrkA or TrkB by their respective ligands are numbered and correspond to the data presented in Table 3.2.

untreated and ligand treated cells are labeled with Cy5 and Cy3, respectively. A pool of all analyzed samples was labeled with Cy2, generating the internal standard. Proteins extracted from neurotrophin-treated and untreated SY5Y–TrkA/B cells were labeled with Cy3 and Cy5, respectively, mixed with Cy2-labeled internal standards (Figure 3.5) and run in one gel. After scanning of gels and manual correction of spot detection, protein spots were matched for statistical analysis and determination of the differentially expressed proteins. In whole cell lysates of SY5Y–TrkA/TrkB, 1700–1900 distinct protein spots were detected by the DeCyder software and subsequent manual correction. The analysis of the expression profiles of SY5Y–TrkB cells resulted in 24 significantly regulated spots induced by BDNF ($p < 0.05$, ratio > 1.5) and 13 regulated spots induced by NGF in SY5Y–TrkA cells ($p < 0.05$, ratio > 1.3) (Table 3.2). A total of 20 spots regulated during the time-course following neurotrophin treatment were specific for SY5Y–TrkB cells, and nine regulated spots were specific for SY5Y–TrkA cells. Four spots were regulated in both cell lines. While three spots demonstrated the same regulation pattern, one was inversely regulated between SY5Y–TrkA and SY5Y–TrkB. The respective protein was identified as tropomyosin-3 by MALDI-PMF. The majority of the proteins differentially expressed upon neurotrophin receptor activation were regulated in the late stimulation phase (10 of 13 protein spots in SY5Y–TrkA and 16 of 24 in SY5Y–TrkB) (Figure 3.6). Most of the proteins identified were assigned to the GeneOntology (GO) class "cytoskeleton organization and biogenesis", but proteins

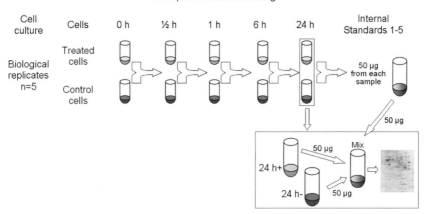

Figure 3.5 Schematic workflow of sample preparation for 2D-DIGE. SY5Y-TrkA and SY5Y-TrkB cells were stimulated by neurotrophins NGF (100 ng/mL) or BDNF (50 ng/mL), respectively ("treated cells"). Controls received fresh medium without neurotrophins at $t = 0$ ("untreated cells"). Samples of the same time point following neurotrophin addition were labeled, mixed and run in one gel with an equal amount of internal standard (B). This standard was pooled from all samples in an experiment. The lower part of the figure exemplarily depicts the procedure for time point $t = 24$ h. Experiments were repeated five times to adjust for inter-experimental variations.

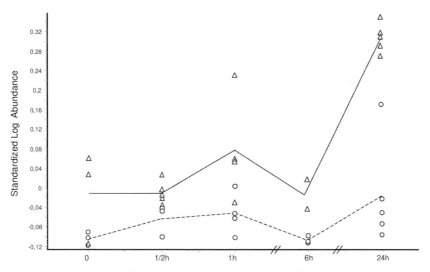

Figure 3.6 Typical kinetics of regulated proteins following neurotrophin treatment of SY5Y-TrkA or SY5Y-TrkB. The majority of proteins like for instance galectin-1 were regulated in the late stimulation phase. Circles (control) and triangles (neurotrophin-treated) represent single standardized abundance values from one gel. Lines connect the average standardized abundance values. Though experiments were repeated five times, the DeCyder software could not detect the here presented protein spot in all gels. Thus, at some time points less than five circles or triangles are represented.

Table 3.2 Summary of significantly regulated proteins following neurotrophin receptor activation in SY5Y-TrkA or SY5Y-TrkB identified by MALDI-MS.

Spot No.	Protein	Theoretical		Experimental		Accession no. NCBI	Regulation/ kinetic group	
		pI	MW [kDa]	pI	MW [kDa]		TrkA	TrkB
1	Dynein	5.0	71.5	5.4	96.9	gi/24307879	none	↑ A
2	Dynein	5.0	71.5	5.5	95.3	gi/24307879	↓ C	↓ A
3	Heat shock 70kDa protein 5	5.0	72.3	5.3	86.8	gi/16507237	↑ C	↑ C
4	Heat shock 70kDa protein 5	5.0	72.3	5.3	86.8	gi/16507237	none	↑ C
5	Lamin A/C isoform 1 precursor	6.8	74.1	6.2	86.6	gi/27436946	none	↑ C
6	Heterogeneous nuclear ribonucleoprotein K isoform a	5.1	51.0	5.4	72.0	gi/14165437	none	↑ B
7	not identified			5.6	67.5		↓ C	↓ B
8	Lamin A/C isoform 2	6.4	65.1	6.0	62.9	gi/27436946	none	↑ C
9	Lamin A/C isoform 2	6.4	65.1	6.1	62.9	gi/27436946	none	↑ C
10	JC5704 protein disulfide-isomerase (EC 5.3.4.1) ER60 precursor	5.9	56.8	5.7	60.9	gi/7437388	none	↑ C
11	JC5704 protein disulfide-isomerase (EC 5.3.4.1) ER60 precursor	5.9	56.8	5.8	60.4	gi/7437388	none	↑ C
12	Dihydrolipoamide dehydrogenase precursor	7.9	54.2	6.3	60.4	gi/4557525	none	↓ B
13	Adenylyl cyclase-associated protein	8.0	51.7	6.3	57.5	gi/5453595	none	↓ B
14	Nuclear matrix protein NMP200 related to splicing factor PRP19	6.1	55.2	6.1	60.5	gi/7657381	none	↓ C
15	TATA binding protein interacting protein 49 kDa	6.0	50.2	6.1	56.6	gi/4506753	none	↓ C
16	ATP synthase, H+ transporting,	9.2	59.8	6.3	55.8	gi/15030240	none	↓ C
17	ATP synthase, H+ transporting,	9.2	59.8	6.2	55.8	gi/15030240	none	↓ C
18	Calumenin	4.4	37.1	5.0	52.0	ggi/4502551	none	↑ C

Table 3.2 (continued)

Spot No.	Protein	Theoretical		Experimental		Accession no. NCBI	Regulation/ kinetic group	
		pI	MW [kDa]	pI	MW [kDa]		TrkA	TrkB
19	A Chain A	8.5	36.7	6.5	35.8	gi/13786849	none	↓ C
20	A27674 tropomyosin 3, fibroblast	4.7	32.9	5.1	34.4	gi/88928	↓ C	↑ C
21	Rho GDP dissociation inhibitor (GDI) alpha	5.0	23.2	5.3	31.2	gi/4757768	none	↓ C
22	AF487339_1 NM23-H1	5.2	19.7	5.8	22.1	gi/29468184	none	↓ B
23	1713410A beta galactoside soluble lectin	5.1	14.6	5.3	12.8	gi/227920	none	↑ C
24	not identified			5.4	24		none	↓ A
25	Lamin A/C isoform 1 precursor	6.8	74.1	6.2	86.6	gi/27436946	↑ C	none
26	Lamin A/C isoform 1 precursor	6.8	74.1	6.2	86.6	gi/27436946	↑ A	none
27	Heterogeneous nuclear ribonucleoprotein K isoform a	5.1	51.0	5.4	70.5	gi/14165437	↓ C	none
28	Lamin B2	5.3	67.7	5.6	71.0	gi/27436951	↑ A	none
29	Vimentin	7.9	54.2	5.2	54.2	gi/4557525	↓ C	none
30	not identified			5.5	49.4		↓ C	none
31	not identified			5.5	49.4		↓ C	none
32	ACTB protein	5.6	40.2	5.5	49.4	gi/15277503	↓ C	none
33	not identified			5.6	26.4		↓ C	none

involved in signal transduction could also be found. Most prominently, SY5Y–TrkB transfected cells up-regulate hnRNP K, which functions as a "RNA silencer" in immature epithelial cells to suppress translation of proteins only needed in differentiated cells. It is intriguing to speculate about a similar function for hnRNP K in preventing expression of proteins found only in differentiated neuronal cells. A recent report indicated that in fact hnRNP K controls the switch from pro-liferation to neuronal differentiation through regulation of p21 (Yano et al., 2005).

In this model system, application of the DIGE system has set the basis for integration of the signaling pathways and functional analysis of effector proteins as well as insights into the underlying courses for the opposing phenotypes caused by activation of different neurotrophin receptors.

3.2.2
Application of Saturation Labeling with CyDye DIGE Saturation Fluors

3.2.2.1 General Procedure
As this type of labeling method aims to label all available cysteines on each protein, many cysteine residues existing as disulfide bonds in proteins must be unfolded and the disulfide bonds broken. This can be achieved under denaturing conditions with a reducing agent such as tris-(carboxyethyl) phosphine hydrochloride (TCEP) (Figure 3.7). In some proteins cysteine residues are buried, and thus the extent of labeling will depend on the accessibility of cysteine within the protein under the reaction conditions used. Also protein quantification in the amounts and concentrations used is not easy to carry out. A labeling optimization in a 2-D titration experiment is therefore obligatory to determine the optimum amount of TCEP and dye required for the protein extract being used. The molar ratio of TCEP : dye should always be kept at 1 : 2 to ensure efficient labeling. Typically, 5 μg of protein lysate requires 2 nmol TCEP and 4 nmol dye for the labeling reaction (assuming an average cysteine content of 2%). If the amount of TCEP/dye is too low, available thiol groups on some proteins will not be labeled and show molecular weight trains in the 2-D image. If the amount of TCEP/dye is too high, nonspecific labeling of the amine groups on lysine residues can occur and have been observed as pI charge trains on the gel. In the standard protocol, proteins are solubilized in lysis buffer (30 mM Tris, 7 M urea, 2 M thiourea, 4% (w/v) CHAPS, pH 8.0). Primary amines or thiols should not be present, because they will compete with the protein for the dye. After lysis, the pH should not deviate

1. Reduction

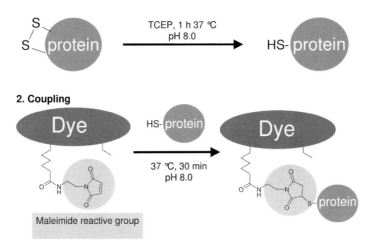

Figure 3.7 Labelling chemistry of CyDye DIGE saturation fluors: dye maleimide group reacts with all available cysteine residues in the protein and due to its net zero charge there is no charge change.

from pH 8; protein concentrations should be between 0.55 and 10 mg/mL. Then cysteine residues are reduced at 37 °C for 1 h by incubating with TCEP (2 mM solution) with a volume determined in the labeling optimization experiment. CyDye DIGE Fluor saturation dye (2 mM working solution in anhydrous dimethyl-formamide) is added for reaction at 37 °C in the dark for a further 30 min. Cy3 is used for the pooled standard and Cy5 for the individual samples. Finally, the reaction is stopped by excess of DTT using 2×lysis buffer (s.a.). 2-DE is according to standard procedure; the alcylation step in the equilibration procedure can be omitted.

3.2.2.2 Example of Use: Analysis of 1000 Microdissected Cells from PanIN Grades for the Identification of a New Molecular Tumor Marker Using CyDye DIGE Saturation Fluors

Pancreatic adenocarcinoma is the fourth leading cause of cancer in the United States (Niederhuber et al., 1995). The main problem of combating pancreatic adenocarcinoma arises from a lack of specific symptoms and limitations in detection methods. The overwhelming majority of pancreatic carcinoma patients are discovered at a late clinical tumor stage. Only 10% of patients show a potentially curable resectable tumor.

In order to identify new molecular markers of the pancreatic tumor progression we established a proteomics approach analyzing 1000 microdissected cells from different pancreatic intraepithelial neoplasias (PanIN) grades (Sitek et al., 2005b). It is believed that PanINs are the precursor lesions of pancreatic ductal adeno-carcinoma (PDAC). PanINs are histologically subdivided into four different grades, namely PanIN-1A/B, PanIN-2 and PanIN-3. All PanINs progress from flat to papillary lesions with increasing degrees of dysplasia (Figure 3.8). To analyze the

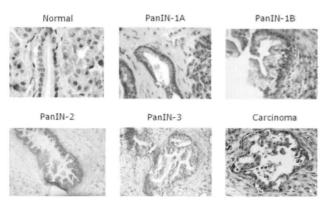

Figure 3.8 Histological images of different PanIN grades stained with H&E. It is believed that pancreatic intraepithelial neoplasias (PanINs) are the precursor lesions of pancreatic ductal adenocarcinoma (PDAC). PanINs are histologically subdivided into four different grades, namely PanIN-1A/B, PanIN-2 and PanIN-3. All PanINs progress from flat to papillary lesions with increasing degrees of dysplasia. The PanIN classification has been established as a common diagnostic criterion at the National Cancer Institute Think Tank (http://pathology.jhu.edu/pancreas_panin).

different PanIN grades we applied microdissection, through which it is possible to analyze the biologically relevant cell type, often in the minority when analyzing precursor lesions. Different techniques have been developed to facilitate micro-dissection (Whetsell et al., 1992; Zhuang et al., 1995; Emmert-Buck et al., 1996; Schütze and Lahr, 1998). To select for PanIN grades, manual microdissection was chosen because we have found it easier and speedier than automatic processing (Sitek et al., 2005b).

Using CyDye DIGE saturation fluors we were able to combine the high resolution power of 2-DE with the high sensitivity of fluorescence imaging for the analysis of the scarce sample amount of 1000 microdissected cells (approximately 2 µg). In this preliminary report we have shown that the established protocol can successfully be applied to the proteome analysis of PanIN-2 grade as well as pancreatic ductal epithelium (Figure 3.9), resulting in the identification of the first molecular markers of PanIN-2 using LC-ESI-MS/MS (Table 3.3). Annexin A2 and annexin A4 have been found to be differentially expressed in PanIN-2 grades. These proteins are members of the annexin family of calcium-dependent phospholipid-binding proteins showing a 45–59% identity with other members of the annexin family (Hauptmann et al., 1989). Although their functions are still not clearly defined, several members of the annexin family have been implicated in membrane-related events along exocytotic and endocytotic pathways (Waisman, 1995). Furthermore, annexin A2 was found to be significantly up-regulated in PanIN-2 grades. It has been shown that the endothelial cell-surface Ca^{2+}-binding

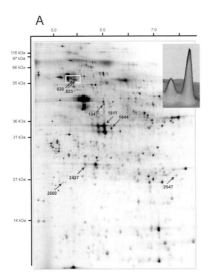
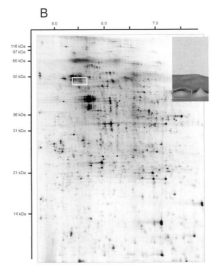
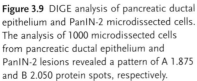

Figure 3.9 DIGE analysis of pancreatic ductal epithelium and PanIN-2 microdissected cells. The analysis of 1000 microdissected cells from pancreatic ductal epithelium and PanIN-2 lesions revealed a pattern of A 1.875 and B 2.050 protein spots, respectively.

Eight significantly regulated proteins are indicated by *arrows* and respective spot number. In the *zoomed sections* two differentially regulated proteins (*white box*) are displayed in 3D view obtained with the DeCyder software.

Table 3.3 First candidate molecular markers for PanIN-2 grade pancreas carcinoma analyzing 1000 microdissected cells.

Spot No.	Spot-name	Accession	Protein	Fold change	T-test	Seq. pI	Seq. MW (kDa)
1	1347	gi|15277503	Actin, gamma 1	4.0	0.017	5.6	40.2
2	2437	gi|18645167	Annexin A2	4.8	0.0008	7.4	38.6
3	1811	gi|4502105	Annexin A4	−4.5	0.0017	5.8	36.1
4	2660	gi|5453740	Myosin regulatory light chain MRCL3	−2.1	0.00031	4.6	19.8
5	820	–	Not identified	−2.65	0.0068	–	–
6	1844	–	Not identified	−13.37	0.0031	–	–
7	2547	gi|48255907	Transgelin	−4.5	0.0017	8.9	22.6
8	823	gi|5030431	Vimentin	−2.5	0.0055	4.8	41.6

protein, annexin A2, activates the t-PA-dependent formation of plasmin from plasminogen and promotes tumor cell invasion (Diaz et al., 2004). In the differential expression profiling of Lu et al., annexin 2A has also been found to be up-regulated in pancreatic carcinoma tissue (Lu et al., 2004).

Furthermore, a group of three actin-associated proteins down-regulated in PanIN-2 grades were found, whereas actin itself was detected as significantly up-regulated. For instance transgelin is an actin-binding protein of unknown function cross-linking actin filaments (Shapland et al., 1993). It has been shown that down-regulation of transgelin may be an important early event in tumor progression and it is considered a diagnostic marker for breast and colon cancer development (Shields et al., 2002). The role of the other actin-associated proteins is at the moment speculative and will be analyzed in more detail when the expression profiles of all PanIN grades and carcinoma samples are available.

Currently, there is very little information available concerning the initiation and prevention of pancreatic cancer. But the candidate protein identified applying DIGE saturation labeling for the analysis of PanIN grades will help to find new diagnostic markers for early detection of pancreatic cancer and to decipher processes within tumor biology. After analyzing all of the PanIN stages a widespread catalogue of protein expression in pancreatic tumor progression will be obtained, helping us to understand pancreatic tumor biology in more detail.

3.2.3
Statistical Aspects of Applying DIGE Proteome Analysis

For the identification of new molecular markers and interaction partners proteome analysis using 2-DE (DIGE) is often applied, but the deduced candidate proteins using the subtractive approach of a differential analysis need further validation.

For reasons of time and cost, the number of candidate proteins is limited and only significant proteins changes can be considered for candidate validation.

A DIGE experiment produces large amounts of data, or more exactly, volume values for thousands of gel spots. In order to make correct inferences from these data, statistical methods are quite important. Many of the statistical methods used in DNA microarray studies can be adapted for analyses of gel data. In this section we focus on the calibration and normalization of protein expression data as well as on the detection of differentially expressed proteins resulting from a DIGE experiment.

3.2.3.1 Calibration and Normalization of Protein Expression Data

From the DeCyder software it is possible to find the background-subtracted spot volumes, i.e., for a single spot the lowest 10th percentile pixel values on the spot boundary have been excluded. Several features of this raw data require calibration and normalization prior to statistical analysis. The volume values are affected by dye-specific system gains and constant additive biases. In Figure 3.10 the Cy5 and Cy3 spot volumes of one gel are plotted against each other. It can clearly be seen that the difference between Cy5 and Cy3 spot volumes increases when the spot volumes increase.

When using m gels for a DIGE study (with internal standard, treatment and control) and regarding each image of a gel as a distinct gel, one receives a data set with $j = 1, ..., 3\ m$ gels. To calibrate the spot volumes Karp et al. (2004) proposed the following statistical model for consideration:

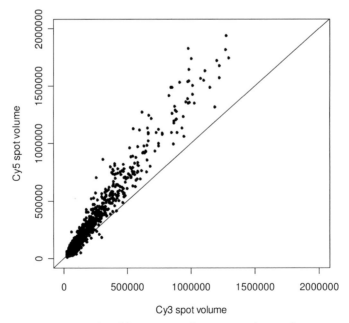

Figure 3.10 Scatterplot of the Cy5 versus the Cy3 spot volumes of a given minimal CyDye gel.

$$\tilde{x}_{ij} = a_j\, x_{ij} + b_{ij} \tag{1}$$

where $\tilde{x}_{ij}$ is the real expression value of the ith spot on the jth gel, x_{ij} is the respective measured spot volume, a_j the scaling factors that adjust for the dye-specific system gain and b_j additive offsets which compensate for the additive biases. These parameters can be estimated by maximum-likelihood estimation. The respective estimation algorithm and the calibration procedure have been implemented within the vsn-package for the open-source software R (available at http://cran.r-project.org) by Huber et al. (2002).

The resulting calibrated spot volumes are lognormal-distributed. Most statistical methods, however, are based on the assumption that the data are normal-distributed. Hence, the calibrated volume values must be normalized. A common method for normalization is to apply the logarithm to the data, but the logarithm has a singularity at zero and causes a bias in the Cy3/Cy5 ratios for small values. The mentioned software package uses the arsinh instead, which for high values is equivalent to the logarithm but has no singularity at zero and its smoothness for small values does not cause biases. The graphs of the logarithm and the arsinh can be compared in Figure 3.11. Applying the logarithm or the arsinh to the data also has the benefit that the variance of the data is stabilized. In the raw data, high volume values usually have a different variance than small volume values. In Figure 3.12 the effect on data normalization using arsinh transformation is shown.

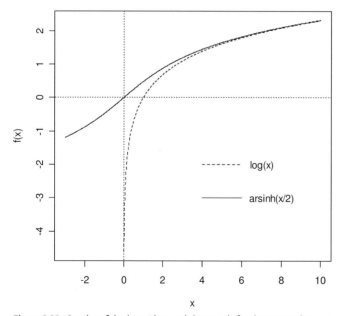

Figure 3.11 Graphs of the logarithm and the arsinh for data normalization.

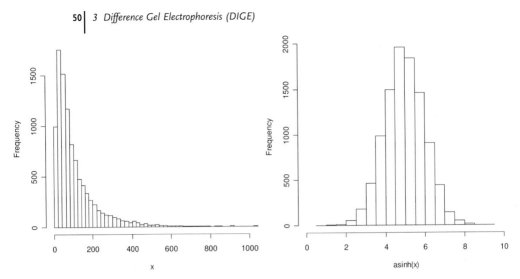

Figure 3.12 Lognormal-distributed volume values before normalization (*left*) and normal-distributed values after the arsinh-transformation (*right*).

3.2.3.2 Detection of Differentially Expressed Proteins

In DIGE experiments it is often of interest to find proteins with altered expression in two different mixtures, e.g., treatment and control. Having transformed the data by the logarithm or the arsinh one has to look for differential expression changes and not for relative changes, because ratios become differences when being transformed by the logarithm or the arsinh, i.e., because $\log(a/b) = \log(a) - \log(b)$.

If the volume values from treatment and control probes come, for example, from the same patient or, as in DIGE experiments, from the same gel, then there is a dependency between the resulting values. Hence, the *t*-test for paired samples, which has even a higher statistical power than the *t*-test for independent samples, can be used here. The term power is explained below. For the paired *t*-test the difference *d* between the treatment and control value is used.

For each single spot the *t*-test for paired samples tests the null hypothesis H_0, that there is no expression change for this spot, versus the alternative hypothesis, H_1 that there is an expression change. Based on the sample, the test decides either to accept H_0 or to reject it and accept H_1.

Be aware that the decision of a statistical test does not supply 100% certainty. A differentiation between the test decision and reality must always be made. In Table 3.4 the two kinds of errors that may occur are shown: a *Type I error* is to reject the null hypothesis when it is true, and a *Type II error* is to accept the null hypothesis when it is not true. The probability α for the Type I error is called the level of the test. The probability β for the Type II error depends on this α-level, the sample size, the expression change to be detected and the variance of the measured values. The probability of rejecting the null hypothesis is called the *power* of the test. It should be very small when the null hypothesis is true and very big when

Table 3.4 Possible results of a statistical test.

		Test decision	
		Protein not differenttially expressed	Protein differentially expressed
Reality	Protein not differentially expressed	Correct decision	Wrong decision (Type I error α)
	Protein differentially expressed	Wrong decision (Type II error β)	Correct decision

the alternative hypothesis is true. The higher the sample size for the experiment the better becomes the power.

Unfortunately it is not possible to keep α and β small at the same time. In practice it is common to specify a small α first, also called the testing level, and determine an appropriate number m of replications to keep β as small as possible afterwards.

Within the t-test the data is summarized in the test-statistic which is t-distributed under the assumption that the data for the treatment and control groups was obtained from normal distributions. The t-test is available in most statistical software packages.

3.2.3.3 Sample Size Determination

It is not easy to directly determine a number m of replicates to use for the t-test. Instead it is more convenient to look how the power of the test behaves when using a specified number of replicates. The power of the paired t-test depends (1) on the probability α for the Type I error, (2) on the standard deviation of the differences between the treatment and control values, (3) on the expression change to be detected and (4) on the number m of replicates which are used. Given these four parameters the power can be plotted as function of the expression change to be detected. In Figure 3.3 the power function is given for different numbers of replicates. On the lower abscissa the relative expression change is given, and on the upper, the differential expression change of the normalized values. From the power one can detect the probability β of the Type II error by the relationship $\beta = 1 -$ power. If in the case of Figure 3.13 one wants to detect a relative expression change of 1.5 then $\beta \approx 0.1$ when using five replicates. Hence, the strategy of finding the correct sample size is to specify the above-named four parameters, plot the power function and read the probability β for the Type II error from the graph. If β is too big, plot the power function for a higher number of replicates. Within the open source software R, the power function for the paired t-test can be plotted.

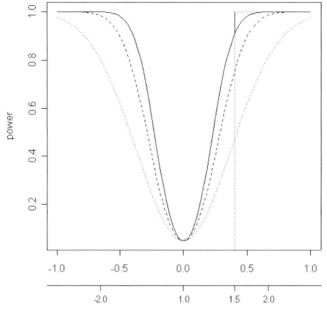

Figure 3.13 Power-function of the paired *t*-test using three replicates (*dotted line*), four replicates (*dashed line*) and five replicates (*solid line*) under the assumption that the standard deviation of the spots differences is s ≈ 0.2.

3.2.3.4 Further Applications

There are many other statistical models which can be used for the evaluation of DIGE studies. Inclusion of not only a group factor, but also a time factor in the experiment methods of the analysis of variance (ANOVA) can be applied to find expression changes within the temporal course of the protein expression or to find interactions between the group and time factor. Several multivariate statistical methods are of use, too. Spots with similar expression profiles can be grouped by cluster analysis or, on the other hand, new spots can be assigned to existing groups by the methods of discriminant analysis.

References

ALBAN, A., DAVID, S. O., BJORKESTEN, L., ANDERSSON, C., SLOGE, E., LEWIS, S., CURRIE, I. (2003). A novel experimental design for comparative two-dimensional gel analysis: two-dimensional difference gel electrophoresis incorporating a pooled internal standard. *Proteomics* **3**, 36–44.

ANDERSON, N. G., ANDERSON, N. L. (1996). Twenty years of two-dimensional electrophoresis: past, present and future. *Electrophoresis* **17**(3), 443–453.

BJELLQVIST, B., EK, K., RIGHETTI, P. G., GIANAZZA, E., GÖRG, A., WESTERMEIER, R., POSTEL, W. (1982). Isoelectric focusing in immobilized pH gradients: principle, methodology and some applications. *J. Biochem. Biophys. Methods* **6**(4), 317–339.

DIAZ, V. M., HURTADO, M., THOMSON, T. M., REVENTOS, J., PACIUCCI, R. (2004). Specific interaction of tissue-type plasminogen activator (t-PA) with annexin II on the membrane of pancreatic cancer cells activates plasminogen and promotes invasion in vitro. *Gut* **53**, 993–1000.

EGGERT, A., IKEGAKI, N., LIU, X.–G., CHOU, T. T., LEE, V. M., TROJANOWSKI, J. Q., BRODEUR, G. M. (**2000**). Molecular dissection of TrkA signal transduction pathways mediating differentiation in human neuroblastoma cells. *Oncogene* **19**, 2043–2051.

EGGERT, A., GROTZER, M. A., IKEGAKI, N., LIU, X. G., EVANS, A. E., BRODEUR, G. M. (**2002**). Expression of the neurotrophin receptor TrkA down-regulates expression and function of angiogenic stimulators in SH-SY5Y neuroblastoma cells. *Cancer Res.* **62**, 1802–1808.

EMMERT-BUCK, M. R., BONNER, R. F., SMITH, P. D., CHUAQUI, R. F., ZHUANG, Z., GOLDSTEIN, S. R., WEISS, R. A., LIOTTA, L. A. (**1996**). Laser capture microdissection. *Science* **274**, 998–1001.

FERNANDEZ-PATRON, C., CASTELLANOS-SERRA, L., HARDY, E., GUERRA, M., ESTEVEZ, E., MEHL, E., FRANK, R. W. (**1998**). Understanding the mechanism of the zinc-ion stains of biomacromolecules in electrophoresis gels: generalization of the reverse-staining technique. *Electrophoresis* **19**, 2398–2406.

GHARBI, S., GAFFNEY, P., YANG, A., ZVELEBIL, M. J., CRAMER, R., WATERFIELD, M. D., TIMMS, J. F. (**2002**). Evaluation of two-dimensional differential gel electrophoresis for proteomic expression analysis of a model breast cancer cell system. *Mol. Cell Proteomics* **1**, 91–98.

GÖRG, A., POSTEL, W., GUNTHER, S. (**1988**). The current state of two-dimensional electrophoresis with immobilized pH gradients. *Electrophoresis* **9**(9), 531–546.

GÖRG, A., OBERMAIER, C., BOGUTH, G., HARDER, A., SCHEIBE, B., WILDGRUBER, R., WEISS, W. (**2000**). The current state of two-dimensional electrophoresis with immobilized pH gradients. *Electrophoresis* **21**(6), 1037–1053.

HAUPTMANN, R., MAURER-FOGY, I., KRYSTEK, E., BODO, G., ANDREE, H., REUTELINGSPERGER, C. P. (**1989**). Vascular anticoagulant beta: a novel human Ca^{2+}/phospholipid binding protein that inhibits coagulation and phospholipase A2 activity. Its molecular cloning, expression and comparison with VAC-alpha. *Eur. J. Biochem.* **185**, 63–71.

HERBERT, B. (**1999**). Advances in protein solubilization for two-dimensional electrophoresis. *Electrophoresis* **20**(4–5), 660–663.

HEUKESHOVEN, J., DERNICK, R. (**1988**). Improved silver staining procedure for fast staining in PhastSystem Development Unit. I. Staining of sodium dodecyl sulfate gels. *Electrophoresis* **9**, 28–32.

HUBER, W., HEYDEBRECK, A., von SÜLTMANN, H., POUSTKA, A., VINGRON, M. (**2002**). Variance stabilization applied to microarray data calibration and to the quantification of differential expression. *Bioinformatics* **18**, S96–S104.

JOHNSTON, R. F., PICKETT, S. C., BARKER, D. L. (**1990**). Autoradiography using storage phosphor technology. *Electrophoresis* **11**, 355–360.

KARP, A. N., KREIL, D. P., LILLEY, K. S. (**2004**). Determining a significant change in protein expression with DeCyder during a pairwise comparison using two-dimensional difference gel electrophoresis. *Proteomics* **4**, 1421–1432.

KLOSE, J. (**1975**). Protein mapping by combined isoelectric focusing and electrophoresis of mouse tissues. A novel approach to testing for induced point mutations in mammals. *Humangenetik* **26**, 231–243.

KLOSE, J., KOBALZ, U. (**1995**). Two-dimensional electrophoresis of proteins: an updated protocol and implications for a functional analysis of the genome. *Electrophoresis* **16**, 1034–1059.

KNOWLES, M. R., CERVINO, S., SKYNNER, H. A., HUNT, S. P., DE FELIPE, C., SALIM, K., MENESES-LORENTE, G., MCALLISTER, G., GUEST, P. C. (**2003**). Multiplex proteomic analysis by two-dimensional differential in-gel electrophoresis. *Proteomics* **3**, 1162–1171.

LOPEZ, M. F., PATTON, W. F. (**1997**). Reproducibility of polypeptide spot positions in two-dimensional gels run using carrier ampholytes in the isoelectric focusing dimension. *Electrophoresis* **18**, 338–343.

Lu, Z., Hu, L., Evers, S., Chen, J., Shen, Y. (2004). Differential expression profiling of human pancreatic adenocarcinoma and healthy pancreatic tissue. *Proteomics* **4**, 3975–3988.

Miller, F. D., Kaplan, D. R. (2001). On Trk for retrograde signaling. *Neuron* **32**, 767–770.

Nakagawara, A., Arima-Nakagawara, M., Scavarda, N. J., Azar, C. G., Cantor, A. B., Brodeur, G. M. (1993). Association between high levels of expression of the TRK gene and favorable outcome in human neuroblastoma. *N. Engl. J. Med.* **328**, 847–854.

Nesterenko, M. V., Tilley, M., Upton, S. J. (1994). A simple modification of Blum's silver stain method allows for 30 minute detection of proteins in polyacrylamide gels. *J. Biochem. Biophys. Methods* **28**, 239–242.

Neuhoff, V., Stamm, R., Pardowitz, I., Arold, N., Ehrhardt, W., Taube, D. (1990). Essential problems in quantification of proteins following colloidal staining with coomassie brilliant blue dyes in polyacrylamide gels, and their solution. *Electrophoresis* **11**, 101–117.

Niederhuber, J. E., Brennan, M. F., Menck, H. R. (1995). The National Cancer Data Base report on pancreatic cancer. *Cancer* **76(9)**, 1671–1677.

O'Farrell, P. H. (1975). High resolution two-dimensional electrophoresis of proteins. *J. Biol. Chem.* **250**, 4007–4021.

Rabilloud, T. (2000). *Proteome research: two-dimensional gel electrophoresis and identification methods.* Springer, Berlin.

Rabilloud, T., Strub, J. M., Luche, S., van Dorsselaer, A., Lunardi, J. (2001). A comparison between Sypro Ruby and ruthenium II tris (bathophenanthroline disulfonate) as fluorescent stains for protein detection in gels. *Proteomics* **1**, 699–704.

Schulte, J. H., Schramm, A., Klein-Hitpass, L., Klenk, M., Wessels, H., Hauffa, B. P., Eils, J., Eils, R., Brodeur, G. M., Schweigerer, L., Havers, W., Eggert, A. (2005). Microarray analysis reveals differential gene expression patterns and regulation of single target genes contributing to the opposing phenotype of TrkA- and TrkB-expressing neuroblastomas. *Oncogene* **24(1)**, 165–177.

Schütze, K., Lahr, G. (1998). Identification of expressed genes by laser-mediated manipulation of single cells. *Nat. Biotechnol.* **16**, 737–742.

Shapland, C., Hsuan, J. J., Totty, N. F., Lawson, D. J. (1993). Purification and properties of transgelin: a transformation and shape change sensitive actin-gelling protein. *Cell Biol.* **121**, 1065–1073.

Shaw, J., Rowlinson, R., Nickson, J., Stone, T., Sweet, A., Williams, K., Tonge, R. (2003). Evaluation of saturation labeling two-dimensional difference gel electrophoresis fluorescent dyes. *Proteomics* **3**, 1181–1195.

Shields, J. M., Rogers-Graham, K., Der, C. J. (2002). Loss of transgelin in breast and colon tumors and in RIE-1 cells by Ras deregulation of gene expression through Raf-independent pathways. *J. Biol. Chem.* **277**, 9790–9799.

Sitek, B., Apostolov, O., Stühler, K., Pfeiffer, K., Meyer, H. E., Eggert, A., Schramm, A. (2005a). Identification of dynamic proteome changes upon ligand activation of Trk-receptors using two-dimensional fluorescence difference gel electrophoresis and mass spectrometry. *Mol. Cell Proteomics* **4**, 291–299.

Sitek, B., Lüttges, J., Marcus, K., Klöppel, G., Schmiegel, W., Meyer, H. E., Hahn, S. A., Stühler, K. (2005b). Application of DIGE saturation labeling for the analysis of microdissected precursor lesions of pancreatic adenocarcinoma. *Proteomics* **5**, in press.

Tonge, R., Shaw, J., Middleton, B., Rowlinson, R., Rayner, S., Young, J., Pognan, F., Hawkins, E., Currie, I., Davison, M. (2001). Validation and development of fluorescence two-dimensional differential gel electrophoresis proteomics technology. *Proteomics* **1**, 377–396.

Unlü, M., Morgan, M. E., Minden, J. S. (1997). Difference gel electrophoresis: a single gel method for detecting changes in protein extracts. *Electrophoresis* **18**, 2071–2077.

WAISMAN, D. M. (1995). Annexin II tetramer: structure and function. *Mol. Cell Biochem.* 149–150, 301–322.

WHETSELL, L., MAW, G., NADON, N., RINGER, D. P., SCHAEFER, F. V. (1992). Polymerase chain reaction microanalysis of tumors from stained histological slides. *Oncogene* 7, 2355–2361.

YAN, J. X., HARRY, R. A., SPIBEY, C., DUNN, M. J. (2000). Postelectrophoretic staining of proteins separated by two-dimensional gel electrophoresis using SYPRO dyes. *Electrophoresis* 21, 3657–3665.

YANO, M., OKANO, H. J., OKANO, H. (2005). Involvement of Hu and hnRNP K in neuronal differentiation through p21 mRNA posttranscriptional regulation. *J. Biol. Chem.* 280, 12690–12699.

ZHUANG, Z., BERTHEAU, P., EMMERT-BUCK, M. R., LIOTTA, L. A., GNARRA, J., LINEHAN, W. M., LUBENSKY, I. A. (1995). A micro-dissection technique for archival DNA analysis of specific cell populations in lesions < 1 mm in size. *Am. J. Pathol.* 146, 620–625.

4
Biological Mass Spectrometry:
Basics and Drug Discovery Related Approaches

Bettina Warscheid

Abstract

Rapid advances in biological mass spectrometry have promoted proteomics to being a key technology in molecular cell biology and biomedical research. Current mass spectrometry-based proteomics provides the capability for the identification of cellular and subcellular proteomes, the study of changes in protein concentrations via stable isotope labeling methods and the mapping of functional protein modules isolated by means of affinity purification. The large potential of proteomics techniques to identify and accurately quantify proteins from complex biological samples has been well recognized in the larger scientific community and it will in all probability have a great impact on future drug research.

4.1
Introduction

In 1981, Barber et al. demonstrated the analysis of a peptide of about 1300 Da in a mass spectrometer. To ionize the native peptide, the molecules were embedded in a glycerol matrix and bombarded with a high energy beam of argon atoms, a technique referred to as fast atom bombardment (FAB). Since ionization of hydrophobic peptides in the FAB process is favored compared to hydrophilic ones, the applicability of this technique to peptide mixture analysis is limited. Yet an avenue toward ionizing large biomolecules was apparent and encouraged the development of new ionization techniques, which also triggered the design of commercial mass spectrometers with expanded mass ranges.

In 1988, Karas and Hillenkamp embedded proteins in a large molar excess of a UV-absorbing crystal matrix and irradiated the sample with a laser beam at suitable wavelengths (Karas and Hillenkamp, 1988). In this process of matrix-assisted laser desorption/ionization (MALDI), large proteins can be transferred into the gas phase as intact molecules and become, for the most part, singly protonated. Since MALDI relies on the use of a pulsed laser, mass analysis is usually performed

Proteomics in Drug Research
Edited by M. Hamacher, K. Marcus, K. Stühler, A. van Hall, B. Warscheid, H. E. Meyer
Copyright © 2006 Wiley-VCH Verlag GmbH & Co. KGaA, Weinheim
ISBN: 3-527-31226-9

with a time-of-flight (TOF) analyzer of theoretically unlimited mass range. Since the mass spectrometer determines the mass-to-charge (m/z) ratio of the molecules, the ability to generate multiply charged ions would allow for the use of instruments with limited mass ranges in peptide and protein analysis. This was realized by Fenn and coworkers in the 1980s when they sprayed a diluted solution of protein in a high voltage electrostatic field gradient, generating multiprotonated species of intact proteins (Meng et al., 1988). Owing to this propensity of the electrospray ionization (ESI) process, quadrupole instruments with a mass range of m/z 2000 (where $z = 1$) could henceforth be used to analyze molecules with masses exceeding the nominal m/z range of the instrument by about 50 times.

The advent of both ESI and MALDI revolutionized the analysis of large biomolecules of low volatility such as peptides and proteins by their capability to form stable ions with little excess energy, enabling the determination of molecular weights even in protein mixtures. To obtain information specific to the primary structure of proteins, however, principles such as the activation of molecules via collisions with small neutral molecules, which have been used in the study of gaseous ion chemistry for decades, had to be adapted and helped to propel mass spectrometry to being of the most important tools in the field of proteomics.

4.2
Ionization Principles

MALDI and ESI represent the predominant ionization techniques in mass spectrometry-based proteomics, as recognized by the Nobel Prize in chemistry in 2002. MALDI is mainly used to volatize and ionize simple polypeptide samples for mass spectrometric (MS) analysis at high speed. The analysis of more complex peptide mixtures is usually conducted via ESI mass spectrometry (ESI MS) coupled online with a high-pressure liquid chromatography (HPLC) system to concentrate and separate peptides prior to MS analysis.

4.2.1
Matrix-Assisted Laser Desorption/Ionization (MALDI)

MALDI sample preparation takes place on a metal target by coprecipitation of the analyte molecules and matrix material present in a molar excess of 1000 to 10 000. Upon drying, the solid analyte-doped matrix sample is irradiated under high vacuum conditions by a nitrogen laser with a wavelength of 337 nm and nanosecond pulse-width. Organic molecules well suited to MALDI show high absorbance at the wavelength of the laser employed, and isolate the analyte molecules in a solid crystalline matrix to reduce their intermolecular interactions. For peptide analysis, matrices of α-cyano-4-hydroxycinnamic acid (HCCA) or 2,5-dihydrobenzoic acid (DHB) are preferred; the latter or sinnapinic acid (SA) are best applicable to protein analysis.

During the MALDI process, matrix crystals are excited and sublimated by laser irradiation. While the matrix plume is expanding, intact analyte molecules and matrix molecules are jointly transferred into the gas phase. The amount of energy transferred to the analyte molecules in this process depends on the physico-chemical properties of the matrix employed and may cause the unimolecular decay of the analyte molecules within a very short time frame. Generally, little internal energy is transferred from DHB to the analyte molecules with increasing energy transfer from SA to HCCA. The latter, which is considered a "hot" matrix, is normally the matrix of choice when analyzing peptides with MALDI–TOF MS on a microsecond time scale. In recent years, MALDI sources have also been successfully coupled to ion-trapping instruments. Extensive collisional cooling of MALDI ions is mandatory for their stabilization in trapping experiments on the millisecond time scale. Higher sensitivity is therefore obtained with the "cooler" DHB matrix than with HCCA.

MALDI favors the formation of singly charged ions, which makes the interpre-tation of mass spectra straightforward. There is evidence that the matrix molecules play a crucial role within the ionization process. Yet the origin of ions produced in MALDI and the mechanisms underlying ionization are still not fully understood. Among various proposed ionization pathways, analyte ionization by gas phase proton transfer reactions in the expanding plume with photoionized matrix molecules has been favored. Ionization efficiencies in MALDI experiments mainly depend on the matrix compound and the sample preparation technique employed. Uniform incorporation of peptides into a monolayer of matrix crystals generally improves ion signal intensities. Increasing analyte concentrations, however, often cause ion signal suppression. Furthermore, the primary structure of a peptide influences its ionization efficiency. Peptides containing arginine at their C-termini, for example, show higher ionization rates than peptides with a lysine residue at the C-terminal position (Krause et al., 1999).

Considering sample concentrations in the picomole to femtomole range on the target and the ablated sample volume during MALDI, it can be estimated that only a few attomoles of the analyte are needed for informative MS analysis. Nonetheless, high concentrations of salt, buffer, and detergents should be avoided as they impede analyte–matrix cocrystallization and cause ion suppression, both effects leading to a significant loss of sensitivity. High sensitivity and high mass resolution are obtained for peptide analysis in the mass range below 2500 Da by MALDI-MS. Mass analysis of peptide ions below 800 Da is impeded by the appearance of matrix ions and their abundant cluster ions in the MALDI spectra. Peak broadening and loss of sensitivity is usually observed for peptides and proteins at higher masses, due to unimolecular decay occurring to some extent during MALDI.

The speed of MALDI analysis depends on the laser pulse rate. With the recent introduction of 200-Hz lasers, samples can be analyzed ten times faster than before. This development is especially advantageous for offline liquid chromato-graphy (LC)/MALDI applications (Ericson et al., 2003). MALDI mass analysis is performed considerably faster than LC separation, allowing for chromatographic separation runs conducted in parallel. In an off-line LC/MALDI-MS experiment, complex peptide samples are separated by LC, and while the different peptide species are eluting from the LC column, they are deposited in discrete fractions of low complexity with the matrix of choice on the MALDI target. The automated sampling of the effluent on the MALDI plate in discrete spots is facilitated by the use of targets containing hydrophilic surface sites, which significantly reduces spot size and also improves spotting precision. Offline LC/MALDI samples can be stored and provide the option to perform multiple analyses at different instrumental parameter settings.

In recent years, two more approaches – MALDI imaging (Chaurand and Caprioli, 2002; Chaurand et al., 2002; Stoeckli et al., 2001) and surface-enhanced laser desorption/ionization (SELDI) (Hutchens, 1993; Petricoin et al., 2002a,b; Tang et al., 2004) – have emerged that focus on the discovery of polypeptide biomarkers abundant in the diseased state and either absent or only a minor presence in the healthy state. While the first method employs MALDI for molecular imaging of tissues or cells, SELDI is preferable to screen body fluids from patients by ion mass profiling combined with advanced statistical analysis. For MALDI imaging, a thin slice of, for example, tumor tissue is placed on the sample plate and coated with matrix. Then the laser scans over the surface to generate an ion map of the tissue. To identify molecular patterns diagnostic of the diseased state, images of ions with specific m/z ratios are reconstructed, revealing their distribution on the analyzed tissue slice. In SELDI, separation chemistry on the target is employed to enrich peptides from urine or serum samples, for example. SELDI has also been coupled to quadrupole TOF (qTOF) instruments, providing superior mass resolution and enabling a better distinction of ions similar in mass (Caputo et al., 2003; Guo et al., 2005; Tang et al., 2004).

Both strategies, however, face the analytical challenge of the high complexity of biological samples, resulting in ion suppression, limited reproducibility in MS data and a strong bias against low-abundance constituents as well as polypeptides with higher molecular masses. Various publications have demonstrated the applicability of these discovery-driven MS methods to screening biological material in short analysis times (Petricoin et al., 2002a; Petricoin et al., 2002b; Yanagisawa et al., 2003). However, a dependable differentiation between the diseased and normal states appears feasible only if a prominent, disease-distinctive set of biomarkers is detected with high reliability and adequate accuracy.

4.2.2
Electrospray Ionization

In ESI, the endergonic transfer of ions from solution into the gas phase is accomplished by desolvation. The electric field penetrates the analyte solution and separates positive and negative ions in an electrophoresis-like process. The positive charges accumulate on the surface of the droplets; when the surface tension is exceeded, the characteristic "Taylor cone" is formed and the microspray occurs (Wilm and Mann, 1994). At this point, the droplets are close to their stability limit (Rayleigh limit), which is determined by the Coulomb forces of the accumulated positive charges and the surface tension of the solvent. When the repelling Coulomb forces reach the cohesion forces due to solvent evaporation, nanometer-sized droplets are formed by a cascade of spontaneous divisions (Coulomb explosion). The ions still present in the liquid phase become completely desolvated during the transfer into the mass spectrometer by further solvent evaporation. If the electric field on the surface of the droplets is high enough, the direct desorption of bare ions occurs. During the whole process, little additional energy is transferred to the ions, which therefore remain stable in the gas phase. In comparison with MALDI, ESI represents an even softer ionization technique and causes no fragmentation of analyte ions.

Peptides and proteins subjected to ESI become protonated in the positive mode, while the number of multiple charges increases with the molecular weight of the biomolecules (Fenn et al., 1989). If the mass spectrometer provides sufficient mass resolution, the charge state of a compound can be directly derived from the difference of m/z values of its isotopes in the mass spectrum (0.5 Da mass differences for doubly charged peptides and about 0.3 Da mass differences for triply charged peptides). A major benefit of the generation of multiply charged ions from polypeptides is that they are measured at a proportional fraction of their molecular weight, and can therefore be readily analyzed with a less sophisticated instrument of limited mass range.

ESI takes place at atmospheric pressure. The ions produced are efficiently transmitted through a small orifice into the ion optics of the mass spectrometer. The desolvation of ions is assisted by nitrogen gas which streams from the interface into the region where ionization occurs. This curtain gas stream also hinders neutral molecules entering the high vacuum region. The ESI source basically consists of a capillary tube which continuously transfers the analyte solution into an electric field. The capillary also acts as counter-electrode to the interface plate; a potential difference of about 2–6 kV is typically applied to generate a stable spray with flow rates in the range of nano- to microliters per minute. Since the ion current of a given compound correlates with its concentration over a wide range and is rather independent of the liquid flow rate, ESI sources operating at nanoliter flow rates (<50 nL/min) have become exceedingly popular, drastically reducing sample consumption with an enhancement of sensitivity. The analytical properties of the nano ESI ion source and its great potential for proteomic research were described by Wilm et al. (Wilm and Mann, 1996; Wilm et al., 1996). This source basically consists of a very small, metal-coated glass capillary with an inner diameter of about 1 μm micrometer at the tip. When loading the capillary with a sample volume of 1 μL dissolved in 50 : 50 MeOH : H_2O, for example, it can be analyzed for more than 20 min without loss of sensitivity; ionization efficiency is increased by a factor of about 100 compared to conventional ESI. This setup enables the MS analysis of an enzymatic protein digest in the low femtomole concentration range (Wilm et al., 1996). Sensitivity in mass analysis can even be increased by coupling a nano ESI source with a high-performance liquid chromatography (HPLC) system operating at a flow rate of about 200 nL/min (Banks, 1996;

Mitulovic et al., 2003). An additional benefit of online HPLC/ESI-MS is that sample cleanup, analyte concentration and separation are accomplished in an automated process. However, polypeptides of high hydrophobicity are often retained on reversed-phase columns using a standard gradient and will therefore not be transferred into the MS device for mass analysis. Additionally, very hydrophobic peptides and proteins (e.g., membrane proteins) tend to precipitate in the small glass capillaries during the electrospray process, which leads to the breakup of the spray.

Another problem concerns the suppression of ion formation from an analyte caused by another analyte or by the presence of another constituent (e.g., buffer) in the solution. Since the MS response significantly depends on solvent and sample composition, ion signal intensities of a given analyte do not necessarily correlate with its concentration in the sample. For the quantitative analysis of an analyte of interest by ESI-MS (the same is true for MALDI-MS), the use of an adequate internal standard is therefore mandatory.

4.3
Mass Spectrometric Instrumentation

The promising use of various MALDI- and ESI-MS approaches in proteome research has accelerated many developments in the instrumental design and concepts of mass spectrometry. Novel MS techniques and intriguing combinations of existing instruments have emerged. Nonetheless, the basic principle of measuring the mass-to-charge ratios of analytes remains.

In general, it is mainly the process of ion formation that determines the mass analyzer used. MALDI is commonly coupled to TOF analyzers, while ESI is coupled to quadrupole or ion-trapping instruments (Jonsson, 2001; Mann et al., 2001). Yet novel MS techniques have been developed for advanced proteomic research which allow for the efficient transfer of MALDI or ESI ions into a mass analyzer almost independent of its working principle. Owing to the soft ionization characteristics of MALDI and ESI, mass-to-charge ratios of intact molecules can be readily measured. To obtain information on the primary structure of poly-peptides by MS, however, more sophisticated experiments within the MS device have to be performed. Amino acid sequence-specific data on polypeptides can be generated with mass spectrometers that allow for ion isolation, gas phase peptide fragmentation, and fragment ion analysis. Sequence-specific fragmentation of isolated peptide ions by collision-induced dissociation (CID) requires the combined use of two mass analyzers with the same or two different ion separation principles, commonly referred to as tandem MS instruments. Exceptions are ion trap instruments and MALDI–TOF MS. The first allows for the performance of multiple consecutive cycles of peptide ion fragmentation (MS^n) in a single-stage device (Jonscher and Yates, 1997), whereas in the latter peptide ions undergo unimolecular decay in the field-free drift tube commonly referred to as postsource decay (PSD) (Spengler, 1997).

In this chapter, a description of the working principles of TOF as well as TOF/TOF and qTOF analyzers is provided. The latter two, illustrated in Figure 4.1, are hybrid instruments and have already had a big impact on proteomic research.

The TOF analyzer is a very robust, simple, and fast instrument, which justifies its widespread use in proteomics research. In this technique, a set of ions is accelerated from the ion source into the field-free drift tube. Since the same constant acceleration voltage is applied, the flight-time of ions down the tube is determined by their kinetic energy and m/z ratio. Generally, the flight-time of ions at a given charge state is inversely proportional to their molecular mass, with smaller ions arriving earlier at the detector than higher mass ions. However, small differences in the kinetic energy of ions with the same m/z ratio (and therefore in the velocity of these ions) can cause a substantial deterioration of mass resolution in TOF MS analysis. To effectively correct for the initial spatial and kinetic energy distribution of MALDI ions (and therefore to improve mass resolution), the concepts of delayed extraction and reflectron TOF have successfully been established in current MALDI–TOF instruments. In proteomics, masses of intact peptides from a protein digest are normally measured by MALDI–TOF to acquire a characteristic peptide mass fingerprint (PMF) spectrum. To obtain sequence information on peptide ions with a single stage TOF analyzer, MALDI–PSD analysis is performed in the reflectron ion mode. To increase the rate of unimolecular decay in PSD experiments, the sample is ionized at increased laser energies. As a direct consequence, a decrease in mass resolution is observed in MALDI fragmentation spectra. Further problems involved in MALDI–PSD analysis relate to the limited ability to select precursor ions within a narrow mass window and to general constraints in peptide ion fragmentation by unimolecular decay.

These limitations were addressed by the development of the MALDI–TOF/TOF mass spectrometer (Medzihradszky et al., 2000) (Figure 4.1a). In this tandem MS instrument, precursor ion selection is performed in the first TOF analyzer, whereas fragment ion mass analysis takes place in the second reflector–TOF instrument. A collision cell is placed between the two mass analyzers and allows for true MS/MS analysis of isolated MALDI ions by CID (Figure 4.1a). To retain the capability of high-throughput peptide mass analysis, ions are generated with a fast pulsing laser. However, MS/MS analysis with a MALDI–TOF/TOF instrument is not significantly faster than with other types of tandem MS instruments such as ion traps.

In addition to this vital improvement in MALDI–TOF MS, the concept of orthogonal ion acceleration had great impact on the development of a novel instrumental configuration, the qTOF analyzer (Krutchinsky et al., 2000; Loboda et al., 2000) (Figure 4.1b). In this instrumental design, the ion source is decoupled from the TOF analyzer by extensive gas collisions in the quadrupole stages and well defined ion packages are accelerated perpendicularly into the TOF analyzer. MALDI and ESI ion sources can be interchangeably coupled to the qTOF instrument. For MS/MS analysis, ions of a particular m/z ratio are selected in the Q1 quadrupole and fragmented by collisional activation with nitrogen or argon atoms in the collision cell (Q2). The masses of fragment ions generated are then

a) MALDI-TOF/TOF

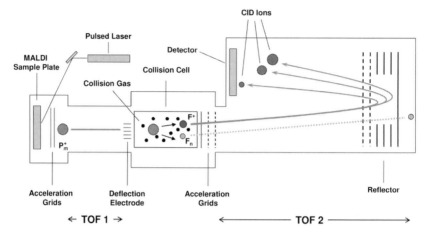

b) MALDI- or ESI-Quadrupole TOF

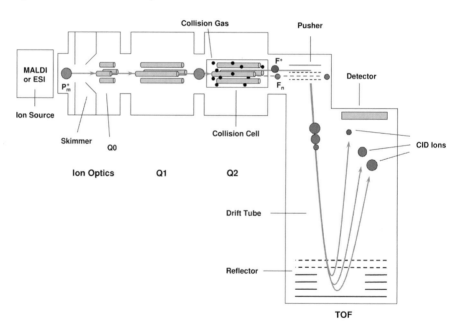

Figure 4.1 Schematic representations of a time-of-flight/time-of-flight (TOF/TOF) instrument with a matrix-assisted laser desorption/ionization (MALDI) source (**a**), and a quadrupole time-of-flight (qTOF) instrument which can be interchangeably coupled with a MALDI or electrospray ionization (ESI) source (**b**). In both instrumental setups, peptide ions can be fragmented via collision-induced dissociation (CID) in order to obtain specific information on peptide amino acid sequences.

analyzed by the reflector TOF analyzer (Figure 4.1b). Peptides at the low femtomole level can be measured with high mass resolution of about 10 000 full width at half maximum (FWHM) and with high mass accuracy of about 10 ppm in both MS and MS/MS modes of this instrument. Its usefulness in proteomic research was early demonstrated (Shevchenko et al., 2000). In current proteomics, the TOF and hybrid TOF instruments as well as three-dimensional ion trap analyzers are predominantly used. Regarding new types of instruments, linear ion traps either as stand-alone devices (Schwartz et al., 2002) or in combination with two quadrupoles (Le Blanc et al., 2003) or Fourier transform ion cyclotron resonance (FTICR)-MS (Wilcox et al., 2002) will certainly take a leading role in future proteomic research.

4.4
Protein Identification Strategies

In proteomics, two main strategies are followed for the identification of proteins in complex samples as illustrated in Figure 4.2. The first employs two-dimensional polyacrylamide gel electrophoresis (2-D PAGE) to greatly reduce the complexity of biological samples prior to peptide MS analysis. In the second strategy, limited protein separation by one-dimensional gel electrophoresis (1-D PAGE) is performed and then combined with online nano HPLC/ESI–tandem MS analysis of peptides in an automated fashion. Alternatively, protein mixtures can be separated by LC [e.g., strong cation exchange (SCX) chromatography]. In both strategic tracks, proteins are converted to a set of peptides by enzymatic digestion prior to MS analysis. Although the sample complexity is increased again by protein digestion, the main reasons to subject peptides instead of proteins to MS analysis are based on the facts that (1) MS sensitivity is best in the nominal mass range below 2500 Da,

---

Figure 4.1

a In the MALDI–TOF/TOF analyzer, ions are accelerated to high kinetic energy and separated according to their mass-to-charge (m/z) ratios along their flight in the second reflectron TOF analyzer (TOF2) and subsequently detected. For mass spectrometric/mass spectrometric (MS/MS) analysis, peptide ions of specific m/z ratios are selected in the first TOF analyzer (TOF1), collisionally activated and fragmented in the collision cell. The fragment ions (F^+) generated are subsequently separated in the TOF2 section followed by detection; neutral fragments (F_n) are lost.

b The qTOF instrument combines three quadrupole sections (Q0, Q1, and Q2) with a reflector TOF analyzer for mass analysis of ESI or MALDI ions. Q0 is part of the ion optics for focusing of ions generated by MALDI or ESI. Peptide ions of interest are isolated in Q1 and fragmented by CID in Q2, which is used as collision cell. Fragment ions (F^+) formed are then perpendicularly accelerated into the reflectron TOF analyzer to acquire the corresponding MS/MS spectrum. For extensive fragmentation of MALDI peptide ions, argon is generally used as collision gas irrespective of the instrumental design. Peptide ions formed by ESI can be readily fragmented with nitrogen or argon as collision gas in qTOF MS/MS experiments.

(2) informative sequence information is most readily obtained from peptides of up to 20 amino acid residues via MS/MS and (3) peptides are more soluble than proteins and therefore easier to handle, to chromatographically separate and to electrospray.

Visualized protein spots in a 2-D gel are cut out and separately digested in-gel with a sequence-specific protease (e.g., trypsin). Subsequently, the molecular masses of the proteolytic peptide mixture generated from a single protein spot are analyzed by MALDI coupled with a TOF, TOF/TOF, qTOF or high-resolution FTICR instrument to acquire the corresponding PMF spectrum (Figure 4.2). In this method (also termed peptide mass mapping), the identity of a precursor protein can be determined using bioinformatic tools by matching the experimental peptide masses with lists of *in-silico* generated peptide masses of proteins cataloged in a database. The specificity and therefore the probability of protein identification with high reliability can be improved by sequence analysis of selected peptides by MALDI-MS/MS with a TOF/TOF or qTOF instrument. This also provides the possibility of coupling LC separation of complex proteolytic peptide mixtures with MALDI–TOF MS analysis (offline LC/MALDI).

If a protein spot cannot successfully be identified by MALDI analysis, peptide sequencing by automated nano HPLC/ESI MS/MS is an alternative. This provides superior sensitivity due to the preconcentration of peptides on trapping columns (Mitulovic et al., 2003) and delivers considerably more sequence-specific information at the peptide level. However, nano HPLC coupled online with ESI-tandem MS analysis is much more time-consuming than MALDI–PMF. Therefore, it is predominantly used to efficiently identify complex protein samples which have only been separated partially, e.g., by 1-D PAGE. This case represents the

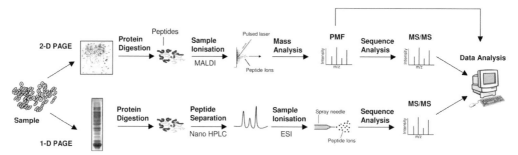

Figure 4.2 Schematic overview of two protein identification strategies commonly followed in proteomics. Protein samples are separated by either two-dimensional (2-D) or one-dimensional (1-D) polyacrylamide gel electrophoresis (PAGE). In both strategic tracks, proteins are converted into a set of peptides by enzymatic digestion (e.g., with trypsin) prior to MS analysis. Peptide mass fingerprinting (PMF) by MALDI MS is predomi-nantly used to identify proteins separated by 2-D PAGE. In contrast, protein mixtures only partially separated by 1-D PAGE are analyzed by nano high-performance liquid chromatography (nano HPLC) coupled online to ESI tandem MS (ESI MS/MS). For protein identification, peptide mass maps or sequence specific data obtained are used to search against databases via bioinformatic tools (see text for further details on both strategies).

second protein identification strategy illustrated in Figure 4.2. Visualized protein bands are excised, followed by protein digestion and separation of the proteolytically generated peptide mixtures by nano HPLC. While peptides are eluting from the chromatographic column, they are directly transferred into the nano ESI ion source via a fused silica capillary to become ionized and analyzed by MS/MS-capable instruments (e.g., ion traps or qTOF). Fragmentation of peptides is typically performed at low energy CID conditions in order to induce cleavages of the amide bonds between the amino acid residues (for a detailed review about peptide sequencing see Steen and Mann, 2004). An alternative strategy, known as the shotgun approach, is to initially digest a protein mixture with a sequence-specific enzyme and subsequently to subject the highly complex proteolytic peptide mixture to 2-D chromatographic separations (e.g., SCX/reversed phase) coupled online to ESI-MS/MS (Link et al., 1999; Opiteck et al., 1997; Washburn et al., 2001; Wolters et al., 2001). In all these cases, algorithms such as Sequest or Mascot are used for large-scale protein identification by searching sequence databases using the MS/MS data obtained in the experiments. However, certain criteria have to be applied to ensure high reliability of the identification of proteins using bioinformatics tools. It should be noted that generally only a fraction of an entire protein sequence is identified, which usually suffices for identification but does not allow the complete characterization of an entire protein. More sophisticated strategies are necessary to detect posttranslationally modified and/or N- or C-terminally processed protein sequences.

4.5
Quantitative Mass Spectrometry for Comparative and Functional Proteomics

Comparative MS-based proteomics has emerged as a promising tool in the field of drug discovery. To reveal potential drug targets or the effects of drugs on biological systems, a set of diseased and control, or perturbed and nonperturbed samples is compared. Quantitative protein analysis represents a targeted proteomic approach as it aims at identifying only those proteins which show altered expression profiles or undergo changes in the relative degree of posttranslational modification. In addition, new strategies have been developed to harness quantitative MS for the study of protein–protein interactions as discussed later in this chapter. Since the latter approach allows the placing of proteins of interest into a functional context, it in particular offers great potential for the discovery of new targets in drug research.

In quantitative proteomics, two alternative strategies have been developed. The first one is based on 2-D PAGE combined with mass spectrometry for protein identification (Haynes and Yates, 2000). This method and current advances in the differential display of proteins by 2-D PAGE are discussed at length in another chapter of this book. The second approach exploits mass spectrometry in combination with stable isotope labeling for gaining accurate quantitative information on proteins. Quantitative MS via stable isotope labeling of proteins

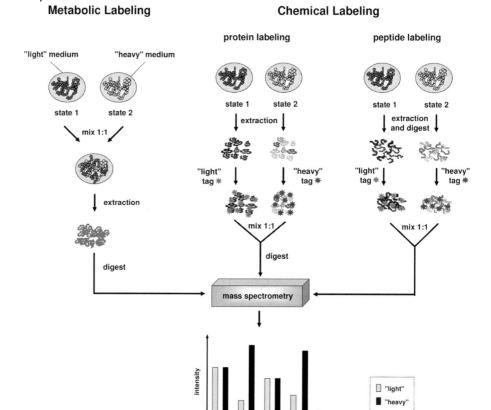

Figure 4.3 Schematic representation of different stable isotope labeling strategies used for determination of protein concentration ratios by MS. Metabolic labeling: *heavy* isotopes are incorporated *in situ* into the proteome of a cell population (state 2) by using cell culture medium enriched in ^{15}N atoms (> 96%) or containing stably isotopically labeled amino acids (e.g., $^{13}C_6$-arginine). A second cell population (state1) is grown in normal medium (*light*), which is used as control for relative quantitation experiments. Light- and heavy-labeled cell populations are immediately mixed in a 1 : 1 ratio after cell harvest. Subsequently, cells are jointly lysed, proteins are extracted and subjected to any further purification and/or separation steps. Alternatively, cells from two conditions can be initially lysed and then combined to be further processed. The proteins are then digested with a sequence-specific protease and proteolytic peptide mixtures generated are subjected to MS analysis. Peptide ratios can be obtained from MS spectra, which eventually allows relative quantification of protein concentrations in two different cell states. Chemical labeling: proteins or peptides extracted from cells of two different states are differentially labeled via chemical reaction with stably isotopically encoded tags (light and heavy). Most tagging reagents specifically react with cysteine residues or primary amino groups in proteins or peptides. Once the label is introduced, samples from two conditions can be mixed in a 1 : 1 ratio, enzymatically digested and relatively quantified by MS analysis. Labeling at the protein level provides the investigator with the possibility of jointly separating differentially labeled samples by gel electrophoresis or liquid chromatography. This greatly reduces the sample complexity prior to digestion, which facilitates relative quantitation via MS analysis.

or peptides represents a universal tool applicable to samples derived from cell lines or tissues. This strategy has also provided shotgun methods capable of quantitation and identification in the same experiment (Wolters et al., 2001; Washburn et al., 2001, 2002, 2003). Typically, a differentially labeled set of samples is combined in a 1 : 1 ratio prior to MS analysis. Relative quantitative information on proteins is obtained from mass spectra of isotope-coded peptides. Since MS is not an absolutely quantitative method, the determination of the absolute amount of an analyte is only achievable with respect to an adequate internal standard. To this end, a stable isotope-coded peptide representing the endogenous peptide is synthesized in its heavy form and used as internal standard that is added to the sample in a defined quantity (Desiderio and Zhu, 1998). MS measurement of the ratio of endogenous to synthetic peptide then allows for absolute quantitation of the former. This method is also applicable to the quantitative determination of the degree of posttranslational modification in a protein of interest (Gerber et al., 2003). Peptide synthesis is time-consuming, though, and instead of studying a protein that is already known, the aim is often to quantify unknown proteins in a complex biological system. This idea triggered the development of several stable isotope labeling techniques for comparative proteomic studies, using ^{2}H, ^{13}C, ^{15}N or ^{18}O isotopes (Aebersold et al., 2000; Flory et al., 2002; Goshe and Smith, 2003; Lill, 2003; Tao and Aebersold, 2003). Differentially labeled samples are simultaneously analyzed by MS and quantitative information is derived by comparing ion signal intensities or peak areas of peptide pairs observed in the corresponding mass spectra. A benefit of using ^{13}C, ^{15}N or ^{18}O isotopes is that the peptides present in the "heavy" isotopic forms do not exhibit significant isotope-dependent chromatographic shifts from their "light" counterparts, thus improving the precision for quantifying peptide abundances by LC/ESI-MS.

Stable isotope-enriched media can be used to directly impart a specific isotope signature to proteins without structural changes. Alternatively, isotope-coded tags can be introduced at the peptide or protein level depending on the labeling method applied. Therefore, current stable isotope labeling methods fall in two major categories: metabolic (*in vivo*) labeling and chemical (*in vitro*) labeling. Figure 4.3 shows an overview of the different approaches to relative quantitative analysis of proteins by stable isotope labeling and MS (see text below for details). Irrespective of the method chosen, the complete labeling of proteins ($\geq$ 95%) is a prerequisite to obtaining quantitative information on proteins with high accuracy.

4.6
Metabolic Labeling Approaches

Metabolic labeling is an *in situ* method which employs ^{15}N or isotope-encoded (^{13}C or ^{2}H) amino acids. Organisms are grown in controlled media under defined conditions in order to ensure translation of mRNA entities to uniformly isotopically encoded proteins. Consequently, the approach is generally restricted to cell culture systems and is not applicable to quantitative MS analysis of proteins derived from body fluids or tissue biopsy. Since the internal standard is generated *in situ*, differentially labeled cells can be pooled directly after harvest, which allows for consistent sample processing, assuring accurate protein quantitation. In addition, metabolic labeling provides the investigator with the possibility of effectively reducing the complexity of biological samples at the protein level via gel electrophoresis or liquid chromatography. Separation prior to enzymatic digestion is advantageous as it generally improves qualitative as well as quantitative information on complex protein mixtures obtained by MS analysis.

4.6.1
^{15}N Labeling

In 1999, the concept of using ^{14}N/^{15}N-containing media for relative quantitative MS analysis of proteins in yeast was introduced (Oda et al., 1999). In this work, wild-type versus mutant yeast cells were compared. Identification and quantitation of individual proteins were performed simultaneously by MS analysis. The reported approach also provided information on the level of phosphorylation of the protein kinase Ste20. In the same year, Pasa-Tolic et al. demonstrated quantitative profiling of *Escherichia coli* proteins on a large scale via *in vivo* stable isotope labeling combined with intact mass tag analysis by high-resolution FTICR-MS (Pasa-Tolic et al., 1999). The idea of using accurate mass tags for high-throughput quantitation was further exploited for the analysis of whole cell lysate from two populations of bacteria, one grown in normal medium and one in medium enriched in ^{15}N (> 96%) (Conrads et al., 2001; Smith et al., 2001, 2002). To simplify quantitative profiling of proteolytic peptide mixtures from ^{14}N/^{15}N-encoded proteins in whole cell lysates, Conrads et al. specifically enriched cysteine-containing peptides via a thiol-reactive affinity tag prior to FTICR-MS analysis (Conrads et al., 2001). However, a general drawback of cysteine tagging methods [such as the isotope-coded affinity tag (ICAT) technology discussed later in this chapter] is that a large protein fraction does not contain cysteine residues at all and often the cysteine content of the remaining fraction is low. The probability of analyzing a suitable set of isotopically encoded peptides to infer accurate quantitative information on precursor proteins is therefore significantly reduced.

With increasing numbers of growth cycles, endogenous proteins become enriched in ^{15}N atoms when cells are cultured in ^{15}N media; eventually all ^{14}N atoms are replaced by ^{15}N. The corresponding mass shift of ^{14}N/^{15}N-labeled proteins or peptides can only be predicted if the amino acid sequences of the proteins are known. This certainly complicates quantitative evaluation of the mass spectral data acquired. Therefore, the use of high-resolution mass spectrometry, such as FTICR, has been recommended (Conrads et al., 2001). However, a suitable algorithm can predict the expected mass shift of peptides or proteins based on the proteome of the organism studied. ^{15}N- and ^{14}N-labeled peptides can easily be distinguished by their isotope distributions observed in mass spectra. Owing to the use of ^{15}N media of > 96% purity and therefore incomplete replacement of ^{14}N, the ^{15}N-labeled peptides exhibit additional isotope peaks in mass spectra.

2-D PAGE is often employed for high-resolution separation of ^{14}N/^{15}N-labeled cell lysates followed by MS-based identification and quantitation of distinct proteins. Alternatively, the applicability of a shotgun approach to determine concentration ratios of ^{14}N/^{15}N-encoded proteins from yeast on a large scale has been demonstrated (Washburn et al., 2002, 2003). To support data analysis in shotgun proteomics, a correlation algorithm for automated protein quantitation has been developed (MacCoss et al., 2003). In another approach for gel-free proteomics, Wang and coworkers reported an inverse labeling strategy to simplify protein quantitation by ^{15}N combined with MS (Wang et al., 2002). In this experimental approach, both perturbed and control samples are independently cultured in ^{14}N- and ^{15}N-enriched media, generating a set of four sample pools as opposed to two in a conventional labeling experiment. Differentially labeled control and perturbed samples are combined, processed, and analyzed by mass spectrometry. Peptide mass spectra obtained were then compared by subtractive analysis to reveal only those peptide pairs which show altered concentrations. This strategy provides confidence in detecting changes in protein concentrations. However, the workload and costs are doubled, so its necessity must be determined on a case by case situation.

In situ ^{15}N labeling has successfully been applied to studying bacteria, yeast, and some mammalian cells at moderate costs. Metabolic labeling of *Caenorhabditis elegans* by feeding ^{15}N-labeled *E. coli* cells (Krijgsveld et al., 2003) or ^{15}N labeling of potato plants for structural protein analysis (Ippel et al., 2004) have also been reported. However, *in vivo* labeling of higher eukaryotic organisms is expensive and may significantly affect growth and protein expression profiles. A detailed study of the effects of ^{15}N labeling on these systems has not yet been reported.

4.6.2
Stable Isotope Labeling by Amino Acids in Cell Culture (SILAC)

Another very elegant *in situ* method for accurate quantitative assessment of protein or peptide abundances between two given cell lines is the stable isotope labeling by amino acids in cell culture (SILAC) (Ong et al., 2002). Mann and coworkers grew mammalian cell lines in media lacking the essential amino acid leucine, but supplemented with its nonradioactive, deuterated form (^{2}H$_3$-Leu) (Ong et al., 2002). Analysis of cell morphology showed that ^{2}H$_3$-Leu caused no differences in the growth of the cells. After five cell cycles, proteins were completely labeled with ^{2}H$_3$-Leu. Jiang and English reported metabolic labeling of a yeast strain auxotroph for leucine that was grown on synthetic complete media containing natural abundance Leu or ^{2}H$_{10}$-Leu (Jiang and English, 2002). Before this concept was harnessed for quantitative proteomics, incorporation of ^{13}C/^{15}N/^{2}H-encoded amino acids during cell culture was utilized to increase confidence in mass fingerprint database searches (Chen et al., 2000).

To prevent chromatographic differences in peptides containing ^{1}H/^{2}H isotopes, the SILAC approach was successfully extended to the use of ^{13}C$_6$-arginine (Ong et al., 2003), improving the accuracy of quantitation. When trypsin is used for protein digestion (note that trypsin cleaves proteins on the carboxy-terminal sides of arginine and lysine residues) in a $^{13}C_6$-arginine experiment, the label is localized at the C-terminus of arginine-containing peptides, while lysine peptides are unlabeled and therefore do not provide quantitative information in peptide mass spectra. Fenselau and coworkers successfully used metabolic labeling with $^{13}C_6$-arginine and $^{13}C_6$-lysine for comparative proteomics in drug-resistant and parental breast cancer cell lines (Gehrmann et al., 2004). This double isotope labeling approach was previously used to quantify changes in protein phosphorylation in resting versus activated human embryonic kidney cells (Ibarrola et al., 2003). Since both $^{13}C_6$-lysine and $^{13}C_6$-arginine were employed during culturing, tryptic digestion of proteins resulted in proteolytic peptide mixtures in which all peptides were labeled at the C-terminus, except for the original C-terminus of the proteins. This is certainly a considerable improvement to the SILAC approach, since it is essential to obtain an adequate set of peptide pairs from each protein for accurate relative quantitation, although there may still be variability from protein to protein.

It has been demonstrated very recently that SILAC is applicable to multiplex experiments in order to measure dynamics of protein abundances in cells in response to stimuli such as growth factors or drugs (Molina et al., 2005). Henrietta Lacks (HeLa) cells were grown in normal medium or in media lacking normal arginine and supplemented with either $^{13}C_6{}^{14}N_4$- or $^{13}C_6{}^{15}N_4$-arginine. After harvesting the cells, differentially labeled cell populations were pooled in a 1 : 1 : 1 ratio for combined sample processing followed by triplicate LC/ESI-MS/MS peptide analyses to test for technical reproducibility. Interestingly, standard deviations obtained in quantitative measurements do vary significantly between proteins, even though the same number of peptide pairs was evaluated. Nevertheless, it was shown that metabolic labeling allows for reliable measurements of alterations in protein abundances of 1.5- to 2-fold in a single biological experiment. To account for biological variations in protein expression profiles, at least three independent biological experiments must be performed.

The SILAC approach has also been used to investigate metastatic prostate cancer development at the protein level (Everley et al., 2004). The fact that proteins showed altered concentration ratios by quantitative MS was confirmed by western blotting. In addition, proteomic approaches for quantitation of protein phosphorylation via SILAC combined with MS analysis have been described (Gruhler et al., 2005; Ibarrola et al., 2003, 2004). A recent study reports on identification as well as relative quantitation of *in vivo* methylation sites of proteins in HeLa cells by stable isotope labeling with $^{13}C^2H_3$-methionine (Ong et al., 2004).

As shown by the examples given above, SILAC is a very flexible, simple, and accurate procedure for relative quantitation that provides deeper insight into biological systems based on cell culture.

4.7
Chemical Labeling Approaches

Chemical reactions are used to introduce isotope-coded tags into specific sites in proteins and peptides after sample collection, i.e., following tissue biopsy or cell harvest. *In vitro* labeling of biomolecules has been achieved in a variety of ways, including enzymatically directed labeling and incorporation of isotope tags prior to or following protein digestion (Figure 4.3).

4.7.1
Chemical Isotope Labeling at the Protein Level

Gygi et al. described an approach for accurate quantitation and concurrent identification of individual proteins within complex mixtures from yeast using chemical reagents termed isotope-coded affinity tags (ICATs) and tandem MS (Gygi et al., 1999). The ICAT reagent consists of a iodoacetyl moiety reacting with cysteine residues, an isotope-encoded linker ($^1H_8$ or $^2H_8$) and a biotin tag. In this method, a set of two samples is differentially tagged and combined in a 1 : 1 ratio, the proteins are separated and then are enzymatically digested. The major innovation here was the selective enrichment of cysteine-containing peptides by the affinity tag, reducing the sample complexity by a factor of about 10 prior to MS analysis. Since the ICAT-encoded peptide pairs differ by at least 8 Da, this labeling method is applicable to MS analysis with low resolution instruments such as ion traps.

The ICAT approach promised to be a widely applicable tool for comparative proteomics of cells and tissues. However, it fails for proteins which are cysteine-free and often the cysteine content of proteins is rather low, rendering them unavailable for accurate quantitative MS analysis. Following this first report on the ICAT strategy (Gygi et al., 1999), a number of studies have further conveyed the ICAT idea in quantitative proteomic research (Griffin et al., 2001; Han et al., 2001; Smolka et al., 2002; Tao and Aebersold, 2003). However, there are several drawbacks in the ICAT strategy: (1) ICAT reagents are relatively large (about 500 Da) and their addition to proteins may be sterically hindered; (2) prolonged reaction times are often needed to achieve complete incorporation of ICAT tags into proteins which may lead to partial derivatization of lysine, histidine, methionine, tryptophan, and tyrosine residues; and (3) the chemical stability of the thioether linkage towards molecular oxygen is low, which may result in uncontrolled partial cleavage of the label via β-elimination. Smolka et al. systematically evaluated the ICAT method and addressed the need for optimization of labeling conditions (Smolka et al., 2001). However, even if complete labeling of samples can be achieved, the molecular masses of peptides bearing the ICAT label are significantly increased and the analysis of the corresponding MS/MS spectra (especially for small peptides of fewer than ten amino acids) is complicated due to additional fragmentation of the affinity tag. Nevertheless, improvements in ICAT technology have been made (Tao and Aebersold, 2003), such as the development of a photocleavable ICAT reagent bound to a solid support, thus eliminating the large biotin tag (Zhou et al., 2002, 2004), and the introduction of a modified version of the ICAT reagent that uses acid-labile isotope-coded extractants (ALICE) (Qiu et al., 2002). Finally, ^{13}C-containing ICAT reagents have been introduced to ensure coelution of peptides during LC separation. This new ICAT generation has been applied to the analysis of a cortical neuron proteome sample to identify proteins regulated by the antitumor drug camptothecin (Yu et al., 2004).

In 2003, a stable isotope labeling similar to ICAT was reported utilizing a membrane-impermeable biotinylating reagent, which contains a cleavable linker and specifically reacts with primary amino groups, i.e., the ε-amino group of lysine residues and the free amino termini of isolated intact proteins (Hoang et al., 2003). The capability of this reagent, available in the light and heavy forms for labeling and enrichment of lysine-containing peptides followed by relative quantitative assessment, was shown for two standard proteins. Labeling of proteins was performed under nondenaturing conditions which resulted in incomplete incorporation of the label. Certainly, this considerably limits the reproducibility and accuracy of the method and therefore its usefulness in quantitative proteomics. However, the authors proposed that the method designed would allow for targeting of cell surface proteins (without the need for subcellular fractionation) which could be useful in drug research (Hoang et al., 2003).

Recently, Lottspeich and coworkers reported an alternative approach, termed isotope-coded protein label (ICPL), which is also specific to primary amino groups (Schmidt et al., 2005). Since ICPL is based on stable isotope tagging at the protein level, it is applicable to any protein sample, including extracts from tissues or body fluids. The ICPL tag is an isotope-coded *N*-nicotinoyloxy-succinimide (Nic-NHS) and different versions such as $^{1}H_{4}/^{2}H_{4}$- and $^{12}C_{6}/^{13}C_{6}$-Nic-NHS have been synthesized for their use in quantitative proteomics on a global scale, the latter ensuring coelution of the isotopic peptide pairs and therefore accurate quantitation by LC/MS analysis. So far, the efficiency of the approach has been demonstrated by comparative analysis of *E. coli* lysates differentially spiked with standard proteins (Schmidt et al., 2005). Advantages of the ICPL method are the generally high abundance of lysine residues in proteins providing multiple labeling sites and an increase in MS sensitivity for labeled peptides which permits accurate quantitation and yields in improved sequence coverage of proteins. However, modification of proteins by Nic induces a strong shift in pH resulting in a drastic change in the migration of labeled proteins during 2-D PAGE with isoelectric focusing in the first dimension (Schmidt et al., 2005). This may be advantageous in the analysis of very basic proteins (e.g., ribosomal proteins) but considerably reduces the resolution of 2-D PAGE for the bulk of proteins.

4.7.2
Stable Isotope Labeling at the Peptide Level

Various alternatives to the ICAT and ICPL technologies have been reported but these methods are consistently based on chemical tagging at the peptide level. In this chapter, only the most common and promising peptide labeling approaches will be discussed (Lill, 2003; Moritz and Meyer, 2003).

Regnier and coworkers coined the term global internal standard technology (GIST) which employs N-acetoxy-[^{2}H$_3$]succinimide or ^{2}H$_4$-succinic anhydride as acetylation reagents to modify all primary amino groups in peptides (Chakraborty and Regnier, 2002; Geng et al., 2000; Ren et al., 2004; Wang et al., 2002; Zhang and Regnier, 2002; Zhang et al., 2001). The applicability of GIST to the quantitation of phosphoproteins enriched via immobilized metal affinity chromatography (IMAC) has recently been demonstrated (Riggs et al., 2005). In this work, relative differences in phosphopeptide concentration between samples were derived from isotope ratio measurements of the peptide isoforms observed in mass spectra. One should bear in mind, however, that the modification of lysine residues reduces the basicity; consequently, peptide ionization efficiencies can be low. In addition, N-terminally blocked peptides containing no further lysine residues are not susceptible to quantitative MS analysis via GIST. To address this issue, a combination of GIST and labeling of the C-terminal carboxyl groups by ^{18}O has been proposed and tested for the comparison of the relative protein expression level in epidermal cells grown in the presence or absence of epidermal growth factor (Liu and Regnier, 2002).

Proteolytic peptide labeling with ^{18}O isotopes was first reported by Schnölzer et al. for its use in *de novo* sequencing experiments. Chemical labeling of the C-termini of peptides with ^{16}O/^{18}O isotopes leads to a doublet signature for the y-ion series in MS/MS spectra (Schnolzer et al., 1996). Since the b-ion series do not contain a label, MS/MS data interpretation is quite straightforward. The usefulness of ^{18}O labeling for *de novo* peptide sequencing was shown with MALDI–TOF (Mo et al., 1997), ESI–qTOF (Shevchenko et al., 1997), ESI–ion trap (Qin et al., 1998), and ESI–FTICR (Kosaka et al., 2000) instruments.

In 2000, Roepstorff and coworkers described ^{18}O labeling for peptide and protein quantitation (Mirgorodskaya et al., 2000). However, the uncontrolled incorporation of one or two ^{18}O atoms into peptides and issues related to the overlap of two isotopic distributions complicated the calculation of peptide ratios from the MS spectra. Fenselau and coworkers successfully harnessed ^{18}O labeling for comparative proteomics using shotgun methods (Yao et al., 2001). Generally, two protein samples are proteolytically digested in parallel, one in H$_2$^{16}O and the other one in H$_2$^{18}O. Alternatively, proteolytic digestion can be conducted prior to ^{18}O labeling, which increases the flexibility of this method (Yao et al., 2003). The incorporation of ^{18}O atoms in the carboxy-termini of peptides is catalyzed by serine proteases (e.g., trypsin) via hydrolysis (Reynolds et al., 2002; Yao et al., 2001, 2003). Double labeling can be obtained under adequate conditions (sufficient digestion times, high enzyme concentration and high enzyme activity) inducing a fixed mass shift of 4 Da into peptides. High as well as low resolution mass spectrometers can be employed for quantitative analysis of $^{16}O/^{18}O$-labeled peptide pairs (Heller et al., 2003; Sakai et al., 2005). ^{18}O labeling as a tool for proteomics was also evaluated by Figeys and coworkers (Stewart et al., 2001) and a number of biologically relevant applications have been reported so far (Berhane and Limbach, 2003; Blonder et al., 2005; Bonenfant et al., 2003; Brown and Fenselau, 2004; Qian et al., 2005; Wang et al., 2001; Zang et al., 2004).

Stable isotope labeling methods commonly impart a mass difference as basis for relative quantitation by evaluation of peptide peaks in MS spectra. These methods are primarily designed for the quantitative analysis of a binary sample set. Even though the feasibility of multiplex protein quantitation from peptide mass spectra via SILAC (Molina et al., 2005) or ICPL (Schmidt et al., 2005) has been demonstrated, the further increase in the overall sample complexity has to be considered. Hence, high-resolution separation prior to MS analysis for the diminution of peak overlapping in peptide mass spectra is a prerequisite for accurate protein quantitation. To address this issue, a new generation of stable isotope tagging reagents, termed iTRAQ, has recently been introduced enabling the simultaneous quantitative comparison of a twofold up to a fourfold sample set (Ross et al., 2004). The iTRAQ reagent is an *N*-hydroxysuccinimide ester which reacts with primary amines in peptides. It links an isotope-encoded tag consisting of a mass balance group (carbonyl) and a reporter group (based on *N*-methyl-pirazine) to peptides via the formation of an amide bond. In the iTRAQ method, the protein samples to be compared are separately digested with a protease, differentially labeled, pooled, and then jointly separated by chromatography followed by tandem MS analysis. Owing to the specific mass design of the ^{13}C, ^{15}N, and ^{18}O-encoded reporter and balance groups, the overall nominal mass of each of the four different iTRAQ reagents is kept constant and differentially labeled peptides therefore appear as single peaks in MS scans. When iTRAQ tagged peptides are subjected to MS/MS analysis, the amide bond of the tag fragments in a way similar to the cleavage of peptide bonds. The mass balancing carbonyl moiety, however, is lost with no retain of charge (neutral fragment) and liberates the reporter group as isotope-encoded fragment entities of 114.1–117.1 Da. The ion fragments (e.g., b- and y-ion series) of modified peptides are isobaric and fragmentation spectra obtained are usually improved with respect to ion signal intensities and sequence coverage. The relative concentration of the peptides can be deduced from the intensity ratios of reporter ions observed in the MS/MS spectra. The applicability of the iTRAQ method to multiplex quantitative protein profiling has been exemplified by simultaneous relative and absolute quantitative analysis of wild-type yeast versus two isogenic mutant strains (Ross et al., 2004).

An inherent drawback of this methodology is that quantitative information is only obtained from those peptides which are subjected to MS/MS analysis. This fact stresses the necessity for efficient peptide separation via multidimensional chromatography (e.g., SCX/reversed-phase HPLC) prior to tandem MS analysis. Moreover, ion trap instruments cannot be employed in quantitative iTRAQ experiments since their peptide fragmentation spectra do not contain information on the low m/z range (low-mass cut off) in which reporter ions would appear (Doroshenko and Cotter, 1995). On the other hand, the use of fast scanning mass analyzers would be an advantage in order to increase the fraction of peptides from a total HPLC run that are subjected to MS/MS analysis. In a recent study, DeSouza et al. employed both iTRAQ and cleavable ICAT reagents in combination with multidimensional LC and tandem MS analysis to discover potential markers for endometrial cancer (EmCa) tissues (DeSouza et al., 2005). As can be expected, the use of ICAT led to identification of a higher proportion of lower abundance signaling proteins while iTRAQ was clearly biased towards abundant sample constituents (i.e., ribosomal proteins and transcription factors). From this set of experiments, a total of nine potential markers for EmCa were discovered. Pyruvate kinase was found to be overexpressed in EmCa tissues using both iTRAQ and ICAT labeling (DeSouza et al., 2005).

In any case, the researcher should prefer stable isotope labeling methods which permit simple, specific, and complete incorporation of isotope tags at the earliest possible stage of sample processing. Furthermore, one should aspire to acquiring resolved peptide profiles and isotope patterns which imply the use of multidimensional separation technologies and high-resolution mass analyzers (e.g., FTICR, TOF/TOF, or qTOF). However, protein quantitation of high accuracy can also be achieved with low resolution mass spectrometers such as ion trap instruments (Heller et al., 2003). In general, qTOF and TOF/TOF analyzers provide more accurate information on peptide abundances than ion trapping instruments due to space–charge limitations of the latter. The improvement in ion statistics and signal-to-noise ratios via prolonged scanning times is particularly beneficial for accurate quantitation of lower abundance constituents. While primarily it is the MS system used that determines accuracy and dynamic range of isotope ratio measurements, reproducibility strongly depends on the entire experimental strategy employed. The latter provides good reason for minimizing the number of separate sample processing steps, and thus for performing stable isotope labeling at the protein level (also in case chemical labeling is required for quantitative analysis of clinical and mammalian specimens). In terms of accuracy, it is essential to perform protein quantitation on the basis of multiple peptide pairs. Comparing stable isotope labeling followed by MS with differential 2-D PAGE using fluorescent dyes (DIGE technology), Heck and coworkers demonstrated that relative protein quantities measured with both methods are generally in good agreement (Kolkman et al., 2005). It should be noted that for verification of altered ratios of protein concentrations in a biological system the analysis of independent repetitions (at least three biological experiments) is a must, irrespective of the quantitation method employed.

4.8
Quantitative MS for Deciphering Protein–Protein Interactions

Stable isotope labeling in combination with mass spectrometry is on its way to revolutionizing the study of protein–protein interactions in cells covering both stable complexes as well as transient interactions. There is a variety of techniques available for the purification of protein complexes. In-depth information on purification strategies as well as protein complex analysis by MS means is provided in the literature (Bauer and Kuster, 2003; Dziembowski and Seraphin, 2004; Fritze and Anderson, 2000; Gingras et al., 2005; Puig et al., 2001; Rigaut et al., 1999; Shevchenko et al., 2002; Terpe, 2003). In a conventional approach, the components of a purified complex and a control are separated by 1-D PAGE, proteins are subsequently stained and band patterns are compared. Distinct bands are cut out from the gel followed by in-gel protein digestion and peptide LC/ESI-MS/MS analysis for protein identification. The lists of the individual protein compositions are then compared to distinguish between background proteins and specific interacting partners in the purified complex. However, due to limitations in peptide MS/MS analyses of complex samples performed separately (e.g., MS/MS analysis of only a subset of peptides in a complex mixture), the nonidentification of proteins in a control does not provide convincing evidence for the specificity of protein–protein interactions and therefore further proof is required, e.g., by biochemical methods such as western blotting.

To address this issue, quantitative MS methods can be utilized to allow for identification of specific interacting partners in protein complexes with high reliability. In a metabolic labeling approach, two distinct cell populations (e.g., one containing the bait protein fused with a tag for affinity purification, the other one representing the control) are differentially labeled with stable isotopes during culturing, and pooled in a 1 : 1 ratio followed by cell lysis and isolation of the protein complex via affinity purification (Figure 4.4). In contrast, in chemical labeling approaches using ICAT or peptide-specific reagents, affinity purification of protein complexes derived from cell cultures or tissues of two distinctive differentiation or developmental states is performed separately followed by stable isotope tagging of proteins or proteolytic peptides. In either case, samples are combined after completion of the labeling step and proteolytic peptide mixtures are jointly analyzed by LC/tandem MS (Figure 4.4). In such experiments, protein identification is obtained on the basis of sequence-specific information in peptide fragmentation spectra. In addition, specific interacting partners can be distinguished from copurifying proteins via ratio measurements of light peptides versus their heavy counterparts. True binding proteins become specifically enriched via affinity purification of complexes, and therefore peptide ratios are significantly greater than should be observed in mass spectra representing these proteins. In contrast, nonspecific binding partners are present in equal amounts, resulting in peptide ratios of about 1 (Figure 4.4). In the case of protein complex versus control, direct discrimination between *bona fide* and nonspecific interacting partners can be achieved by comparing ratios of peptide pairs.

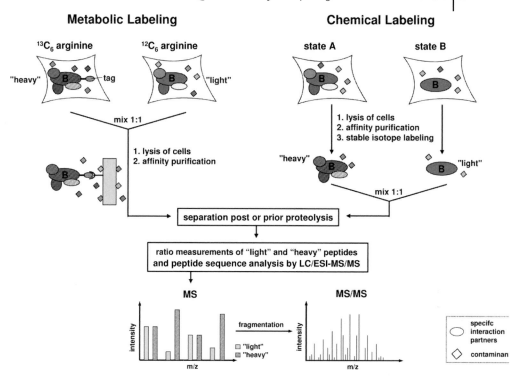

Figure 4.4 Schematic representation of the strategies followed for accurate mapping of protein complexes by stable isotope labeling combined with MS. Relative quantitative MS allows for discrimination between specific interacting partners and contaminants which are often copurified with the protein complex of interest. In the metabolic labeling approach, two cell populations are differentially labeled with stably isotopically encoded amino acids (e.g., $^{13}C_6$-/$^{12}C_6$-arginine) during culturing. Following the labeling experiment, the heavy cell population, which contains the bait protein (termed *B*) fused with a tag for affinity purification, and the light cell population which is used as control are mixed in a 1 : 1 ratio followed by cell lysis and protein complex isolation. In the chemical labeling approach, cells from cultures or tissues of two distinctive differentiation or developmental states (termed states *A* and *B*) are separately lysed followed by affinity purification of protein complexes and stable isotope tagging of proteins. Subsequently, the isotopically tagged protein samples can be pooled in a 1 : 1 ratio and further processed. In both approaches, proteolytic peptide mixtures are generated and then separated by liquid chromatography (LC) followed by qualitative as well as relative quantitative analysis using MS/MS. Specific interacting partners can be differentiated from copurifying proteins based on ion signal intensities of light peptides versus their heavy counterparts in MS spectra. Peptide ratios significantly greater than 1 point to the presence of specific interacting partners; peptides ratios of about 1 are usually distinctive for nonspecific interacting partners. Protein identification is performed on the basis of sequence-specific information observed in MS/MS spectra.

A major promise of this approach is that protein complex purification protocols can be simplified and mild washing conditions used; thus, the ability to screen for transient or weak interaction partners is improved. This has been demonstrated by Aebersold and coworkers who used ICAT technology combined with MS analysis to identify specific interacting partners of a large RNA polymerase II (Pol II) preinitiation complex (PIC) only partially purified from nuclear extracts by a single-step promoter DNA affinity procedure (Ranish et al., 2003).

Comparative analysis of two or more isolated protein complexes looks for differences in protein composition. Using ICAT, dynamic changes in the composition and abundance of STE12 complexes immunopurified from yeast cells in different states (Ranish et al., 2003) as well as in protein complexes associating with the MafK transcription factor upon erythroid differentiation in MEL cells (Brand et al., 2004) were detected. Pandey and coworkers have recently provided detailed instructions for using the SILAC approach to study protein complexes, protein–protein interactions, and the dynamics of protein abundance and posttranslational modifications (Amanchy et al., 2005). This *in situ* labeling approach combined with affinity pull-down experiments was shown to be useful to study (1) signaling complexes involved in the epidermal growth factor receptor (EGFR) pathway of EGF-stimulated versus unstimulated HeLa cells (Blagoev et al., 2003); (2) peptide–protein interactions in EGFR signaling (Schulze and Mann, 2004); and (3) early events in EGFR signaling in a time-course experiment (Blagoev et al., 2004). Hochleitner et al. have recently reported the determination of absolute protein quantities of the human U1 small nuclear ribonuclearprotein (snRNP) complex by using synthetic peptides and stable isotope peptide labeling combined with MS analyses (Hochleitner et al., 2005).

The examples given above clearly demonstrate that quantitative MS-based proteomics is a powerful tool for shedding light on the stoichiometry, dynamics and assembly processes of functional protein modules in biological systems. Since drugs can equally be used as affinity baits, this methodology could also provide valuable information on cellular target proteins in drug research.

4.9
Conclusions

Biological mass spectrometry represents the core technology in proteome research. Instrumental designs and fragmentation techniques are still advancing at an expeditious pace. Novel proteomic strategies, with MS, biochemical as well as molecular cell biological techniques acting in concert, promise to dramatically extend our current knowledge of biological systems in the near future. In quantitative proteomics, it is crucial to consider the variability of biological systems in order to deliver meaningful results regarding protein quantities. This biological variability should not be underestimated but rather be addressed by independent repetitions and advanced statistical data analysis. Certainly, evaluation of large quantitative data sets is very laborious and only feasible using bioinformatics.

In our laboratory, the current lack of suitable software tools for accurate large-scale determination of protein concentration ratios deduced from MS or MS/MS spectra was addressed by in-house development of an efficient software package supporting quantitative analysis of data obtained in any experiment employing stable isotope labeling. Quantitative MS in combination with epitope tagging opens up an additional major area for future proteomic research: the mapping and characterization of functional protein modules on a large scale. This adds to the great promise of proteomics in drug research. Yet, the final impact of proteomics in the discovery of new drug targets cannot easily be foreseen.

Acknowledgements

My special thanks go to Dr. Silke Oeljeklaus for helpful discussion and critical reading of the manuscript and to her and Sebastian Wiese for assistance in the design of the figures presented in this chapter.

References

AEBERSOLD, R., RIST, B., GYGI, S. P. (2000). Quantitative proteome analysis: methods and applications. *Ann. N.Y. Acad. Sci.* **919**, 33–47.

AMANCHY, R., KALUME, D. E., PANDEY, A. (2005). Stable isotope labeling with amino acids in cell culture (SILAC) for studying dynamics of protein abundance and posttranslational modifications. *Sci. STKE* **267**, 12.

BANKS, J. F. JR. (1996). High-sensitivity peptide mapping using packed-capillary liquid chromatography and electrospray ionization mass spectrometry. *J. Chrom. A* **743**, 99–104.

BARBER, M., BORDOLI, R. S., SEDGEWICK, R. D. AND TYLER, A. N. (1981). Fast-atom bombardment of solids (F. A. B.): a new ion source for mass spectrometry. *J. Chem. Soc. Chem. Commun.* **293**, 325–327.

BAUER, A., KUSTER, B. (2003). Affinity purification–mass spectrometry. Powerful tools for the characterization of protein complexes. *Eur. J. Biochem.* **270**, 570–578.

BERHANE, B. T., LIMBACH, P. A. (2003). Stable isotope labeling for matrix-assisted laser desorption/ionization mass spectrometry and post-source

decay analysis of ribonucleic acids. *J. Mass Spectrom.* **38**, 872–878.

BLAGOEV, B., KRATCHMAROVA, I., ONG, S. E., NIELSEN, M., FOSTER, L. J., MANN, M. (2003). A proteomics strategy to elucidate functional protein–protein interactions applied to EGF signaling. *Nat. Biotechnol.* **21**, 315–318.

BLAGOEV, B., ONG, S. E., KRATCHMAROVA, I., MANN, M. (2004). Temporal analysis of phosphotyrosine-dependent signaling networks by quantitative proteomics. *Nat. Biotechnol.* **22**, 1139–1145.

BLONDER, J., HALE, M. L., CHAN, K. C., YU, L. R., LUCAS, D. A., CONRADS, T. P., ZHOU, M., POPOFF, M. R., ISSAQ, H. J., STILES, B. G., VEENSTRA, T. D. (2005). Quantitative profiling of the detergent-resistant membrane proteome of iota-b toxin induced vero cells. *J. Proteome Res.* **4**, 523–531.

BONENFANT, D., SCHMELZLE, T., JACINTO, E., CRESPO, J. L., MINI, T., HALL, M. N., JENOE, P. (2003). Quantification of changes in protein phosphorylation: a simple method based on stable isotope labeling and mass spectrometry. *Proc. Natl. Acad. Sci. USA* **100**, 880–885.

BRAND, M., RANISH, J. A., KUMMER, N. T., HAMILTON, J., IGARASHI, K.,

FRANCASTEL, C., CHI, T. H., CRABTREE, G. R., AEBERSOLD, R., GROUDINE, M. (2004). Dynamic changes in transcription factor complexes during erythroid differentiation revealed by quantitative proteomics. *Nat. Struct. Mol. Biol.* **11**, 73–80.

BROWN, K. J., FENSELAU, C. (2004). Investigation of doxorubicin resistance in MCF-7 breast cancer cells using shot-gun comparative proteomics with proteolytic 18O labeling. *J. Proteome Res.* **3**, 455–462.

CAPUTO, E., MOHARRAM, R., MARTIN, B. M. (2003). Methods for on-chip protein analysis. *Anal. Biochem.* **321**, 116–124.

CHAKRABORTY, A., REGNIER, F. E. (2002). Global internal standard technology for comparative proteomics. *J. Chromatogr. A* **949**, 173–184.

CHAURAND, P., CAPRIOLI, R. M. (2002). Direct profiling and imaging of peptides and proteins from mammalian cells and tissue sections by mass spectrometry. *Electrophoresis* **23**, 3125–3135.

CHAURAND, P., SCHWARTZ, S. A., CAPRIOLI, R. M. (2002). Imaging mass spectrometry: a new tool to investigate the spatial organization of peptides and proteins in mammalian tissue sections. *Curr. Opin. Chem. Biol.* **6**, 676–681.

CHEN, X., SMITH, L. M., BRADBURY, E. M. (2000). Site-specific mass tagging with stable isotopes in proteins for accurate and efficient protein identification. *Anal. Chem.* **72**, 1134–1143.

CONRADS, T. P., ALVING, K., VEENSTRA, T. D., BELOV, M. E., ANDERSON, G. A., ANDERSON, D. J., LIPTON, M. S., PASA-TOLIC, L., UDSETH, H. R., CHRISLER, W. B., et al. (2001). Quantitative analysis of bacterial and mammalian proteomes using a combination of cysteine affinity tags and 15N-metabolic labeling. *Anal. Chem.* **73**, 2132–2139.

DESIDERIO, D. M., ZHU, X. (1998). Quantitative analysis of methionine enkephalin and beta-endorphin in the pituitary by liquid secondary ion mass spectrometry and tandem mass spectrometry. *J. Chromatogr. A* **794**, 85–96.

DESOUZA, L., DIEHL, G., RODRIGUES, M. J., GUO, J., ROMASCHIN, A. D., COLGAN, T. J., SIU, K. W. (2005). Search for cancer markers from endometrial tissues using differentially labeled tags iTRAQ and cICAT with multidimensional liquid chromatography and tandem mass spectrometry. *J. Proteome Res.* **4**, 377–386.

DOROSHENKO, V. M., COTTER, R. J. (1995). High-performance collision-induced dissociation of peptide ions formed by matrix-assisted laser desorption/ ionization in a quadrupole ion trap mass spectrometer. *Anal. Chem.* **67**, 2180–2187.

DZIEMBOWSKI, A., SERAPHIN, B. (2004). Recent developments in the analysis of protein complexes. *FEBS Lett.* **556**, 1–6.

ERICSON, C., PHUNG, Q. T., HORN, D. M., PETERS, E. C., FITCHETT, J. R., FICARRO, S. B., SALOMON, A. R., BRILL, L. M., BROCK, A. (2003). An automated noncontact deposition interface for liquid chromatography matrix-assisted laser desorption/ ionization mass spectrometry. *Anal. Chem.* **75**, 2309–2315.

EVERLEY, P. A., KRIJGSVELD, J., ZETTER, B. R., GYGI, S. P. (2004). Quantitative cancer proteomics: stable isotope labeling with amino acids in cell culture (SILAC) as a tool for prostate cancer research. *Mol. Cell Proteomics* **3**, 729–735.

FENN, J. B., MANN, M., MENG, C. K., WONG, S. F., WHITEHOUSE, C. M. (1989). Electrospray ionization for mass spectrometry of large biomolecules. *Science* **246**, 64–71.

FLORY, M. R., GRIFFIN, T. J., MARTIN, D., AEBERSOLD, R. (2002). Advances in quantitative proteomics using stable isotope tags. *Trends Biotechnol.* **20**, 23–29.

FRITZE, C. E., ANDERSON, T. R. (2000). Epitope tagging: general method for tracking recombinant proteins. *Methods Enzymol.* **327**, 3–16.

GEHRMANN, M. L., HATHOUT, Y., FENSELAU, C. (2004). Evaluation of metabolic labeling for comparative proteomics in breast cancer cells. *J. Proteome Res.* **3**, 1063–1068.

GENG, M., JI, J., REGNIER, F. E. (2000). Signature-peptide approach to detecting proteins in complex mixtures. *J. Chromatogr. A* **870**, 295–313.

GERBER, S. A., RUSH, J., STEMMAN, O., KIRSCHNER, M. W., GYGI, S. P. (2003). Absolute quantification of proteins and phosphoproteins from cell lysates by tandem MS. *Proc. Natl. Acad. Sci. USA* **100**, 6940–6945.

GINGRAS, A. C., AEBERSOLD, R., RAUGHT, B. (2005). Advances in protein complex analysis using mass spectrometry. *J. Physiol.* **563**, 11–21.

GOSHE, M. B., SMITH, R. D. (2003). Stable isotope-coded proteomic mass spectrometry. *Curr. Opin. Biotechnol.* **14**, 101–109.

GRIFFIN, T. J., HAN, D. K., GYGI, S. P., RIST, B., LEE, H., AEBERSOLD, R., PARKER, K. C. (2001). Toward a high-throughput approach to quantitative proteomic analysis: expression-dependent protein identification by mass spectrometry. *J. Am. Soc. Mass Spectrom.* **12**, 1238–1246.

GRUHLER, A., OLSEN, J. V., MOHAMMED, S., MORTENSEN, P., FAERGEMAN, N. J., MANN, M., JENSEN, O. N. (2005). Quantitative phosphoproteomics applied to the yeast pheromone signaling pathway. *Mol. Cell Proteomics* **4**, 310–327.

GUO, J., YANG, E. C., DESOUZA, L., DIEHL, G., RODRIGUES, M. J., ROMASCHIN, A. D., COLGAN, T. J., SIU, K. W. (2005). A strategy for high-resolution protein identification in surface-enhanced laser desorption/ionization mass spectrometry: Calgranulin A and chaperonin 10 as protein markers for endometrial carcinoma. *Proteomics* **5**, 1953–1966.

GYGI, S. P., RIST, B., GERBER, S. A., TURECEK, F., GELB, M. H., AEBERSOLD, R. (1999). Quantitative analysis of complex protein mixtures using isotope-coded affinity tags. *Nat. Biotechnol.* **17**, 994–999.

HAN, D. K., ENG, J., ZHOU, H., AEBERSOLD, R. (2001). Quantitative profiling of differentiation-induced microsomal proteins using isotope-coded affinity tags and mass spectrometry. *Nat. Biotechnol.* **19**, 946–951.

HAYNES, P. A., YATES III, J. R., (2000). Proteome profiling-pitfalls and progress. *Yeast* **17**, 81–87.

HELLER, M., MATTOU, H., MENZEL, C., YAO, X. (2003). Trypsin catalyzed 16O-to-18O exchange for comparative proteomics: tandem mass spectrometry comparison using MALDI–TOF, ESI–QTOF, and ESI–ion trap mass spectrometers. *J. Am. Soc. Mass Spectrom.* **14**, 704–718.

HOANG, V. M., CONRADS, T. P., VEENSTRA, T. D., BLONDER, J., TERUNUMA, A., VOGEL, J. C., FISHER, R. J. (2003). Quantitative proteomics employing primary amine affinity tags. *J. Biomol. Tech.* **14**, 216–223.

HOCHLEITNER, E. O., KASTNER, B., FROHLICH, T., SCHMIDT, A., LUHRMANN, R., ARNOLD, G., LOTTSPEICH, F. (2005). Protein stoichiometry of a multiprotein complex, the human spliceosomal U1 small nuclear ribonucleoprotein: absolute quantification using isotope-coded tags and mass spectrometry. *J. Biol. Chem.* **280**, 2536–2542.

HUTCHENS, T. W., YIP, T. T. (1993). New desorption strategies for the mass spectrometric analysis of macromolecules. *Rapid Commun. Mass Spectrom.* **7**, 576–580.

IBARROLA, N., KALUME, D. E., GRONBORG, M., IWAHORI, A., PANDEY, A. (2003). A proteomic approach for quantification of phosphorylation using stable isotope labeling in cell culture. *Anal. Chem.* **75**, 6043–6049.

IBARROLA, N., MOLINA, H., IWAHORI, A., PANDEY, A. (2004). A novel proteomic approach for specific identification of tyrosine kinase substrates using [13C]tyrosine. *J. Biol. Chem.* **279**, 15805–15813.

IPPEL, J. H., POUVREAU, L., KROEF, T., GRUPPEN, H., VERSTEEG, G., VAN DEN PUTTEN, P., STRUIK, P. C., VAN MIERLO, C. P. (2004). In vivo uniform (15)N-isotope labelling of plants: using the greenhouse for structural proteomics. *Proteomics* **4**, 226–234.

JIANG, H., ENGLISH, A. M. (2002). Quantitative analysis of the yeast proteome by incorporation of isotopically labeled leucine. *J. Proteome Res.* **1**, 345–350.

JONSCHER, K. R., YATES III, J. R. (1997). The quadrupole ion trap mass spectrometer – a small solution to a big challenge. *Anal. Biochem.* **244**, 1–15.

Jonsson, A. P. (**2001**). Mass spectrometry for protein and peptide characterisation. *Cell Mol. Life Sci.* **58**, 868–884.

Karas, M., Hillenkamp, F. (**1988**). Laser desorption ionization of proteins with molecular masses exceeding 10,000 daltons. *Anal. Chem.* **60**, 2299–2301.

Kolkman, A., Dirksen, E. H., Slijper, M., Heck, A. J. (**2005**). Double standards in quantitative proteomics: direct comparative assessment of difference in gel electrophoresis and metabolic stable isotope labeling. *Mol. Cell Proteomics* **4**, 255–266.

Kosaka, T., Takazawa, T., Nakamura, T. (**2000**). Identification and C-terminal characterization of proteins from two-dimensional polyacrylamide gels by a combination of isotopic labeling and nanoelectrospray Fourier transform ion cyclotron resonance mass spectrometry. *Anal. Chem.* **72**, 1179–1185.

Krause, E., Wenschuh, H., Jungblut, P. R. (**1999**). The dominance of arginine-containing peptides in MALDI-derived tryptic mass fingerprints of proteins. *Anal. Chem.* **71**, 4160–4165.

Krijgsveld, J., Ketting, R. F., Mahmoudi, T., Johansen, J., Artal-Sanz, M., Verrijzer, C. P., Plasterk, R. H., Heck, A. J. (**2003**). Metabolic labeling of *C. elegans* and *D. melanogaster* for quantitative proteomics. *Nat. Biotechnol.* **21**, 927–931.

Krutchinsky, A. N., Zhang, W., Chait, B. T. (**2000**). Rapidly switchable matrix-assisted laser desorption/ionization and electrospray quadrupole-time-of-flight mass spectrometry for protein identification. *J. Am. Soc. Mass Spectrom.* **11**, 493–504.

Le Blanc, J. C., Hager, J. W., Ilisiu, A. M., Hunter, C., Zhong, F., Chu, I. (**2003**). Unique scanning capabilities of a new hybrid linear ion trap mass spectrometer (Q TRAP) used for high sensitivity proteomics applications. *Proteomics* **3**, 859–869.

Lill, J. (**2003**). Proteomic tools for quantification by mass spectrometry. *Mass Spectrom. Rev.* **22**, 182–194.

Link, A. J., Eng, J., Schieltz, D. M., Carmack, E., Mize, G. J., Morris, D. R., Garvik, B. M., Yates III, J. R. (**1999**).

Direct analysis of protein complexes using mass spectrometry. *Nat. Biotechnol.* **17**, 676–682.

Liu, P., Regnier, F. E. (**2002**). An isotope coding strategy for proteomics involving both amine and carboxyl group labeling. *J. Proteome Res.* **1**, 443–450.

Loboda, A. V., Krutchinsky, A. N., Bromirski, M., Ens, W., Standing, K. G. (**2000**). A tandem quadrupole/time-of-flight mass spectrometer with a matrix-assisted laser desorption/ionization source: design and performance. *Rapid Commun. Mass Spectrom.* **14**, 1047–1057.

MacCoss, M. J., Wu, C. C., Liu, H., Sadygov, R., Yates III, J. R. (**2003**). A correlation algorithm for the automated quantitative analysis of shotgun proteomics data. *Anal. Chem.* **75**, 6912–6921.

Mann, M., Hendrickson, R. C., Pandey, A. (**2001**). Analysis of proteins and proteomes by mass spectrometry. *Annu. Rev. Biochem.* **70**, 437–473.

Medzihradszky, K. F., Campbell, J. M., Baldwin, M. A., Falick, A. M., Juhasz, P., Vestal, M. L., Burlingame, A. L. (**2000**). The characteristics of peptide collision-induced dissociation using a high-performance MALDI–TOF/TOF tandem mass spectrometer. *Anal. Chem.* **72**, 552–558.

Meng, C. K., Mann, M., Fenn, J. B. (**1988**). On protons of proteins – a beams a beam for a that. *Z. Phys. D* **10**, 361–368.

Mirgorodskaya, O. A., Kozmin, Y. P., Titov, M. I., Korner, R., Sonksen, C. P., Roepstorff, P. (**2000**). Quantification of peptides and proteins by matrix-assisted laser desorption/ionization mass spectrometry using (18)O-labeled internal standards. *Rapid Commun. Mass Spectrom.* **14**, 1226–1232.

Mitulovic, G., Smoluch, M., Chervet, J. P., Steinmacher, I., Kungl, A., Mechtler, K. (**2003**). An improved method for tracking and reducing the void volume in nano HPLC–MS with micro trapping columns. *Anal. Bioanal. Chem.* **376**, 946–951.

Mo, W., Takao, T., Shimonishi, Y. (**1997**). Accurate peptide sequencing by post-source decay matrix-assisted laser desorption/ionization time-of-flight mass spectrometry. *Rapid Commun. Mass Spectrom.* **11**, 1829–1834.

MOLINA, H., PARMIGIANI, G., PANDEY, A. (2005). Assessing reproducibility of a protein dynamics study using in vivo labeling and liquid chromatography tandem mass spectrometry. *Anal. Chem.* **77**, 2739–2744.

MORITZ, B., MEYER, H. E. (2003). Approaches for the quantification of protein concentration ratios. *Proteomics* **3**, 2208–2220.

ODA, Y., HUANG, K., CROSS, F. R., COWBURN, D., CHAIT, B. T. (1999). Accurate quantification of protein expression and site-specific phosphorylation. *Proc. Natl. Acad. Sci. USA* **96**, 6591–6596.

ONG, S. E., BLAGOEV, B., KRATCHMAROVA, I., KRISTENSEN, D. B., STEEN, H., PANDEY, A., MANN, M. (2002). Stable isotope labeling by amino acids in cell culture, SILAC, as a simple and accurate approach to expression proteomics. *Mol. Cell Proteomics* **1**, 376–386.

ONG, S. E., KRATCHMAROVA, I., MANN, M. (2003). Properties of 13C-substituted arginine in stable isotope labeling by amino acids in cell culture (SILAC) *J. Proteome Res.* **2**, 173–181.

ONG, S. E., MITTLER, G., MANN, M. (2004). Identifying and quantifying in vivo methylation sites by heavy methyl SILAC. *Nat. Methods* **1**, 119–126.

OPITECK, G. J., LEWIS, K. C., JORGENSON, J. W., ANDEREGG, R. J. (1997). Comprehensive on-line LC/LC/MS of proteins. *Anal. Chem.* **69**, 1518–1524.

PASA-TOLIC, L., JENSEN, P. K., ANDERSON, G. A., LIPTON, M. S., PEDEN, K. K., MARTINOVIC, S., TOLIC, N., BRUCE, J. E., SMITH, S. R. (1999). High throughput proteome-wide precision measurements of protein expression using mass spectrometry. *J. Am. Chem. Soc.* **121**, 7949–7950.

PETRICOIN III, E. F., ORNSTEIN, D. K., PAWELETZ, C. P., ARDEKANI, A., HACKETT, P. S., HITT, B. A., VELASSCO, A., TRUCCO, C., WIEGAND, L., WOOD, K., et al. (2002a). Serum proteomic patterns for detection of prostate cancer. *J. Natl. Cancer Inst.* **94**, 1576–1578.

PETRICOIN, E. F., ARDEKANI, A. M., HITT, B. A., LEVINE, P. J., FUSARO, V. A., STEINBERG, S. M., MILLS, G. B., SIMONE, C., FISHMAN, D. A., KOHN, E. C., LIOTTA, L. A. (2002b). Use of proteomic patterns in serum to identify ovarian cancer. *Lancet* **359**, 572–577.

PUIG, O., CASPARY, F., RIGAUT, G., RUTZ, B., BOUVERET, E., BRAGADO-NILSSON, E., WILM, M., SERAPHIN, B. (2001). The tandem affinity purification (TAP) method: a general procedure of protein complex purification. *Methods* **24**, 218–229.

QIAN, W. J., MONROE, M. E., LIU, T., JACOBS, J. M., ANDERSON, G. A., SHEN, Y., MOORE, R. J., ANDERSON, D. J., ZHANG, R., CALVANO, S. E., et al. (2005). Quantitative proteome analysis of human plasma following in vivo lipopolysaccharide administration using 16O/18O labeling and the accurate mass and time tag approach. *Mol. Cell Proteomics* **4**, 700–709.

QIN, J., HERRING, C. J., ZHANG, X. (1998). De novo peptide sequencing in an ion trap mass spectrometer with 18O labeling. *Rapid Commun. Mass Spectrom.* **12**, 209–216.

QIU, Y., SOUSA, E. A., HEWICK, R. M., WANG, J. H. (2002). Acid-labile isotope-coded extractants: a class of reagents for quantitative mass spectrometric analysis of complex protein mixtures. *Anal. Chem.* **74**, 4969–4979.

RANISH, J. A., YI, E. C., LESLIE, D. M., PURVINE, S. O., GOODLETT, D. R., ENG, J., AEBERSOLD, R. (2003). The study of macromolecular complexes by quantitative proteomics. *Nat. Genet.* **33**, 349–355.

REN, D., PENNER, N. A., SLENTZ, B. E., REGNIER, F. E. (2004). Histidine-rich peptide selection and quantification in targeted proteomics. *J. Proteome Res.* **3**, 37–45.

REYNOLDS, K. J., YAO, X., FENSELAU, C. (2002). Proteolytic 18O labeling for comparative proteomics: evaluation of endoprotease Glu-C as the catalytic agent. *J. Proteome Res.* **1**, 27–33.

RIGAUT, G., SHEVCHENKO, A., RUTZ, B., WILM, M., MANN, M., SERAPHIN, B.

(1999). A generic protein purification method for protein complex characterization and proteome exploration. *Nat. Biotechnol.* **17**, 1030–1032.

RIGGS, L., SEELEY, E. H., REGNIER, F. E. (2005). Quantification of phosphoproteins with global internal standard technology. *J. Chromatogr. B Anal. Technol. Biomed. Life Sci.* **817**, 89–96.

ROSS, P. L., HUANG, Y. N., MARCHESE, J. N., WILLIAMSON, B., PARKER, K., HATTAN, S., KHAINOVSKI, N., PILLAI, S., DEY, S., DANIELS, S., et al. (2004). Multiplexed protein quantification in *Saccharomyces cerevisiae* using amine-reactive isobaric tagging reagents. *Mol. Cell Proteomics* **3**, 1154–1169.

SAKAI, J., KOJIMA, S., YANAGI, K., KANAOKA, M. (2005). 18O-labeling quantitative proteomics using an ion trap mass spectrometer. *Proteomics* **5**, 16–23.

SCHMIDT, A., KELLERMANN, J., LOTTSPEICH, F. (2005). A novel strategy for quantitative proteomics using isotope-coded protein labels. *Proteomics* **5**, 4–15.

SCHNOLZER, M., JEDRZEJEWSKI, P., LEHMANN, W. D. (1996). Protease-catalyzed incorporation of 18O into peptide fragments and its application for protein sequencing by electrospray and matrix-assisted laser desorption/ionization mass spectrometry. *Electrophoresis* **17**, 945–953.

SCHULZE, W. X., MANN, M. (2004). A novel proteomic screen for peptide–protein interactions. *J. Biol. Chem.* **279**, 10756–10764.

SCHWARTZ, J. C., SENKO, M. W., SYKA, J. E. (2002). A two-dimensional quadrupole ion trap mass spectrometer. *J. Am. Soc. Mass Spectrom.* **13**, 659–669.

SHEVCHENKO, A., CHERNUSHEVICH, I., ENS, W., STANDING, K. G., THOMSON, B., WILM, M., MANN, M. (1997). Rapid 'de novo' peptide sequencing by a combination of nanoelectrospray, isotopic labeling and a quadrupole/time-of-flight mass spectrometer. *Rapid Commun. Mass Spectrom.* **11**, 1015–1024.

SHEVCHENKO, A., LOBODA, A., SHEVCHENKO, A., ENS, W., STANDING, K. G. (2000).

MALDI quadrupole time-of-flight mass spectrometry: a powerful tool for proteomic research. *Anal. Chem.* **72**, 2132–2141.

SHEVCHENKO, A., SCHAFT, D., ROGUEV, A., PIJNAPPEL, W. W., STEWART, A. F., SHEVCHENKO, A. (2002). Deciphering protein complexes and protein interaction networks by tandem affinity purification and mass spectrometry: analytical perspective. *Mol. Cell Proteomics* **1**, 204–212.

SMITH, R. D., ANDERSON, G. A., LIPTON, M. S., MASSELON, C., PASA-TOLIC, L., SHEN, Y., UDSETH, H. R. (2002). The use of accurate mass tags for high-throughput microbial proteomics. *Omics* **6**, 61–90.

SMITH, R. D., PASA-TOLIC, L., LIPTON, M. S., JENSEN, P. K., ANDERSON, G. A., SHEN, Y., CONRADS, T. P., UDSETH, H. R., HARKEWICZ, R., BELOV, M. E., et al. (2001). Rapid quantitative measurements of proteomes by Fourier transform ion cyclotron resonance mass spectrometry. *Electrophoresis* **22**, 1652–1668.

SMOLKA, M., ZHOU, H., AEBERSOLD, R. (2002). Quantitative protein profiling using two-dimensional gel electrophoresis, isotope-coded affinity tag labeling, and mass spectrometry. *Mol. Cell Proteomics* **1**, 19–29.

SMOLKA, M. B., ZHOU, H., PURKAYASTHA, S., AEBERSOLD, R. (2001). Optimization of the isotope-coded affinity tag-labeling procedure for quantitative proteome analysis. *Anal. Biochem.* **297**, 25–31.

SPENGLER, B. (1997). Post-source decay analysis in matrix-assisted laser desorption/ionization mass spectrometry of biomolecules. *J. Mass Spectrom.* **32**, 1019–1036.

STEEN, H., MANN, M. (2004). The ABC's (and XYZ's) of peptide sequencing. *Nat. Rev. Mol. Cell Biol.* **5**, 699–711.

STEWART, I. I., THOMSON, T., FIGEYS, D. (2001). 18O labeling: a tool for proteomics. *Rapid Commun. Mass Spectrom.* **15**, 2456–2465.

STOECKLI, M., CHAURAND, P., HALLAHAN, D. E., CAPRIOLI, R. M. (2001). Imaging mass spectrometry: a new technology for the analysis of protein expression in mammalian tissues. *Nat. Med.* **7**, 493–496.

TANG, N., TORNATORE, P., WEINBERGER, S. R. (**2004**). Current developments in SELDI affinity technology. *Mass Spectrom. Rev.* **23**, 34–44.

TAO, W. A., AEBERSOLD, R. (**2003**). Advances in quantitative proteomics via stable isotope tagging and mass spectrometry. *Curr. Opin. Biotechnol.* **14**, 110–118.

TERPE, K. (**2003**). Overview of tag protein fusions: from molecular and biochemical fundamentals to commercial systems. *Appl. Microbiol. Biotechnol.* **60**, 523–533.

WANG, S., ZHANG, X., REGNIER, F. E. (**2002**). Quantitative proteomics strategy involving the selection of peptides containing both cysteine and histidine from tryptic digests of cell lysates. *J. Chromatogr. A* **949**, 153–162.

WANG, Y. K., MA, Z., QUINN, D. F., FU, E. W. (**2001**). Inverse 18O labeling mass spectrometry for the rapid identification of marker/target proteins. *Anal. Chem.* **73**, 3742–3750.

WANG, Y. K., MA, Z., QUINN, D. F., FU, E. W. (**2002**). Inverse 15N-metabolic labeling/ mass spectrometry for comparative proteomics and rapid identification of protein markers/targets. *Rapid Commun. Mass Spectrom.* **16**, 1389–1397.

WASHBURN, M. P., ULASZEK, R., DECIU, C., SCHIELTZ, D. M., YATES III, J. R. (**2002**). Analysis of quantitative proteomic data generated via multidimensional protein identification technology. *Anal. Chem.* **74**, 1650–1657.

WASHBURN, M. P., ULASZEK, R. R., YATES III, J. R. (**2003**). Reproducibility of quantitative proteomic analyses of complex biological mixtures by multidimensional protein identification technology. *Anal. Chem.* **75**, 5054–5061.

WASHBURN, M. P., WOLTERS, D., YATES III, J. R. (**2001**). Large-scale analysis of the yeast proteome by multidimensional protein identification technology. *Nat. Biotechnol.* **19**, 242–247.

WILCOX, B. E., HENDRICKSON, C. L., MARSHALL, A. G. (**2002**). Improved ion extraction from a linear octopole ion trap: SIMION analysis and experimental demonstration. *J. Am. Soc. Mass Spectrom.* **13**, 1304–1312.

WILM, M., MANN, M. (**1996**). Analytical properties of the nanoelectrospray ion source. *Anal. Chem.* **68**, 1–8.

WILM, M., SHEVCHENKO, A., HOUTHAEVE, T., BREIT, S., SCHWEIGERER, L., FOTSIS, T., MANN, M. (**1996**). Femtomole sequencing of proteins from polyacrylamide gels by nano-electrospray mass spectrometry. *Nature* **379**, 466–469.

WILM, M., MANN, M. (**1994**). Electrospray and Taylor-Cone theory, Dole's beam of macromolecules at last? *Int. J. Mass Spectrom. Ion Process* **136**, 167–180.

WOLTERS, D. A., WASHBURN, M. P., YATES III, J. R. (**2001**). An automated multidimensional protein identification technology for shotgun proteomics. *Anal. Chem.* **73**, 5683–5690.

YANAGISAWA, K., SHYR, Y., XU, B. J., MASSION, P. P., LARSEN, P. H., WHITE, B. C., ROBERTS, J. R., EDGERTON, M., GONZALEZ, A., NADAF, S., et al. (**2003**). Proteomic patterns of tumour subsets in non-small-cell lung cancer. *Lancet* **362**, 433–439.

YAO, X., AFONSO, C., FENSELAU, C. (**2003**). Dissection of proteolytic 18O labeling: endoprotease-catalyzed 16O-to-18O exchange of truncated peptide substrates. *J. Proteome Res.* **2**, 147–152.

YAO, X., FREAS, A., RAMIREZ, J., DEMIREV, P. A., FENSELAU, C. (**2001**). Proteolytic 18O labeling for comparative proteomics: model studies with two serotypes of adenovirus. *Anal. Chem.* **73**, 2836–2842.

YU, L. R., CONRADS, T. P., UO, T., ISSAQ, H. J., MORRISON, R. S., VEENSTRA, T. D. (**2004**). Evaluation of the acid-cleavable isotope-coded affinity tag reagents: application to camptothecin-reated cortical neurons. *J. Proteome Res.* **3**, 469–477.

ZANG, L., PALMER TOY, D., HANCOCK, W. S., SGROI, D. C., KARGER, B. L. (**2004**). Proteomic analysis of ductal carcinoma of the breast using laser capture micro dissection, LC–MS, and 16O/18O isotopic labeling. *J. Proteome Res.* **3**, 604–612.

ZHANG, R., REGNIER, F. E. (**2002**). Minimizing resolution of isotopically coded peptides in comparative proteomics. *J. Proteome Res.* **1**, 139–147.

ZHANG, R., SIOMA, C. S., WANG, S., REGNIER, F. E. (2001). Fractionation of isotopically labeled peptides in quantitative proteomics. *Anal. Chem.* **73**, 5142–5149.

ZHOU, H., BOYLE, R., AEBERSOLD, R. (2004). Quantitative protein analysis by solid phase isotope tagging and mass spectrometry. *Methods Mol. Biol.* **261**, 511–518.

ZHOU, H., RANISH, J. A., WATTS, J. D., AEBERSOLD, R. (2002). Quantitative proteome analysis by solid-phase isotope tagging and mass spectrometry. *Nat. Biotechnol.* **20**, 512–515.

5
Multidimensional Column Liquid Chromatography (LC) in Proteomics – Where Are We Now?

Egidijus Machtejevas, Klaus K. Unger and Reinhard Ditz

Abstract

Multidimensional liquid chromatography (MD-LC) is developing into a powerful separation platform in proteomics research. So far, it is being applied as nano-reversed phase LC in the capillary format to resolve peptides from protein digests after 2D gel electrophoresis. As a front-end technique for mass spectrometry (MS) in proteomics, MD-LC offers a number of advantages such as integrated sample clean up, a wide range of selective filters as separation modes, and high and tuneable column loadability, and is an automated, error-tolerant and easy-to-use system to couple offline or online to MS. Improvements in sample preparation, resolution, and data analysis are, however, necessary before the full potential of MD-LC for the study of proteomes can be revealed. Furthermore, the potential of combining LC with electrically driven modes such as isotachophoresis (ITP), isoelectrofocusing (IEF), capillary electrophoresis (CE) and capillary electrochromatography (CEC) is not yet being fully exploited in proteomics. While LC instruments already show a high degree of flexibility and robustness in use, the design, development and manufacture of appropriate packings and stationary phases for protein separations with respect to low mass transfer resistance and high selectivity is still in its infancy. Novel restricted access material with strong cation exchange (RAM-SCX) functionality has shown high efficiency in sample clean-up after direct biofluid injections. Using the restricted access material combined with an MD-LC platform, peptide maps having more than 800 peptides with molecular weights ranging from 1000 to 3500 Da were obtained, allowing the implementation of biomarker screening as a routine procedure. This fully automated, easy-to-handle analytical separation system reproducibly provides peptide maps for different patients and evidently has great potential in the field of differential proteomics.

Proteomics in Drug Research
Edited by M. Hamacher, K. Marcus, K. Stühler, A. van Hall, B. Warscheid, H. E. Meyer
Copyright © 2006 Wiley-VCH Verlag GmbH & Co. KGaA, Weinheim
ISBN: 3-527-31226-9

5.1
Introduction

At the beginning of 2005, it seems useful to recapitulate the state-of-the-art in the proteomics field:

- The focus in proteomics is still on protein discovery and protein structure analysis rather than on dynamic proteomics and protein function.
- 2D Gel electrophoresis/mass spectrometry (MS) remains the dominant technology platform.
- 2D Gel electrophoresis/MS has not delivered the contributions in drug discovery and in biomarker search expected by the pharmaceutical industry.
- The major application of liquid chromatography (LC) thus far is as nano-LC after digestion of proteins into peptides.
- Multidimensional liquid chromatography (MD-LC) has not yet developed as a front-end technique in proteomics due to the lack of fundamental understanding on how to handle complex sample mixtures with an enormously wide abundance and dynamic range.
- LC instrumentation has reached a mature enough stage to be applied in proteomics on a routine basis.
- The development of novel supports and stationary phases, in particular for protein separation, and the understanding of the separation mechanisms are still in their infancy; further improvements must be made.

Today, proteomics focuses on a top-down approach aiming at the analysis of peptides with numbers of 10 000 and greater. It remains rather questionable whether this processing strategy will provide the necessary answers. Further, the general trend is directed towards the comprehensive and global approach. The situation outlined by the paradigm "if you don't know where you are going, any road will do" should preferably be avoided. It remains to be seen whether a targeted approach will be more successful, but it will obviously be economically more attractive. Another challenge is the reception of representative and focused data.

The major targets in proteomics are in broad-based diagnostic and therapeutic applications. This requires error-avoidance and easy-to-use systems with significant reductions in system complexity. There are a number of essential features of MD-LC that will enable its development into a powerful technology for proteomics: LC can provide a considerable number of selective filters to be combined with a powerful MD-LC system: size exclusion (SEC), reversed phase (RP), hydrophobic interaction (HIC), hydrophilic interaction (HILIC), ion exchange (IEX), immobilized metal affinity and a range of group-specific, biomimetic and specific affinity chromatography (AF) modes. The mass loadability of LC columns is much higher than for 2D gel electrophoresis systems and can be tuned to the requirements of a MD system. LC modes can be incorporated into the sample clean-up which in return becomes more selective, robust and reproducible, thus enhancing the quality of the final data. The most important feature is however, that MD-LC can be automated with a high degree of robustness and reproducibility. MD-LC can also be coupled with electrically driven separation modes such as capillary electrophoresis (CE) and capillary electrochromatography (CEC), which significantly enhances the separation power capabilities. These potentials have neither been thoroughly discussed in the proteomics community, nor evaluated in their fundamental terms.

5.2
Why Do We Need MD-LC/MS Methods?

The general strategy in proteomic research includes sample preparation, protein or peptide separation, their identification, and data interpretation. There are several challenges to be addressed in separation technologies. First, the separation of a complex protein mixture is certainly not an easy task because human body fluids, such as urine or blood, contain several thousands of proteins and peptides. Recent publications (reviewed in [1,2]) lead to the impression that the separation of such complex samples – containing at least several thousand different polypeptides – is not possible in a single liquid-chromatographic run due to the lack of resolving power of 1D LC and limitations in the dynamic range of MS. In addition, larger polypeptides frequently cannot be separated by HPLC. The initial problem is that a vast number of biomolecules are usually present in a biological sample. Depending on the genome, one may be dealing with the some 25 000 proteins. However, the expected number of proteins can even be much higher than the number of genes, as a single gene can give rise to multiple proteins due to co- and posttranslational modifications, degradation intermediates, and alternative splicing products. The difficulty of analyzing the proteome of a cell or serum is that a human cell type, for example, may express up to 20 000 proteins at any time [3] with a predicted dynamic range of up to five orders of magnitude [4]. Complicating proteome analysis is the fact that a given proteome is not static but rather dynamic, being defined by a combination of the cell's genome, environment, and history [5]. Existing methodologies are not adequate to resolve and detect these large numbers of proteins present at such different levels of concentration. In addition, analyzing peptidomics, some tissues or body fluids contain only very low concentrations of peptide hormones, requiring a preconcentration prior to analysis [6].

2D polyacrylamide gel electrophoresis (2D PAGE) and MS are well-established and the most commonly employed techniques in proteomics today. 2D PAGE, however, provides limited information of the total amount of proteins. Low-abundance proteins and small peptides are not detected [1]. Additional methodologies and techniques in sample preparation, selective enrichment, high resolution separation, and detection need to be developed which would allow even higher resolution than 2D PAGE. Acceptable sensitivity to detect the low-abundance proteins is also still an issue. LC can address some of the above-mentioned problems. In comparison with gel-based separation methods, sample handling and preparation are simplified and automated. MD-LC has a number of advantages, such as higher sensitivity, faster analysis time, variable sample size (preconcentration of the target substances is possible), a large number of separating mechanisms, and, most importantly, it is amenable to automation. However, because of the wide dynamic range, no single chromatographic or electrophoretic procedure is likely to resolve a complex mixture of cell or tissue proteins and peptides.

The quantitative dynamic range currently accessible, for example in the plasma proteome, can be assessed by plotting the clinical "reference interval" associated with a sampling of proteins measured in plasma by validated diagnostic assays in commercial clinical laboratories. At the high-abundance end, serum albumin (normal concentration range 35–50 mg mL^{-1}); at the low-abundance end, interleukin 6 (normal concentration range 0–5 pg mL^{-1}) are measured as a sensitive indicator of inflammation or infection [7]. These two clinically useful proteins differ in plasma abundance by a factor of 10^{10}. The fact that both can be measured in hospital laboratories is a demonstration of the power of current immunoassay technology. Finding a molecule of an analyte present at 10 pg mL^{-1} among the albumin molecules present at 55 mg mL^{-1} is very challenging. It is sobering to recognize that this immense dynamic range is achieved only by technologies that presuppose the identity of the analyte: available methods for unbiased protein discovery, such as 2D PAGE or LC MS, have typical dynamic ranges of only 10^2–10^4, which are substantially too low for comprehensive proteome mapping by some 6–8 orders of magnitude. There is currently no single experiment or platform that permits the analysis of all these sample constituents on a proteome-wide scale. Therefore, a number of technologies and platforms have been developed to study specific subsets of components to overcome these limitations.

A 2D or 3D approach employing two or more orthogonal separation techniques or separation methods with different mechanisms of separation will significantly improve the chances of resolving a complex mixture of cell proteins into their individual components. MD peptide separation will play an increasingly important role on the road to identifying and quantifying the proteome.

5.3
Basic Aspects of Developing a MD-LC/MS Method

5.3.1
General

MD (multistage, multicolumn) chromatography was early discovered to be a powerful tool for separating complex mixtures [8]. Two of the protagonists were J. C. Giddings [9,10] and J. F. K. Huber [11]. MD-LC is based on coupling columns in an online or offline mode, which are operated in an orthogonal mode, i.e., separate the sample mixture by different separation mechanisms. The sample separated on the first column (first dimension) is separated into fractions that can then be further treated independently of each other. The practical consequence is an enormous gain in peak capacity (number of peaks resolved at a given resolution) and the potential for independent optimization of the separation conditions for each fraction. Simultaneously, there is the option of relative enrichment/depletion and peak compression by fractionation.

In proteomics MD-LC, columns should be combined with a MD system, which allows the separation of proteins under the following conditions: large diversity and abundance of components in chemical structure and composition, small differences in chemical composition, large differences in molecular size and mass, an extremely large abundance ratio of $1:10^8$, and a wide concentration range, arising from the fact that the number of constituents increases exponentially with decreasing concentration.

To achieve these goals in the desired combination requires a rational design of a separation system with the following selection criteria. First, the choice and sequence of the LC phase system (stationary phase, mobile phase) needs to be tuned to achieve a high peak capacity and optimum resolution, to design a system which effectively depletes the high-abundance constituents and enriches the target compounds, and fractionates the remaining depleted fractions with a high selectivity. In an online mode, the speed of analysis within the dimensions has to be adjusted in such a way that the resolution on the first column is achieved at a low speed. A high-speed analysis in the second dimension is required to resolve the fractions from the first dimension [12]. As a consequence of the high abundance ratio the high-abundance proteins must be depleted and removed most effectively at the beginning of the MD separation to avoid an overload of the subsequent columns, which would compromise their selectivity and reproducibility. In other words, the mass loadability of columns in the MD column train plays a significant role, otherwise displacement phenomena and unwanted protein–protein interactions will take place, which may change the downstream composition of the individual fractions in an irreproducible way [13].

5.3.2
Issues to be Considered

Before the development of a separation strategy, it is necessary to ask an important question: what is the desired outcome? By manipulating a complex protein mixture in one way or another, one could selectively enhance trapping of proteins of interest, and exclude unwanted ones. It is of the greatest importance to make the right choice for the order of consecutive selective filters to approach target substances with high yields and recoveries (Figure 5.1). The term "selective filter" includes sample clean-up procedures and strategies, and chromatographic separations.

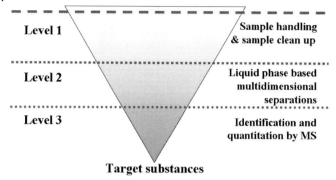

Figure 5.1 Selective filters in LC.
Each level could be reached employing different techniques, e.g.,
level 1, SPE, SEC, IEF or ITP; level 2, SEC, RPC, HIC, HILIC, IMAC or AC;
level 3, MS, MS/MS.

5.3.3
Sample Clean-up

This is the most important step of the whole process. It should be selective with respect to depletion of highly abundant species and in addition with respect to the enrichment of the target compounds. Sample clean-up can be performed either by an electric field through isoelectric focusing (IEF) and isotachophoresis (ITP), or chromatographically by employing solid phase extraction (SPE). Further developments of SPE are columns with restricted access materials (RAM), which combine two chromatographic modes for selective depletion and enrichment. There are several affinity-based columns on the market to efficiently remove albumin, imunoglobulins, and some other high-abundance compounds. Manual operations should be avoided, because of the possible errors and limited throughput. A fully automated online approach with integrated sample clean up is the method of choice.

Mass loadability of SPE and RAM columns play a key role in executing the sample clean-up. It is advisable to work below the overload regime of the column. Otherwise, displacement effects and other phenomena such as secondary interaction by adsorbed species might take place, which will lead to nonreproducible results. This last statement is particularly important when the task is to monitor medium-to-low-abundance proteins; large sample volumes in the milliliter range are therefore usually applied.

5.3.4
Choice of Phase Systems in MD-LC

Separation methods exploiting different physical properties of proteins have been combined with varying degrees of success. The ultimate goal is a rapid separation strategy that can be coupled with detection methods, such as MS, to provide comprehensive monitoring of the concentration differences, interactions, and structures of proteins in the proteome. MD separation typically relies on using two or more independent physical properties of the proteins. Physical properties commonly exploited include size (Stokes' radius), charge, hydrophobicity and biological interaction or affinity. When the targeted properties are independent entities, the separation methods are considered orthogonal.

Ideally, components that are not separated in the first separation step are resolved in the second. Peak capacity is the number of individual components that can be resolved by a separation method. A mathematical model shows that if the MD separations are orthogonal, then the total peak capacity is the product of the individual peak capacities of each dimension [14]. Load capacity is defined as the maximum amount of material that can be run in a separation while maintaining chromatographic resolution. MD separations can be designed to significantly increase the load capacity in a first dimension to achieve enrichment of low-abundance or trace components in a peptide mixture, while the necessary peak capacity may be obtained in the second separation dimension [15].

The primary criteria for the choice of a separation phase system are selectivity and orthogonality, mass loadability, and biocompatibility (in case of quantification). As a rule of thumb, the first dimension should possess a high mass loadability combined with sufficient selectivity and maintenance of bioactivity. Ion exchange chromatography (IEC) is therefore the method of choice offering charge selectivity. In principle, there are two options in IEC, to employ either a cation or an anion exchanger, which in turn influences the pH working range. Note that either cationic or anionic species are resolved, i.e., only a limited number of species from the whole spectrum. The IEC columns are operated via salt gradients with increasing ionic strength. Consequently, the salt load must be removed before the fractions are transferred to the second dimension column.

It is most common to use reversed phase chromatography (RPC) as the second dimension. The term RPC stands for a number of columns with different degrees of hydrophobicity. The most commonly applied phases are n-octadecyl-bonded silicas (RP-18 columns). An intrinsic feature of RP columns is their desalting property: Salts are eluted at the front of the chromatogram when running a gradient elution with an acidic buffer/acetonitrile mobile phase with increasing acetonitrile content. The hydrophobic surface of the RP packing and the hydrophobic eluent are not favorable with respect to providing a biocompatible environment for proteins: they may change their conformation or denature which may be seen by the appearance of broad peaks, splitting of peaks, etc. RP columns possess a much lower mass loadability than IEC columns (10 mg of protein per gram of packing as compared to 100 mg in IEC). An advantage of RPC is the fact that the eluents are MS-compatible, provided volatile buffers such as ammonium acetate are employed.

In the case of an online MD-LC system, the speed of analysis in the second dimension should be as high as possible [12]. This, however, conflicts with the requirement of high resolution or high peak capacity. The highest peak capacity in gradient elution RPC is obtained with a shallow gradient at relatively low flow-

Table 5.1 Comparison of performance parameters of MD-LC systems at different dimensions. Positive aspects are represented with + signs, while negative aspects represented with – signs.

	1D	2D	3D	> 3D
Selectivity (peak cap.)	+	++	+++	++++
Loading Capacity	–	+	++	+++
Sensitivity	++	+	+	+
Throughput	++	+	–	– –
Efficiency	++	++	++	++
Automation possibility	+++	++	+	+
Robustness	++	+	–	– –
Cost	+	–	– –	– – –
Operator skill level	++	+	–	– –

rate. Thus, a compromise between the desired peak capacity and the gradient time is inevitable. Often gradient times of several hours are applied for the analysis of peptides from protein digests.

A question often arising is: how many dimensions do we need in MD-LC? Table 5.1 surveys the performance parameters of MD-LC systems for different dimensions. It becomes obvious that as the number of dimensions increases, the peak capacity will increase. As mentioned earlier, in an ideal case the total peak capacity of the MD-LC system is equal to the product of the individual orthogonal dimensions! At the same time, more than two dimensions in an online MD-LC system results in a very sophisticated instrumental setup that may be difficult to control. The major goal in proteomics for the common user is to design a highly efficient, error-minimizing and easy-to-handle system. Reduction of the system complexity is the major demand. In the system presented here, analyzing endogenous peptides we were able to work with one sample clean-up column (RAM-SCX) with analytical dimensions ($L = 25$ mm, I. D. = 4 mm) with elution performed by five salt steps. The fractions were transferred via a valve to a RP trap column and then further resolved on a 100 mm I. D. Chromolith Performance RP-18 monolithic column. Using this platform, we were able to analyze 2000–4000 endogenous peptides in ~100 µL of human sputum with a MW of between 500 and 4000 in 10 h. Thus, a 2D-LC system with tailor-made stationary phases under optimized operation conditions is sufficient to resolve even highly complex matrices.

McDonald et al. [16] reported the use of a third dimension/phase adding an additional section of RP material behind the first RP and SCX. Although, this extra RP dimension of separation indeed allowed the identification of even more proteins, the whole data collection time had to be significantly extended.

5.3.5
Operational Aspects

In order to end up with enough material for MS-based identification, in most cases one has to start with high sample amounts or large volumes. In MD-LC this means one employs large I. D. columns initially and conclude with capillary columns of 50–100 μm I. D. Corresponding to the column I. D. the volume flow-rate changes from milliliters per minute to nanoliters per minute. At constant column length, the amount of packing in the columns decreases. For example, a 10 cm, 4 mm I. D. column contains approximately 1 g of packing; a 100 mm × 100 μm capillary column contains approximately 10 μg of packing.

Column miniaturization from millimeters to micrometers I. D. has two consequences apart from flow-rate reduction: the injection volume decreases as well as the mass loadability. The diminution of flow-rate and injection volume in relation to column I. D. and an example of the mass loadability for peptides of RP columns of gradated I. D. is displayed in Table 2.

In MD-LC all columns except SEC columns are operated under gradient elution conditions. Usually one starts with eluent A and adds increasing amounts of eluent B with a stronger eluting solvent (linear gradient). Other gradient operations are step gradient or pulse injection of the stronger eluent. Depending on the mode of gradient operation one will obtain different results at equal sampling rate for the fractionation on the first dimension column. Usually approximately 5–40 fractions of equal volume are collected from the first dimension. Each fraction generates approximately another 50 fractions on the second dimension column. The optimal number of fractions from the second column is determined by the time to be invested analyzing fractions by MS and investigating spectra, and the

Table 5.2 Estimation of HPLC column mass loadability.
Assumptions: column dimensions, $L = 100$ mm, total column porosity $\varepsilon_t = 0.7$, skeleton porosity $\varepsilon_s = 0.3$, silica skeleton density 2.2 g/cm^3.

Column I. D. (μm)	Flow rate (μL min^{-1})	Column volume (μL)	Mass of silica per column (mg)	Mass loadability per column (μg)
4600	1000	1660	1100	110
4000	760	1260	830	83
2000	190	310	210	21
1000	47	80	50	5
300	4.3	7	5	0.5
100	0.5	0.8	0.5	0.05
50	0.125	0.2	0.13	0.01
10	0.005	0.08	0.05	0.005

number of peptides one would like to see. A compromise must be sought in order to obtain the required amount of useful data in a reasonable time frame. Note that the evaluation of MS data, and the system automation are also still under development [17].

Either the fractions from the first dimension can be stored on a trap column of suitable dimensions or they can directly be deposited on top of the second dimension column. The direct deposition approach is simpler, and requires less equipment; however, usually the process time is drastically increased. Furthermore, this approach is not always applicable, for example when SEC is used in the first dimension. The use of trap columns is mandatory when the flow-rates in the first and second dimension columns differ by an order of magnitude (see Table 2).

It is essential to select a minimum number of dimensions to handle complex separations. By increasing the number of dimensions one will encounter certain advantages as well as limitations (see Table 1). While increasing the number of separation dimensions can continually increase the peak capacity, it is important to remember that the overall approach must strive to meet the demands of proteomic analysis. That means that the techniques should be robust, should be automated, and should have high throughput capabilities.

5.3.6
State-of-the-Art – Digestion Strategy Included

5.3.6.1 Multidimensional LC MS Approaches

While MD-LC MS has found widespread use in the analysis of peptides from natural sources or generated by proteolytic digestion of larger proteins, the method is not suitable for analyzing proteins directly. First, proteins tend to denature under reversed-phase conditions either by stationary phase or mobile phase induced effects (strongly hydrophobic surfaces, low pH and high organic solvent concentrations) making their quantitative elution rather difficult. Also, measuring the molecular mass of a protein by MS is not sufficient for its unambiguous identification. To circumvent these obstacles the proteins are digested and the separation is performed at the peptide level. One can distinguish two approaches (1) proteins are separated and then digested (top-down proteomics [4]), amd (2) in "shotgun" proteomics a complex protein mixture is first digested (see Figure 5.2a) and peptides are then chromatographically resolved (bottom-up proteomics (see Figure 5.2b) [5]). In both cases, separation technologies play a critical role in protein identification and analysis.

Even though in the shotgun approach, sample complexity is vastly increased, there is an increasing number of reports on the comprehensive analysis of human proteomes using this strategy. The advantage is that after digestion, the peptides obtained are more easily separated than large proteins [16,18–22]. RP-LC is the method of choice as the second dimension in MD-LC due to its high resolving power, high speed of analysis and its desalting capability [23]. The disadvantage of this approach is that one ends up with an extremely large number of peptides to be resolved. Direct analysis of biofluids without prior digestion is a definitive

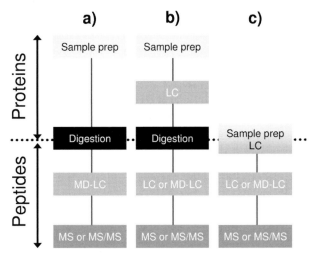

Figure 5.2 LC workflow strategy options in proteomics.
a Bottom-up approach.
b Top-down approach.
c Selective sample clean-up directly combined with chromatographic separation ("digestion-free" strategy).

option in biomarker discovery. Prior digestion gives access to the higher molecular weight proteins, but at the expense of rendering the mixture much more complex. Assuming that a given biofluid contains 1000 proteins and that each protein will generate approximately 50 proteolytic fragments, we are talking about 50 000 or more peptides to be resolved. This task can only be approached by MD protein identification technologies [16–22,24,25]. It should be emphasized that in complex mixtures of peptides, the higher incidence of coelution is the limiting factor for the number of peptides that are identified by MS. Since sensitivity in electrospray MS, for example, is concentration-dependent rather than flow-rate-dependent, the ability to vary the flow in nanoscale LC is advantageous when analyzing more complex or more dilute mixtures of compounds. This technique, known as "peak parking" has already shown its potential in the LC MS analysis of enzymatic protein digests resulting from a proteomic experiment [26]. By reducing the flow-rate on the fly during the separation, elution of peaks can be slowed down, thereby increasing the time for analysis of the mass spectra. In this way, more peptides can be fragmented or a single low-abundance peptide can be fragmented for a longer time.

Another attractive approach is to separate proteins first by IEC according to charge and charge distribution under soft (biocompatible) conditions and collecting fractions. The fractions are subjected to digestion and consecutive reinjection on to a RP column is performed, whereby the separation is based on the hydrophobicity. This is particularly favorable since the mobile phase in the second dimension (RPC) is compatible with the solvent requirements of MS. The restrictions associated with this method lie in the limited size of proteins that can be investigated (MW < 20 000 Da) and the insolubility or incomplete separation of very hydrophobic peptides. All peptide-containing fractions are then investigated by MS to generate a peptide map [27]. This approach has already been found to be sufficient to deal with smaller subsets of the proteome (i.e., several hundred proteins) [17]. These studies also clearly demonstrate that this methodology is not yet suitable for the analysis of a whole proteome due to its enormous complexity. Therefore, pre-selection of the protein from a given tissue or a pre-separation seems mandatory. For example, for the analysis of human urine, SPE (C-18 packings) to trap peptides, followed by IEX chromatography in the first dimension collecting 30 fractions and analysis of the collected fractions by RP-LC (C-18) in the second dimension [28] was successfully employed. A similar procedure was used for the separation of proteins and peptides in human plasma filtrate [29], plasma [30], and blood ultrafiltrate [31].

5.4
Applications of MD-LC Separation in Proteomics – a Brief Survey

In the 1980s, attempts were made to use MD-LC for the resolution of peptide fragments derived from large target proteins for peptide mapping. Putnam's group combined cation and anion exchange chromatography with RP-LC for peptide mapping of high molecular weight proteins [32,33]. Matsuoka et al. demonstrated a similar approach using anion exchange chromatography and RP-LC to fractionate peptides from a brain extract [34]. In these studies, IEX chromatography first resolved peptides based on their charge and then RP-LC further fractionated the peptides based on hydrophobicity. Peptide mixtures were loaded onto an IEX column and eluted step-wise. Fractions were transferred to a RP column. A linear gradient was used to elute the peptide mixture from the RP column and fractions were analyzed by UV detection. Using RP chromatography as the second dimension allows desalting of the peptide mixtures before analysis. Owing to this desalting ability, RP chromatography is preferred as the final dimension in most MD-LC separations. These early attempts foreshadowed experiments in which MD-LC separations were coupled to MS for identifying the peptides.

MD-LC MS is a versatile combination of a commonly used separation technology and MS. Success in proteomics analysis relies on MS; however, the high identification scores obtained by MS are dependent on the resolution power for the highly complex protein mixtures. It is obvious that detection of one particular molecular species depends on the minimum number of those molecules required for the mass spectrometer to be able to detect them. For high-quality data generation, precise chromatographic separation is therefore mandatory. MS separation according to mass adds an additional dimension to a separation, but is acceptable only for screening. MS is often used as a detector or as a separator; it is not always a good idea to mix these features. As MS has limitations with respect to sensitivity, only sufficient concentrations can be identified, so large amounts of sample should be applied. That is how one may possibly solve the dilemma of how much should be injected into the system. If the detection limit of the MS system is known, as is an idea of the concentration range of our target compounds in the sample, the total mass load to apply to the system to achieve the desired sensitivity can be calculated. On the other hand, nonspecific protein loss in a system or inefficient separation strategies may increase the losses and further reduce the overall sensitivity. To reduce complexity and to avoid overloading of the separation system, some approaches deliberately eliminate most of the unwanted proteins by, for example, focusing on the low molecular weight fraction.

One important issue is to find an elegant way to transfer fractions resulting from the first column to the consecutive columns. The simplest approach is to collect fractions from the first dimension, store them and later re-inject them onto the second dimension. However, exposing the fractionated samples to additional surfaces may cause sample losses due to nonspecific adsorption of low-abundance peptides. A more promising strategy involves direct switching of columns by employing valves. Fractions resulting from the first dimension are then trapped or parked until second dimension separation occurs. In these comprehensive procedures, the second separation must be designed to be much faster than the first dimension. This permits the analysis of all fractions eluting from the first dimension column within the period provided by the first dimension separation.

Several comprehensive systems have been developed for the MD-LC fractionation of protein mixtures. Strong cation-exchange or SEC has been coupled with RP-LC to fractionate proteins based on charge or size and then on hydrophobicity. Fractions from the RP column were analyzed by UV detection and MS.

Gygi et al. claimed an offline SCX fractionation coupled with RP and mass spectrometric analysis to be optimal, as opposed to the in-line SCX-RP biphasic methodology used in MD protein identification technology (MudPIT) [35]. The authors argued that the offline SCX approach provides increased loading capacity, improved resolution, greater flexibility and repeated sample analysis. However, MudPIT's comprehensive analysis of complex peptide mixtures avoids the need for complicated switching valves and minimizes sample loss.

For MD separations of peptides using IEC, cation-exchange chromatography has been the method of choice for the first dimension. High loadablity in combination with relatively mild separation conditions is particularly attractive for protein separations. High mass loadability may cause some danger of displacement effects occurring [13]. At a pH of less than 3, negative charges at carboxyl groups and the C-terminus are neutralized as a result of the complete protonation, leaving arginine, lysine and histidine residues plus the N-terminus to contribute to a net positive charge of the peptide. The fully protonated peptides can be fractionated by SCX-LC. For anion exchange–LC, the situation of completely deprotonated basic residues requires a pH greater than 12. This extremely alkaline pH is not compatible with LC packings made of silica because of their chemical instability under these conditions. Polymer-based anion exchangers do not solve the problem completely because the highly alkaline pH causes the functional group on the resin itself to become deprotonated. Although in the IEX mode the major property governing peptide retention results from electrostatic interactions, most ion exchangers also display some hydrophobic properties [36]. This effect partially explains why peptides with the same number of positively charged residues can be resolved by IEX–LC.

One- or two-dimensional LC approaches such as cation exchange followed by online RP-LC MS have demonstrated their resolution power to identify proteins in complex mixtures after tryptic digestion. Adkins et al. [37] performed a proteomic analysis with submilliliter volumes of serum and increased the measurable concentration range for blood proteins. Immunoglobulins were removed from serum with protein A/G and the remaining proteins were digested with trypsin. In serum, 490 proteins were detected by online RP-capillary LC coupled to an ion-trap mass spectrometer. Some low-abundance serum proteins in the nanogram per milliliter range were found, including human growth hormone and interleukins. Chong et al. [38] detected several hundred proteins in cell lysates of human breast cancer cell lines using online nonporous RP-LC–ESI–TOF MS. RP-LC was performed on nonporous C18-coated 2 μm size silica beads, with a separation range of 5–90 kDa. Several hundred proteins were detected and the authors claim that 75–80 proteins were expressed at a higher level in cancer cell lines than in normal human breast cell lines. An elegant approach to identifying urinary polypeptides was published by Spahr et al. [39]. The authors used LC coupled to MS MS to analyze tryptic digests of pooled human urinary proteins. They identified more than 100 polypeptides. Raida et al. employed SCX- and RP-LC coupled to MS to assess the masses of over 3000 peptides from a human hemofiltrate [29]. Although the two LC steps were performed in separate experiments, this study demonstrated the power of MD-LC and MS to fractionate and analyze a staggering number of peptides. However, most of the separated peptides were not identified by MS.

Washburn et al. [22] and Wolters et al. [4] optimized and automated the MudPIT approach. The authors used volatile salts to elute peptides from the SCX column part and computerized LC control for fully automated MD proteomic analysis in a single biphasic column. Biphasic means that the column contains two types of stationary phases: a cation exchanger and a *n*-octadecyl-bonded silica. A discrete fraction of peptides is displaced from the SCX column directly onto a RP column and is then separated and eluted from the RP column into the mass spectrometer. This iterative process is repeated 12 times using increasing salt-gradient elution for the SCX column and an increasing organic concentration for the RP column [22]. The SCX and RP columns are packed at opposing ends of a single capillary column, minimizing the amount of sample loss between the two separation dimensions. A major advantage of this separation technology is that the entire system is coupled directly online with MS, enabling a large number of peptides to be directly identified in a high-throughput manner. The excessive loading capacity of the SCX stationary phase of the biphasic column and the sensitivity of the MS enabled the identification of low-abundance proteins such as transcription factors and protein kinases, typically absent in the 2D PAGE analysis of whole-cell lysates. Link et al. [24] have combined multidimensional SCX- and RP-LC with tandem MS to identify proteins in biological complexes, a method initially called direct analysis of large protein complexes. It was demonstrated that the 2D-LC separation could be performed in a single, biphasic column, which exhibited improved resolution and loading capacity compared to a single-dimension column. In this approach, the SCX column acts as a large peptide reservoir. With at least four times the loading capacity of RP column, SCX-LC greatly increases the number of digested proteins that can be analyzed and therefore allows detection of low-abundance proteins in the mixture. Distinct fractions of peptides are released from the SCX column onto the RP column with increasing salt concentrations. The RP column resolves the peptides in each SCX fraction before tandem MS analysis. Protein fragments fractionated on the SCX–RP column were selected for tandem MS analysis and the proteins were identified using genome-correlated computer analysis. In this study, 75 of the 80 known components of the eukaryotic ribosome, as well as a novel component, were identified [24]. This approach has been shown to be a powerful tool for directly identifying components in protein complexes. It should be noted that long gradient times combined with a large number of fractions make these MD approaches very time-consuming. For example, approximately 83 h was required for MS analysis to collect the data required for the identification of 1484 proteins [22].

One may argue that the number of 1000 proteins in a biofluid sample is still far too small. Comprehensive methods, such as 2D PAGE or MD-LC MS are generally known to be rather time-consuming, making them more suitable for initial discovery efforts than for larger clinical validation studies. In the end, it may not be necessary to visualize every low-abundance protein in order to find significant differences that lead to novel markers.

Opiteck and colleagues combined SEC and RPC to a MD separation of peptides based on size and hydrophobicity [40,41]. The SEC was conducted under either denaturing or nondenaturing conditions. Peptides were loaded onto an SEC column and the effluent was directed to the RP column using a novel computer-controlled interface that directly coupled the two chromatographic separations. Peaks eluting from the first dimension were automatically subjected to RP-LC to separate similarly sized proteins on the basis of their hydrophobicity. The RP-LC system also served to desalt the samples, allowing for their detection by UV absorption at 215 nm, regardless of the SEC mobile phase used. The system had an estimated peak capacity of 495 peptides. However, poor resolution during the SEC step limited its usefulness to the analysis of peptide mixtures from proteolytic digests of single proteins.

Using SEC, most accomplishments of the commonly used RP-LC technique can be employed, such as various detector options, high sample loading capacity, variability in stationary phases and up-scaling option. Often in SEC, non-size-exclusion effects such as electrostatic and hydrophobic interactions between the analyte and stationary phase may be observed. The separation efficiency can be improved by optimizing the mobile phase, flow rate, column length, and sample volume. Practical guidelines for SEC method development have been described [42].

In general, the separation efficiency of SEC is rather low. The efficiency in SEC can be optimized following established procedures. The number of theoretical plates increased from 7100 to 14 000 when the column length was doubled by placing several columns in series [42]. The drawbacks of this approach are the simultaneously increased column backpressure and the analysis time. Decreasing the flow rate from 0.5 to 0.25 mL min^{-1} increased the resolution by more than 50% from 1.4 to 2.3. This contribution to the resolution is more pronounced for larger molecules like proteins because diffusion into and out of the pores is very slow [42]. The resolution can also be increased by decreasing the sample volume. For sample volumes of 200 and 2 μL, resolutions of 0.7 and 1.3 respectively were obtained [42].

Using the isoelectric point (pI) and hydrophobicity as discriminating properties, Wall et al. [43] generated a high peak capacity by coupling liquid-phase IEF with fast RP-LC to resolve proteins from mammalian cell lysates. They achieved a peak capacity of 1000 plus an improved ability to identify low molecular weight and basic proteins as compared to 2D gel electrophoresis. Lubmann et al. [44] and Kachmann et al. [45] used 2D-LC separation and ESI MS to map the protein content of ovarian surface epithelial cells and an ovarian carcinoma-derived cell line. As the first dimension, IEF was employed; the second dimension was nonporous RP chromatography with columns packed with nonporous 2 μm RP-18 particles followed by online coupling to an ESI–TOF MS. The result was a 2D-map of pI versus relative molecular mass (M_r) in analogy to 2D PAGE. Three pI ranges were studied; each containing more than 50 proteins, and about 40% of these proteins were identified by database search. Using a combination of IEX and RP chromatography, termed 2D-LC, complex peptide mixtures, but not protein mixtures have been successfully investigated. In the first dimension, peptides are separated utilizing IEC. Each fraction was further separated in the second RP dimension. The collected fractions were finally investigated by MS to constitute a complete peptide map. These peptide patterns were further employed to identify differences between normal control samples and patient samples to reveal potential biomarkers [27,46].

5.5
Sample Clean-Up: Ways to Overcome the "Bottleneck" in Proteome Analysis

A proteomic analysis of a sample usually consists of four steps. These are extraction of the proteins from the sample, their separation, detection, and finally identification/analysis of the individual separated proteins. Major attention must be paid to the sample processing, sample handling, and the sample clean-up since any error or sample loss during this stage influences the final result.

Most biofluids contain large amounts of well-known proteins such as albumin and IgGs, which overwhelm the separation system and make the detection of low-abundance proteins and peptides very difficult. It is thus advantageous to remove these proteins prior to digestion and separation. Besides the already described approaches that are based on size exclusion fractionation, there are alternative ways of reducing the overall protein load by specific adsorption of albumin and IgG to affinity matrices [47–51]. While usually an affinity matrix is generally highly specific, in high-content samples the specificity of the affinity ligand is limited. There are degrees of specificity between highly selective immunoaffinity matrices and less selective but more robust affinity supports using synthetic ligands. In an effort to reduce the amount of albumin from human serum, a number of affinity matrices have been evaluated based on antibodies or dye ligands. Antibody-mediated albumin removal was efficient and selective. Dye ligand chromatography, a technique that is extensively used in protein chromatography was surprisingly effective [52], in particular with regard to high binding capacities and a long column lifetime, although this was at the expense of selectivity.

Often some classes of target molecules are present in very small amounts and need to be selectively isolated or enriched before identification. Affinity chromatography, which selectively retains proteins or peptides based on specific interactions, has been employed in several MD separation methods for selective trapping of proteins of interest. Phosphorylated, glycosylated or derivatized amino acids are often targeted by affinity chromatography in the first dimension. Immobilized metal-affinity chromatography (IMAC) using Fe^{3+} or Ga^{3+} enriches for phosphorylated peptides [53,54]. Several groups have reported online combination of IMAC-Fe^{3+} and RP chromatography for identifying phosphorylated peptides from 2D peptide maps or synthetic peptide mixtures [55,56].

Recently the group of A. Heck reported a novel analytical procedure for phosphopeptide enrichment employing a developed nanoflow 2D-LC MS/MS setup for characterization of both nonphosphorylated and phosphorylated peptides in two separate measurements [57]. The procedure relies on the unique ion exchange properties of Titansphere, a new type of column material that consists of spherical particles of titanium oxide. In comparison with IMAC, this chromatographic approach has the advantage that fewer column handling steps are required, and therefore, it seems to be a more robust enrichment procedure for the selective enrichment and characterization of phosphopeptides from complex mixtures.

Peptides containing different carbohydrate modifications can be enriched based on their affinity for specific lectins. Focusing on glycosylated peptides using the lectin ConA, Regnier's group has coupled lectin affinity chromatography with RP chromatography [58]. One of the most publicized MD separation techniques that employs an affinity chromatography step is based on the use of isotope-coded affinity tags (ICAT) [59]. In this strategy, cysteinyl residues within proteins are modified with a thiol-reactive reagent that contains a biotin moiety. The proteins are enzymatically digested and the modified peptides recovered using immobilized avidin chromatography. The main purpose of affinity chromatography isolating only the cysteinyl containing peptides is to reduce the complexity of the sample; however, since the ICAT-based strategy is designed for global proteomic studies, the postaffinity chromatography sample is still quite complex.

Approaches using affinity peptide capture and RP-HPLC separations have been applied with varying degrees of success. Overall, MD affinity chromatography is likely to play an increasingly important role in the context of sample clean-up, and identification and characterization of specific classes of peptides and proteins in the proteome.

As many reports have shown, there remains a tremendous amount of information in the low-MW fraction of biological samples such as serum or urine. Serum or plasma may be divided into a high-MW and a low-MW fraction by ultrafiltration or SEC. Nevertheless, even ultrafiltration at a cut-off of 10 kDa still leaves a considerable amount of albumin in urine or in ultrafiltered serum, since the MW cut-off is not sharp as may be related to the distribution of the pore sizes.

Naturally occurring peptides, typically below the kidney size cut-off and, hence, usually collected from urine or from blood hemodialysate, provide a complementary picture of many events at the low-mass end of the plasma proteome. It provides a large source of proteins and peptides below 45 kDa. Such material has been analyzed by combined chromatography and MS approaches to resolve ~5000 different peptides, including fragments of 75 different proteins [60]. Of the fragments, 55% percent were derived from plasma proteins, and 7% of the entries represented peptide hormones, growth factors, and cytokines.

In recent years, special SPE supports possessing restricted access properties have been developed [61–65] to allow the direct injection of untreated biological samples into online SPE–LC systems.

In 1985, Hagestam and Pinkerton [61] published the first restricted access material (RAM). RAM exerts a specific feature by restricting the access of large molecular weight constituents in the sample mixture to the internal surface due to size and shape barriers and favoring the trapping of smaller analytes which pass the barrier provided by the outer surface to access the internal surface. The types of RAMs reviewed in [66] according to the structure of adsorbent surface and type of the barrier can be divided into mixed-functional or dual zone phases, shielded hydrophobic phases, semi-permeable surfaces and internal surface RP supports [67–69]. The low-molecular-mass analytes are retained on the RAM columns by conventional retention mechanisms such as hydrophobic, ionic, or affinity interactions. At the same time, the interaction of large molecules is restricted by a diffusion barrier in SEC-based materials. The limitations occur depending on the pore diameter of the support (ChromSper 5 Biomatrix from Chrompack, LiChrospher ADS from Merck) or a chemical barrier provided via a polymer at the outer surface of the particles (Hisep from Supelco, Ultrabiosep from Hypersil, Biotrap 500 from Chromtech, Capcell Pak from Shiseido).

The direct injection of biological samples onto the chromatographic column without any sample preparation is in most cases highly problematic and may lead to an irreversible contamination of the separation columns, which reduce selectivity and column performance. A powerful asset to circumvent all the named problems is the implementation of RAM for sample preparation. RAMs, a combination of SEC with adsorptive chromatography, have found widespread use in the analysis of drug metabolites and other low-MW compounds, but have only recently been applied to proteomics applications [69]. The RAM is based on a designed surface topography of porous silica particles of 20–40 μm size. The outer surface of the particles possesses electroneutral hydrophilic propoxydiol ligands which show minimal adsorption towards biopolymers, while the pore inside surface is covered with C-18, C-8 and C-4 ligands respectively or with cation and anion exchange groups e.g., with sulfonic acid ligands. An advantage of these adsorbents relies on the simultaneous occurrence of two chromatographic separation mechanisms: selective RP or IEC of low-molecular-mass analytes and SEC for the macromolecular sample constituents. By regulating the pore size of the particles, the MW exclusion can be varied as well as the MW fractionation range, which allows certain analytes to be trapped at the internal surface. In this case, only proteins and peptides below a certain molecular shape and size have access to the inner pore surface of the RAM, and are thus retained while the larger proteins encounter only the hydrophilic, nonadsorptive outer surface, and will be flushed out in the following washing step. Of the RAM, the strong cation exchanger with sulfonic ligands (RAM–SCX) was employed in the sample clean-up of proteins, and proved to show an acceptable capacity towards positively charged peptides and proteins. The features described above, when elegantly combined with column switching, become a powerful tool for direct analysis in the profiling of endogeneous peptides in a fully automated, MD-LC platform.

Applying RAM–SCX to the analysis of hemofiltrate, it was shown that the sample preparation step could be performed online with the subsequent chromatographic separations [12]. Wagner et al. [12] demonstrated a rapid automated online MD-LC system for protein and peptide mapping with integrated sample preparation for proteins of MW less than 20 kDa from a human hemofiltrate. The size-selective fractionation step was followed by anion- or cation exchange chromatography as the first dimension. A column-switching technique, including four RP columns was employed for the second dimension for online fractionation and separation. The system was applied to protein mapping of biological samples of human hemofiltrate as well as of cell lysates originating from a human fetal fibroblast cell line. The separation was performed on the protein level. With this system the analysis of approximately 1000 peaks within an astonishing time frame of less than 2 h was performed. In combination with a 2D chromatographic separation system, this allowed the attainment of a peak capacity of approximately 5000, a number that is comparable to the number of spots obtained on a highly optimized 2D PAGE for proteins. Such high-resolution MD systems have the potential to provide the technical platform for biomarker discovery in combination with modern mass spectrometers and data analysis software.

Recently an extremely simple 2D-LC separation system with integrated sample clean-up has been developed for profiling of endogenous peptides in sputum as a biofluid. After loading the RAM–SCX column of 30 × 15 mm I. D., washing was accomplished and the trapped peptides were eluted from RAM-SCX column by five salt steps with increasing salt concentrations. Each of the five fractions were transferred and desalted directly on a monolithic 4.6 mm I. D. RP column, and further separated into 20 fractions according to peptide hydrophobicity [70].

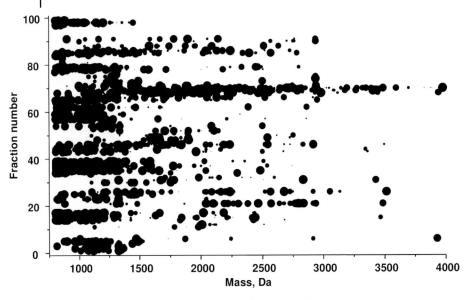

Figure 5.3 The peptide and small protein map from 100 μL human sputum injection.
Columns: sample preparation, RAM-SCX; analytical column, Chromolith Performance
RP-18, 100 × 4.6 mm I. D. Minute fractions analyzed using MALDI–TOF MS.
Fraction numbers corresponds to time scale.
Mark size related to signal intensity.

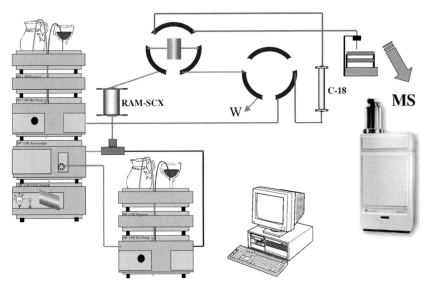

Figure 5.4 MD chromatographic platform configuration:
RAM-SCX column, analytical RP column, MALDI–TOF MS.

In total, a 100-position MALDI plate was filled from every human sputum sample injection and analyzed using MS. Obtained mass spectra contained approximately 1500 signals in a mass range from 800 to 4000 Da. Analyzing the obtained data it was concluded that peptides from the RAM-SCX fractions separated according to mass located at random (Figure 5.3) without obvious repetition. Surprisingly, there was hardly any overlapping for closely located fractions; each of them provided a particular unique peptide profile.

In a search for novel column design for proteomics with enhanced mass-transfer properties, monolithic columns have been developed and manufactured which show lower pressure drop than comparable particulate columns and improved column performance at the same time [71].

The fully automated 2D-LC system (Figure 5.4) performing an effective fractionation combined with offline MALDI–TOF MS offers an enormous potential for human peptidomics screening on a daily basis. By analyzing preliminary results, one should be able to generate peptide maps injecting as little as 150 μL of human sputum to obtain about 1500 signals in the MS. The system offers a high degree of flexibility to be optimized for effective analysis of other biofluids.

5.6
Summary

Proteomics is the large-scale study of protein expression levels and insight into the activity state of all proteins of a proteome or subsystem. MD-LC and 2D-LC followed by tandem MS are fast and accurate technologies for protein identification and characterization in proteomics. Many research groups invest their best efforts; however, most of the LC designs appear to be more suitable for initial discovery than for large clinical validation studies.

Novel RAM with strong cation exchange functionality has shown high efficiency in sample clean-up after direct online biofluid injections. Integration of strong cation exchange RAM columns for automated sample clean up into a 2D-LC system, offline coupled to MALDI–TOF–TOF MS/MS, has demonstrated the potential. This technology platform substantially enhanced selectivity, sensitivity, throughput and robustness for the identification of (target) peptides from human biofluids. Using this platform, peptide maps with more than 800 peptides with molecular weights ranging from 1000 to 3500 Da are obtainable, allowing the implementation of biomarker screening as a routine procedure. This fully automated, easy-to-handle analytical separation system reproducibly provides peptide maps for different patients and evidently has great potential in the field of differential peptidomics.

In the future, proteomics will play a major role in drug discovery, accelerating the various steps involved – target identification, target validation, and drug discovery (efficacy, selectivity and mode of action). Operated on a routine basis, MD-LC may provide the desired data, and after interconnection with the biology, solutions could be found.

References

1 ISSAQ, H. J. (2001). *Electrophoresis* 22, 3629–3638.
2 AEBERSOLD, R., GOODLETT, D. R. (2001). *Chem. Rev.* 101, 269–295.
3 CELIS, E., GROMOV, P. (1999). *Electrophoresis* 20, 16–21.
4 WOLTERS, D. A., WASHBURN, M. P., YATES III, J. R. (2001). *Anal. Chem.* 73, 5683–5690.
5 REGNIER, F., AMINI, A., CHAKRABORTY, A., GENG, M., JI, J., RIGGS, L., SIOMA, C., WANG, S., ZHANG, X. (2001). *LC–GC* 19, 200–213.
6 OOSTERKAMP, A. J., GELPI, E., ABIAN, J. (1998). *J. Mass Spectrom.* 33, 976–983.
7 ANDERSON, N. L., ANDERSON, N. G. (2002). *Mol. Cell Proteomics* 1.11, 845–867.
8 CORTES, H. J. (1990). *Multidimensional chromatography, techniques and applications.* Marcel Dekker, New York.
9 GIDDINGS, J. C. (1984). *Anal. Chem.* 56, 1258–1270.
10 GIDDINGS, J. C. (1995). *J. Chromatogr. A* 703, 3–15.
11 HUBER, J. F. K., LAMPRECHT, G. (1995). *J. Chromatogr. B* 666, 223–232.
12 WAGNER, K., MILIOTIS, T., MARKO-VARGA, G., BISCHOFF, R., UNGER, K. K. (2002). *Anal. Chem.* 74, 809–820.
13 WILLEMSEN, O., MACHTEJEVAS, E., UNGER, K. K. (2004). *J. Chromatogr. A* 1025, 209–216.
14 GIDDINGS, J. C. (1987). *J. High Resolut. Chromatogr.* 10, 319–323.
15 LUNDELL, N., MARKIDES, K. (1992). *Chromatographia* 34, 369–375.
16 MCDONALD, W. H., OHI, R., MIYAMOTO, D. T., MITCHISON, T. J., YATES III, J. R. (2002). *Int. J. Mass Spectrom.* 219, 245–251.
17 HILLE, J. M., FREED, A. L., WÄTZIG, H. (2001). Review, *Electrophoresis* 22, 4035–4052.
18 GEVAERT, K., VAN DAMME, J., GOETHALS, M., THOMAS, G. R., HOORELBEKE, B., DEMOL, H., MARTENS, L., PUYPE, M., STAES, A., VANDEKERCKHOVE, J. (2002). *Mol. Cell Proteomics* 1, 896–903.
19 GRIFFIN, T. J., GYGI, S. P., IDEKER, T., RIST, B., ENG, J., HOOD, L.,

AEBERSOLD, R. (2002). *Mol. Cell Proteomics* 1, 323–333.
20 MACCOSS, M. J., MCDONALD, W. H., SARAF, A., SADYGOV, R., CLARK, J. M., TASTO, J. J., GOULD, K. L., WOLTERS, D., WASHBURN, M., WEISS, A., CLARK, J. I., YATES III, J. R. (2002). *Proc. Natl. Acad. Sci.* 99, 7900–7905.
21 MCDONALD, W. H., YATES III, J. R. (2002). *Dis. Markers* 18, 99–105.
22 WASHBURN, M. P., WOLTERS, D., YATES III, J. R. (2001). *Nat. Biotechnol.* 19, 242–247.
23 LIU, H., LIN, D., YATES III, J. R. (2002). *Biotechniques* 32, 898–902.
24 LINK, A. J., ENG, J., SCHIELTZ, D. M., CARMACK, E., MIZE, G. J., MORRIS, D. R., GARVIK, B. M., YATES III, J. R. (1999). *Nat. Biotechnol.* 17, 676–682.
25 PANG, J. X., GINNANNI, N., DONGE, A. R., HEFTA, S. A., OPITECK, G. J. (2002). *J. Proteome Res.* 1, 161–169.
26 MOSELEY, M. A., BLACKBURN, K., BURKHART, W., VISSERS, H., BORDOLI, R. (2000). Presented at the 48th American Society for Mass Spectrometry Conference, Long Beach, CA, June 2000.
27 SCHULZ-KNAPPE, P., ZUCHT, H.-D., HEINE, G., JÜRGENS, M., HESS, R., SCHRADER, M. (2001). *Combinatorial Chem. High Throughput Screening* 4, 207–217.
28 HEINE, G., RAIDA, M., FORSSMANN, W.-G. (1997). *J. Chromatogr. A* 766, 117–124.
29 RAIDA, M., SCHULZ-KNAPPE, P., HEINE, G., FORSSMANN, W.-G. (1999). *J. Am. Mass Spectrom.* 10, 45–54.
30 RICHTER, R., SCHULZ-KNAPPE, P., SCHRADER, M., STANDKER, L., JURGENS, M., TAMMEN, H., FORSSMANN, W.-G. (1999). *J. Chromatogr. B* 726, 25–35.
31 SCHRADER, M., JURGENS, M., HESS, R., SCHULZ-KNAPPE, P., RAIDA, M., FORSSMANN, W.-G. (1997). *J. Chromatogr. A* 766, 139–145.
32 TAKAHASHI, N., TAKAHASHI, Y., PUTNAM, F. W. (1983). *J. Chromatogr.* 266, 511–522.
33 TAKAHASHI, N., ISHIOKA, N., TAKAHASHI, Y., PUTNAM, F. W. (1985). *J. Chromatogr.* 326, 407–418.

34 Matsuoka, K., Taoka, M., Isobe, T., Okuyama, T., Kato, Y. (1990). *J. Chromatogr.* 515, 313–320.

35 Gygi, S. P., Rist, B., Griffin, T. J., Eng, J., Aebersold, R. (2002). *J. Proteome Res.* 1, 47–54.

36 Zhu, B.-Y., Mant, C. T., Hodges, R. S. (1992). *J. Chromatogr.* 594, 75–86.

37 Adkins, J. N., Varnum, S. M., Auberry, K. J., Moore, R. J., Angell, N. H., Smith, R. D., Springer, D. L., Pounds, J. G. (2002). *Mol. Cell Proteome* 1, 947–955.

38 Chong, B. E., Hamler, R. L., Lubmann, D. M., Ethier, S. P., Rosenspire, A. J., Miller, F. R. (2001). *Anal. Chem.* 73, 1219–1227.

39 Spahr, C. S., Davis, M. T., McGinley, M. D. (2001). *Proteomics* 1, 93–107.

40 Opiteck, G. J., Jorgenson, J. W. (1997). *Anal. Chem.* 69, 2283–2291.

41 Opiteck, G. J., Ramirez, S. M., Jorgenson, J. W., Moseley, M. A. (1998). *Anal. Biochem.* 258, 349–361.

42 Ricker, R. D., Sandoval, L. A. (1996). *J. Chromatogr. A* 743, 43–50.

43 Wall, D. B., Kachman, M. T., Gong, S., Hinderer, R., Parus, S., Miek, D. E., Hanash, S. M., Lubman, D. M. (2000). *Anal. Chem.* 72, 1099–1111.

44 Lubmann, D. M., Kachmann, M. T., Wang, H., Gong, S., Yan, F., Hamler, R. L., O'Neil, K. A., Zhu, K., Buchanan, N. S. Barder, T. J. (2002). *J. Chromatogr. B* 782, 183–196.

45 Kachmann, M. T., Wang, H., Schwartz, D. R., Cho, K. R., Lubmann, D. M. (2002). *Anal. Chem.* 74, 1779–1791.

46 Schrader, M., Schulz-Knappe, P. (2001). *Trends Biotechnol.* 19/S1, S55–S60.

47 Georgiou, H. M., Rice, G. E., Baker, M. S. (2001). *Proteomics* 1, 1503–1506.

48 Kassab, A., Yavuz, H., Odabasi, M., Denizli, A. (2000). *J. Chromatogr. B* 746, 123–132.

49 Nakamura, K., Suzuki, T., Kamichika, T., Hasegawa, M., Kato, Y., Sasaki, H., Inouye, K. (2002). *J. Chromatogr. A* 972, 21–25.

50 Wang, Y. Y., Cheng, P. C., Chan, D. W. (2003). *Proteomics* 3, 243–248.

51 Govorukhina, N. I., Keizer-Gunnink, A., van der Zee, A. G. J., de Jong, S., de Bruijn, H. W. A., Bischoff, R. (2003). *J. Chromatogr. A* 1009, 171–178.

52 Andrecht, S., Anders, J., Hendriks, R., Machtejevas, E., Unger, K. (2004). *Laborwelt* 5/2, 4–7.

53 Andersson, L., Porath, J. (1986). *Anal. Biochem.* 154, 250–254.

54 Posewitz, M. C., Tempst, P. (1999). *Anal. Chem.* 71, 2883–2892.

55 Watts, J. D., Affolter, M., Krebs, D. L., Wange, R. L., Samelson, L. E., Aebersold, R. (1994). *J. Biol. Chem.* 269, 29520–29529.

56 Li, S., Dass, C. (1999). *Anal. Biochem.* 270, 9–14.

57 Pinkse, M. W. H., Uitto, P. M., Hilhorst, M. J., Ooms, B., Heck, A. J. R. (2004). *Anal. Chem.* 76, 3935–3943.

58 Geng, M., Ji, J., Regnier, F. E. (2000). *J. Chromatogr.* 870, 295–313.

59 Gygi, S. P., Rist, B., Gerber, S. A., Turecek, F., Gelb, M. H., Aebersold, R. (1999). *Nat. Biotechnol.* 17, 994–951.

60 Schepky, A. G., Bensch, K. W., Schulz-Knappe, P., Forssmann, W. G. (1994). *Biomed. Chromatogr.* 8, 90–94.

61 Hagestam, I. H., Pinkerton, T. C. (1985). *Anal. Chem.* 57, 1757–1763.

62 Gurley, B. J., Marx, M., Olsen, K. (1995). *J. Chromatogr. B* 670, 358–364.

63 Rudolphi, A., Boos, K. S., Seidel, D. (1995). *Chromatographia* 41, 645.

64 Yu, Z., Westerlund, D., Boos, K. S. (1997). *J. Chromatogr. B* 689, 379–386.

65 Boos, K. S., Rudolphi, A. (1997). *LC–GC Int.* 15, 602–611.

66 Pinkerton, T. C. (1991). *J. Chromatogr.* 544, 13–23.

67 Pinkerton, T. C. (1989). *Eur. Clin. Lab.* 8, 11.

68 Boos, K. S., Rudolphi, A., Vielhauer, S., Walfort, A., Lubda, D., Eisenbeiss, F., Fresenius, J. (1995). *Anal. Chem.* 352, 684–690.

69 Boos, K. S., Grimm, C. H. (1999). *Trends Anal. Chem.* 18, 175–180.

70 Unger, K. K., Machtejevas, E., Hennessy, T. P., Ditz, R. (2004). *Laborwelt* 5/4, 4–10.

71 Nakanishi, K., Soga, N. (1992). *J. Non Cryst. Solids* 139, 1–24.

6

Peptidomics Technologies and Applications in Drug Research

Michael Schrader, Petra Budde, Horst Rose, Norbert Lamping, Peter Schulz-Knappe and Hans-Dieter Zucht

Abstract

The road from target discovery to a marketable drug is prolonged and cost-intensive. Although sequencing of the human genome has identified several potential disease genes, the number of successful drugs against novel targets decreased. Various factors contribute to the high attrition rate: one reason is related to the shortcomings of gene expression approaches to prediction of biological function and pathways of gene products. While approximately 20 000 genes serve as the blueprint of a life, the number of proteins is calculated to be at least one order of magnitude larger. The comprehensive analysis of dynamic alterations of the proteome is now regarded as a new paradigm to improve understanding of the molecular complexity of diseases. Because most drugs target aberrant functions of proteins, the industry increasingly integrates proteomics techniques into the drug development process. The proteome is remarkable dynamic and different mechanisms of posttranslation control exist. An important irreversible control level is proteolytic processing of proteins that releases an even larger number of peptides. At present, approximately 500 proteases have been identified in the human genome that are either involved in the degradation of proteins or liberate novel bioactive peptides. Peptides control important cellular and physiological functions and possess enormous potential in treating diabetes, multiple sclerosis and osteoporosis. Spurred by the use of insulin as an archetype therapeutic peptide, peptide formulation and delivery technologies have been improved in recent decades, and revenues from therapeutic peptides are therefore expected to increase tremendously in the future. Peptidomics is a novel approach to the comprehensive display and identification of dynamic changes in all endogenous peptides and small proteins (peptidome) in response to physiological or pathophysiological conditions, or to therapeutic interventions. The peptidomics technology platform integrates purification of peptides and small proteins (< 20 kDa) from complex protein mixtures by high-performance separation via HPLC, with mass spectrometry analysis of fractionated peptides. This approach is particularly useful for

Proteomics in Drug Research
Edited by M. Hamacher, K. Marcus, K. Stühler, A. van Hall, B. Warscheid, H. E. Meyer
Copyright © 2006 Wiley-VCH Verlag GmbH & Co. KGaA, Weinheim
ISBN: 3-527-31226-9

identifying novel peptide drug candidates and peptide biomarkers that are coupled to the molecular events associated with disease pathomechanisms. These biomarkers are increasingly used in all phases of the drug development process and provide a measure for drug safety, efficacy, toxicity and side effects. Combined with genomics and proteomics, peptidomics enables the identification and development of peptide biomarkers that, knowing the amino acid sequence, can be translated into clinical assays such as radioimmunoassays (RIA) or enzyme-linked immunosorbent assays (ELISA).

6.1
Introduction

Peptides are present in large numbers and in varying amounts in all human body fluids, cells and within tissues. They have many physiological functions, for example, as hormonal messengers, cytokines, antimicrobial agents and protease inhibitors. They also serve as information carriers, reflecting the status of the entire organism since they are generated as a result of metabolism, especially proteolytic degradation of larger proteins. Peptidomics, a novel paradigm in the proteomics research field, is based on the comprehensive characterization of native peptides in complex mixtures (Gänshirt et al., 2005). Research in peptidomics is aimed at understanding the biological and medical relevance of peptides by analyzing their topological distribution between sites, their concentration and structure, and their functional relevance. For drug research and development, peptidomics offers several important and attractive research opportunities, such as the identification of novel peptide drug candidates or drug targets. Identification of peptides as biomarkers is used to explain drug effects and enables differentiation between patient groups (e.g., responder vs. nonresponder). Peptides are also important since they reflect the activity of proteases. Proteases are at the center of medical indications with paramount socioeconomic importance, where they serve as drug targets.

Numerous areas of medical research, such as diabetes, cardiovascular disease, and degenerative diseases will benefit from the novel opportunities offered by peptidomics technologies.

6.2
Peptides in Drug Research

6.2.1
History of Peptide Research

Proteins (from *proteios,* Greek for primary) are considered *the* primary material of life, and the relevance and ubiquitous role of peptides (from *peptein,* Greek for digestion) have been investigated in biochemistry and drug research since their beginning. Early breakthroughs in peptide chemistry have resulted in an exciting history of discoveries of bioactive peptides, particularly peptide hormones with biologically relevant activities. Peptide chemistry has always been a technology-driven area. Pioneering work was related to the development of separation technologies, of analytics and of synthesis strategies for peptides. Drug research in the field of peptide hormones experienced a pivotal breakthrough in the development of insulin therapy. The basis of this success story was the research performed by Banting and Best in Toronto and their first therapeutic treatment of diabetic patients using pancreas extracts (Banting et al., 1991 and Rosenfeld, 2002). It took decades until the primary structure of this peptide hormone was finally resolved by Sanger (1959) and the peptide was developed to its current use as a drug. In the second half of the last century, peptide research evolved extensively, and various peptides of physiological importance were identified. From the early days of peptide hormone discovery, prominent examples are the hypophyseal/ hypothalamic peptides such as the blood-pressure-regulating vasopressin, which was copurified with oxytocin (Acher and Chauvet, 1953 and Acher et al., 1958), the neuropeptides TSH-releasing factor (TRH) (Burgus et al., 1970), luteinizing hormone–releasing hormone (LH–RH), and the growth hormone somatostatin, as well as the gastrointestinal hormones secretin, gastrin and cholecystokinin (for review see Sewald and Jakubke 2002 and Wieland and Bodanszky, 1991). Later, in the 1980s, growth factors and cytokines were discovered and many more bioactive compounds were detected, paralleled by the development of improved analytical methods (Rehfeld, 1998). The ever-increasing number of peptide hormones isolated allows more and more insight into the biochemistry of almost every important physiological function of the human body.

The progress of peptide research was expedited by the completion of the Human Genome Project. The availability of DNA sequence information has made it possible to rapidly discover new peptides by screening DNA and expression libraries. Using bioinformatic tools (Antelmann et al., 2001 and Chen et al., 2003), numerous as yet unknown putative polypeptides have been predicted as candidates with secretory properties or as proteases.

The term peptidomics was first introduced by us in February 2000 at the ABRF conference "From Singular to Global Analyses of Biological Systems", organized by R. Aebersold in Bellevue, USA (Schrader and Schulz-Knappe, 2001). Its genesis was as a short version of peptide proteomics, and meant the comprehensive qualitative and quantitative analysis of all peptides and small proteins of a given biological system. A second research group headed by L. Schoofs at the University of Leuven, Belgium, independently introduced the term in parallel (Clynen et al., 2001), and, in 2001, a couple of further publications already included the term peptidomics in their title (Bergquist and Ekman, 2001; Ramstrom and Bergquist, 2004; Schrader and Schulz-Knappe, 2001; Schulz-Knappe et al., 2001, Verhaert et al., 2001). The list of publications in the field of peptidomics has grown slowly but steadily over the last few years. Another usage of the term peptidomics, not discussed in this chapter, is based on screening antibody peptide interactions for protein expression profiling (Scrivener et al., 2003).

6.2.2
Brief Biochemistry of Peptides

Peptides are polymers of amino acids covalently linked by peptide bonds. There is no official, clear-cut definition to differentiate between peptides and proteins (IUPAC–IUB Joint Commission on Biochemical Nomenclature). Historically, amino acid sequences of up to 10–20 residues are called oligopeptides. Compounds containing more than 50 amino acid residues are called proteins. Here, we use the term peptide for peptides and small proteins in the range from dipeptides to molecules of about 20 kDa molecular weight, corresponding to approximately 150 amino acid residues. This choice is based mainly on the differences between the physicochemical properties of peptides and proteins. Native peptides are usually more hydrophilic than proteins and they tend to dissolve easily in aqueous solutions. Peptides contain secondary structure elements, but do not form rigid three-dimensional structures and, consequently, their folding behavior is mostly reversible. The stability of peptides, for example at higher temperatures, is thus higher than that of proteins, which are denatured easily, a process that is usually irreversible.

With the exception of prokaryotes and lower eukaryotes the synthesis of peptides starts with ribosomal synthesis of larger protein precursors, which are subsequently processed proteolytically. These peptides are therefore indirect gene products being tailored by a series of processing steps. Precise proteolytic processing and degradation is considered to be the most frequent posttranslational modification. In some cases, a single protein precursor is cleaved into a multiplicity of peptides. Prominent examples are the numerous peptide hormones resulting from proteolytic processing of the pro-opio-melanocortin (POMC) hormone precursor in the brain or the glucagon precursor in the pancreas and gut. Such processing of secreted peptide hormones often takes place at dibasic cleavage sites and is catalyzed by specific prohormone convertases within intracellular vesicles (Seidah and Chretien, 1999 and Steiner, 2002). Based on the prohormone conversion cocktail present in a certain tissue, the outcome of processing of a prohormone can be tissue-specific and rather diverse. Posttranslational modifications such as glycosylation and phosphorylation are less frequent in peptides. Disulfide bonds which stabilize a certain spatial three-dimensional structure are found more often in peptides (Sillard et al., 1993), for instance in peptide hormones (e.g., insulin, endothelin, natriuretic peptide family) or peptide antibiotics such as defensins (Hancock and Lehrer 1998; Rehfeld, 1998). Disulfide bonds can lead to rigid stabilizing structures. They, as well as N-terminal pyroglutamylation or C-terminal amidation (Harris, 1989) prevent peptides from premature proteolytic degradation and thus improve the half-life of many peptides with biological function.

6.2.3
Peptides as Drugs

Peptides are very potent compounds which have a promising potential for use as drugs. Examples of peptides currently used as drugs are peptide hormones such as insulin, parathyroid hormone (PTH), brain natriuretic peptide (BNP) and calcitonin (Table 6.1). Other examples are antimicrobial peptides (Ganz, 2004; Hancock and Lehrer, 1998; and Zasloff, 1987) protease inhibitors, and growth factors. However, the usage of peptide drugs has one important drawback: oral application of peptide drugs is difficult, as peptides are usually rapidly degraded by proteolysis after ingestion. This is a major reason why peptide therapeutic drugs tend to have a negative image in the industry compared to other classes of compounds, although it is often difficult to find adequate molecular substitutes

Table 6.1 Peptide-based pharmaceuticals.

Trade name	Corresponding peptide or small protein	Molecular mass (kDa)	Company	Indication	Sales (2002) million US$
Humulin	Insulin	5.7	Eli Lilly	Diabetes	1004
Humalog	Insulin	5.7	Eli Lilly	Diabetes	834
Betaseron	Interferon-β	23	Schering	Multiple Sclerosis	740
Glucagon	Glucagon	3.5	Novo Nordisk	Hypo-glycaemia	110
Forteo	Teriparatide [human parathyroid hormone (1–34)]	6	Eli Lilly	Osteoporosis	> 2000
Miacalcin	Calcitonin	3.4	Novartis	Osteoporosis	435
Genotropin/ Somatropine	Human growth hormone	22	Pfizer	Dwarfism	551
Sandimmune	Cyclosporine	1.2	Novartis	Transplant rejection	1037
Refludan	Lepirudin (recombinant hirudin)	6.98	Schering	Homeostasis	20
Octreotide Sandostatin	Somatostatin (octapeptide)	1	Novartis	Acromegaly	314
Natrecor	Nesiritide [synthetic ana-logue of human B-type natriuretic peptide (BNP)]	3.5	Scios	Congestive heart failure	107

for potential peptide drugs. Insulin has been the most important peptide drug on the market with an annual turnover of several billion Euros for a number of decades. Despite huge investments made by the pharmaceutical industry, there is no adequate small molecule known to replace it. Peptide drugs usually have to be administered by injection. To overcome this limitation, current developments aim at other pharmaceutical formulations such as products suitable for intranasal, enteral or inhalational administration (Kochendoerfer, 2003; Niwa et al., 1995 and Owens, 2002). Many biopharmaceutical products currently in late-stage clinical development or under approval for therapeutic applications are proteins with a molecular mass of less than 40 kDa. We expect that the development of convenient delivery methods will facilitate the acceptance of future peptide drugs.

The characterization of peptides with peptidomics may also be indirectly used for drug development: peptide identification leads to information regarding new target enzymes or receptors. In these cases, the resulting products to be developed are most likely enzyme inhibitors, such as angiotensin-converting enzyme inhibitors or angiotensin receptor antagonists which serve as antihypertensive drugs, or agonists such as LH–RH agonists for the treatment of prostate cancer.

6.2.4
Peptides as Biomarkers

Physiological as well as pathological changes are reflected by the synthesis and the metabolism of proteins and peptides. A biomarker is defined as: "A characteristic that can be measured and evaluated as an indicator of normal biologic processes, pathologic processes or pharmacologic responses to therapeutic intervention" (NIH Biomarker Definitions Working Group, 1998). Biomarker changes are detectable in tissues, and in extracellular fluids, which represent the major link between cells, tissues and organs of an organism as a consequence of secretion, transport or diffusion and further processing of proteins and peptides. Biomarkers are important tools for clinical research and diagnostic use (Aebersold and Mann, 2003; Davidsson et al., 2003; Ilyin et al., 2004 and Zolg and Langen, 2004). Body fluids are the most important specimens in medical research and diagnostics. Of particular interest for diagnostic purposes are readily accessible clinical sample sources such as blood, plasma, serum, urine, cerebrospinal fluid, and, to a lesser extent, lacrimal fluid (tears) or saliva. Very often, a crucial factor is not the detection technology, but the type, quality and quantity of the clinical sample tested.

6.2.5
Clinical Peptidomics

At first glance, blood is the ideal sample from which to obtain a representative clinical sample suitable for peptide research. However, its complexity and the high abundance of certain proteins, such as albumin, fibrinogen, and immunoglobulins, render blood probably the most difficult of all proteomic samples

(Anderson and Anderson, 2002). Moreover, blood consists of a variety of blood cells, all capable of releasing peptides or proteases, which may unpredictably influence the peptide composition of the sample. Finally, coagulation is a process heavily influencing peptide content due to the action of a complex cascade of proteases such as thrombin or factor Xa. Different strategies have been applied to address this issue. These include removal of highly abundant proteins from the sample or fractionation of the sample according to the molecular masses of the analytes of interest (Rose et al., 2004; Schrader and Schulz-Knappe, 2001 (Tammen et al., 2005) and Villanueva et al., 2004). Because of the intrinsic complexity of blood, the analysis of blood peptides remains a challenge, but will certainly be solved, step by step, due to its great importance in clinical diagnostics and research. In addition to the efforts in the biotech and pharmaceutical industries, a comprehensive, multinational approach to unraveling this biological source has already started with the emergence of the Human Plasma Proteome Project.

Peptides from the insulin precursor are good examples for peptide biomarkers present in blood. Insulin itself is used as a therapeutic compound but, at the same time, it is also used as a biomarker to diagnose diabetes (Table 6.2). In addition, byproducts released during hormone processing are used as diagnostics. For example, the connecting peptide (C-peptide) is analyzed as a diabetes biomarker and compared to insulin. It is easier to measure than insulin as its biological half-life is longer, serving as a long-term indicator of insulin concentration. The insulin precursor itself is the third derivative from the insulin gene which is currently evaluated to serve as an indicator for beta-cell failure (Pfützner et al. 2004). Another prominent example of a peptide biomarker is procalcitonin, which is the current gold standard for diagnosis of sepsis (Table 6.2). A further very recent example of

Table 6.2 Peptide-based diagnostics and biomarkers.

Peptide	Molecular mass (kDa)	Disease
Insulin	5.7	Diabetes
C-peptide	3.1	Diabetes
Proinsulin	9.5	Diabetes
Brain natriuretic peptide (BNP)	3.5	Congestive heart failure
N-terminal-pro BNP (NT-proBNP)	8.5	
Amyloid Aβ 1–40	4.1	Alzheimer's disease
Amyloid Aβ 1–42	4.3	Alzheimer's disease
Trypsinogen-activating peptide (TAP)	0.9	Pancreatitis
Procalcitonin (PCT)	14	Bacterial infection/sepsis
β-2 Microglobulin	12	Renal failure
Osteocalcin	5.8	Osteoporosis
Cystatin C	13	Renal failure

a peptide used as a biomarker for the diagnosis of cardiovascular diseases such as heart failure is brain natriuretic peptide (BNP). Whereas the C-terminal bioactive BNP is used for treatment of acute heart failure, an assay specific for the N-terminal propiece of 8.5 kDa (NT-proBNP) (Hobbs, 2003; Karl et al. 1999; Wei et al., 1993 and Yamamoto et al., 1996) is now applied for the evaluation of chronic heart failure in the clinic with great success (Table 6.2). In addition to diagnostic uses, such molecular parameters or biomarkers may prove useful for monitoring the success of a therapy if other clinical parameters are not available. Such biomarkers are needed to monitor degenerative diseases with slow progression. Beta-amyloid peptide is a well-known example of such a biomarker, indicating the status of amyloid and plaque formation in the cerebrospinal fluid of patients suffering from Alzheimer's disease (Aebersold and Mann, 2003 and Blennow and Hampel, 2003) (Table 6.2). Expression profiling of RNAs or proteins has become a promising means to investigate the heterogeneity of cancer. Several studies have already demonstrated the suitability of peptidomic profiling of cancer cell lines (Sasaki et al., 2002; Sato et al., 2002 and Tammen et al., 2003). For example, peptide-based profiling of normal human mammary epithelial cells and the breast cancer cell line MCF-7 revealed complex peptide patterns comprised of up to 2300 peptides. Most of these peptides were common to both cell lines, although about 8% differed in their abundance (Tammen et al., 2003). Some of these differences were shown to originate from proteins supposably involved in tumor formation, progression and metastasis, and may serve as future biomarkers. With better molecular parameters it will be possible to accelerate drug development and improve treatment in such diseases. During the development of novel peptide biomarkers, validation strategies are required and, finally, the transformation of an analytical result into an applicable assay format which can be used in a clinical setting. Immunoassays such as radioimmunoassay (RIA) or enzyme-linked immunosorbent assay (ELISA) are the most widespread analytical methods used in clinical routine and drug development and are the most likely assay systems for novel peptide biomarkers.

6.3
Development of Peptidomics Technologies

6.3.1
Evolution of Peptide Analytical Methods

Bioactive peptides usually occur in small quantities and concentrations within complex biological matrices. This is for two reasons: firstly, bioactive peptides interact with a cell-surface-bound receptor to convey a message to the cell that is further amplified by second messengers or other signal transduction cascades. Therefore, they are usually produced in low concentrations along with other peptides. Secondly, because the bioactive peptides usually affect carefully balanced biological regulatory systems, their localization and half-life are precisely controlled by (proteolytic) inactivation or renal excretion. Therefore, looking for bioactive peptides is akin to "finding unknown needles in a haystack" (Mutt, 1992). As a result, peptide research is substantially driven by innovations in analytical chemistry. Owing to the differences between peptides and proteins regarding physicochemical properties, as already mentioned, different analytical strategies have to be applied to peptides than to larger proteins. Two major developments have had a very stimulating effect on peptide research. First, the development of novel separation methods since the 1960s, including high-performance liquid chromatography (HPLC) instrumentation in parallel with the development of a large variety of high-performance chromatographic media (e.g., reversed phase), enabling purification of low-concentration compounds from complex matrices (Andersson et al., 2000; Pearson et al., 1982 and Schulz-Knappe et al., 2001). In particular, the impressive technological boost in mass spectrometry (MS) over the last 20 years (Aebersold and Mann, 2003 and Roepstorff, 1997) has improved peptide analysis by several orders of magnitude in terms of sensitivity, specificity and speed. A major breakthrough is being able to softly ionize intact peptide molecules and accelerate them in a vacuum without substantial loss of peptide integrity and signal intensity, since fragmentation of peptides could now be prevented. The two major ionization techniques used for peptides are electrospray ionization (ESI) developed by J. B. Fenn and coworkers (Fenn et al., 1989) and matrix-assisted laser desorption/ionization (MALDI) invented by M. Karas and F. Hillenkamp (Karas et al., 1987; Karas, 1996). The technology was awarded the 2002 Nobel Prize for Chemistry, shared by John B. Fenn and Koichi Tanaka.

The high sensitivity of MALDI–MS and the low interference by salts and other components allows the detection of peptide hormones from complex biological mixtures with reduced sample preparation efforts. It is now possible to analyze tissue preparations, whole animal slices, and single cells. A review by Garden et al. (Li et al., 2000) describes this exciting phase of the development of MALDI–time-of flight (TOF)–MS. Other applications include the comprehensive analysis of large natural peptide libraries from blood filtrate with a molecular cut-off at about 20 kDa (Schulz-Knappe et al., 1997) obtained from patients with end-stage renal disease. Large-scale purification of known peptide hormones in milligram amounts (Schrader et al., 1997) as well as large-scale sequencing of blood plasma peptides have also been performed (Raida et al., 1999 and Richter et al., 1999). The latter concept was directly applied to blood plasma, confirming many of the prior peptide identifications (Rose et al., 2004).

6.3.2
Peptidomic Profiling

Transcriptomic expression analysis is a well established high-throughput techno-logy for profiling gene expression in a certain cell or tissue type. It is used to classify compounds and profile a drug's action on a global scale, displaying the changes in a subset of genes. It should be noted that protein expression may not correlate well enough with mRNA levels, since not only the rate of biosynthesis, but also the speed of the degradation processes of mRNAs and proteins determine the abundance of a given protein (Anderson and Seilhamer, 1997; Gygi et al., 1999). Furthermore, posttranslational modifications alter the properties of proteins. This cannot be detected at the mRNA level (Rehfeld and Goetze, 2003).

The overwhelming diversity of proteomes hinders the identification of relevant candidate molecules for commercial purposes. At the inception of this technology, the massive parallel sequencing approach used for the genome identification was considered feasible for proteomics as well. We note that today the strategy has changed toward sequence analysis of more restricted subproteomes, generally in combination with a comparative approach aiming for differentially expressed proteins. Thus, differential display proteomics (Pandey and Mann, 2000) starting with a well defined biological hypothesis, a carefully chosen sample number and a technology able to filter large data sets is favored. Semi-quantitative two-dimensional gel electrophoresis is still the state-of-the-art technology in proteomics. To perform differential peptide profiling, MS has to be applied in a quantitative way. This is still a major challenge concerning the robustness and standardization of sample preparation and measurement routines.

Quantitative methods for protein or peptide analysis using MS have been under development for several years (Pandey and Mann, 2000). Most of these approaches relied on liquid chromatography (LC) coupled to MS (LC–MS) as the analytical method (John and Standker, 2004 and Wang et al., 2003). As a strategy for (absolute) quantification, a chemical label is attached to the analytes (peptides) of a sample, allowing relative quantification of many analytes (Sechi and Oda, 2003), whereas for single analytes, standard calibration curves have to be determined individually (Bucknall et al., 2002). For MALDI–TOF–MS in particular, it was discussed at length whether it would ever become feasible for quantification purposes. A first basic differential display was performed employing a few different spectra from rat neurointermediate lobes (Jimenez et al., 1997) and the hemolymph of *Drosophila* (Uttenweiler-Joseph et al., 1998), respectively. From this it was anticipated that applications in drug research using complex biological samples and sophisticated model systems was possible, if these basic research methods could be translated into robust and standardized routine protocols.

Combination of MS with LC led to the development of platform technologies such as ESI–LC–MS and LC–MALDI–MS. All these technologies share in principle the comprehensiveness for peptide detection and the ability to cope with the complexity of biological samples. A semi-quantitative peptidomic profiling of extracts of peptides to analyze complex biological systems in a comparative manner was described and patented by us (Forssmann et al., 1996 and Schrader and Schulz-Knappe, 2001). Owing to optimized sample preparation, separation techniques and MS, all implemented within a special bioinformatic data analysis software package (Spectromania) (Lamerz et al., 2005), powerful peptidomic profiling technologies have further matured and are now applied in biomarker and drug discovery as well as in human epidemiological and clinical studies (Sasaki et al., 2002; Schrader and Schulz-Knappe, 2001 and Stoeckli et al., 2001). Peptidomic profiling will add further strategic value to conventional protein and mRNA expression profiling studies.

Peptide profiling can be performed at different levels of resolution and sensitivity. A typical example for a less comprehensive one-dimensional approach is the surface-enhanced laser desorption/ionization (SELDI) platform marketed by Ciphergen Biosystems. This technology is related to MALDI, and uses a chemically modified chip surface to selectively capture and enrich peptides and proteins present in a sample for subsequent mass spectrometric analysis. Quite a number of SELDI–MS biomarker analyses have been published, mainly addressing peptides and small proteins as biomarker candidates from human serum samples (Chen et al., 2004; Carrette et al., 2003 and Petricoin and Liotta, 2004). Other technologies are currently evolving using magnetic beads or batch extraction steps for peptide extraction from body fluids (Pusch et al., 2003).

6.3.3
Top-Down Identification of Endogenous Peptides

As a result of a differential peptide profiling, certain peptides that differ significantly between the compared groups are selected for subsequent sequence determination. This is a prerequisite for meaningful interpretation of the biological context, for supporting the validation process and enabling development and application of the findings. At the present time, most peptide sequence information is generated by MS methods combined with database comparison. This reflects the high-throughput capabilities of the technique compared to earlier methods such as Edman sequencing or amino acid analysis. In conventional proteomics, protein sequences are identified by trypsin digestion of proteins to produce characteristic peptide mass fingerprints, which are compared with *in silico* digests of large databases. As native peptides usually contain only a few proteolytic cleavage sites, this approach is not applicable to peptide structure determination. The identification of the amino acid sequence of peptides and small proteins is achieved by employing a so-called top-down approach, which involves high-resolution mass spectrometric measurement and fragmentation of intact, ionized molecules in the gas phase (Kelleher, 2004). The lack of practical experience in top-down identification of endogenous peptides from complex mixtures has recently been overcome by the development of optimized experimental protocols, and it seems to be a promising strategy for future studies (Baggerman et al., 2004; Möhring et al., 2005).

The top-down methodology is generally applicable for peptides from body fluids, tissues or other biological sources. This approach also allows the identification of posttranslational modifications. From our experience, we would recommend adjusting the type of instrumentation according to the maximal molecular mass of the peptides and proteins under investigation. In the molecular mass range of up to 3 kDa, MALDI–TOF–TOF–MS/MS is readily applicable (for example Ultraflex, Bruker, D; 4700 proteomics analyzer, Applied Biosystems, USA). Above that range, hybrid ESI–TOF instrumentation (for example Q–ToF, Micromass, UK; QSTAR, Applied Biosystems) is recommended. If the molecular mass range of the proteins of interest clearly exceeds 10 kDa, the application of either classical peptide mapping (Aebersold and Mann, 2003) or application of top-down Fourier transform ion cyclotron resonance (FT–ICR–MS) should be favored (Kelleher, 2004). The latter has the advantage of potential prepurification of the molecules in the mass spectrometer, but requires even more sophisticated instrumentation.

6.4
Applications of Differential Display Peptidomics

6.4.1
Peptidomics in Drug Development

The development of new therapeutic drugs is a risky, expensive and lengthy process. A recent survey (DiMasi et al., 2003) estimated that the average costs of bringing a new drug or new chemical entity (NCE) to market is US$ 800 million. The drug development process, which is the time from discovery to market, can take 10–15 years. It is divided into a preclinical development phase of 2–4 years, a clinical development phase of 3–6 years and additional time for regulatory affairs. Only approximately 10% of NCEs entering development make it to market. While the productivity of drug discovery and early development has increased over the years, the number of NCEs submitted to the FDA to enter phase I clinical trials has decreased by 50% in 2002–2003 compared to 1996–1997. Approximately 58% of the overall costs for drug development are spent on clinical trials; however, ~50% of new drugs fail in late-stage phase III clinical trials. Major causes for terminating drug development are concerns regarding safety and toxicity as well as lack of efficacy. Further costs are generated postapproval because the pharmaceutical industry has to monitor the feedback from healthcare professionals to identify long-term side effects, risk factors and interactions with other drugs. The growing economic pressures in pharmaceutical markets carry the risk that pharmaceutical companies will be forced to favor the development of drugs with the potential for lucrative markets of > US$ 1 billion, but will discontinue the development of new therapeutic drugs for orphan diseases with lower expected revenues.

In order to increase productivity and success, it is essential to accelerate drug development times and to reduce attrition rates in cost-intensive phase III clinical trials (DiMasi, 2002). What is needed to improve the success rates is to profile NCEs more rigorously at the preclinical stage. Lead compounds that fail earlier in development would substantially reduce attrition and costs in clinical trials. Most pharmaceutical companies have now recognized the need to implement biomarker discovery strategies as early as possible to advance the drug development process (Richardson, 2002 and Walgren and Thompson, 2004). It is expected that biomarkers can predict the outcome of late-stage clinical trials with a high degree of certainty early in the process. In the future, biomarkers will play an important role in drug safety and efficacy measurements as well as in the identification of interindividual variations in drug responses (Frank and Hargreaves, 2003).

The biomarker definition proposed by the NIH group is not conceptually restricted to proteins as indicators of a biological or pathobiological process (Lesko and Atkinson, 2001). However, as most drugs directly target proteins or affect their expression, the biomarker initiatives increasingly utilize proteomics and, in particular, peptidomics technologies (Richardson, 2002; Zolg and Langen, 2004). Technically, proteomics and peptidomics are complementary approaches to monitor alterations in the protein and peptide composition within an organism. A major goal of proteomics is to characterize protein expression (in a mass range of 10–200 kDa) on a global scale. However, the proteome is not static, but of a highly dynamic nature. Many proteins and enzymes are regulated by posttranslational modifications that alter their activity but not necessarily their abundance. Furthermore, proteolytic processing is an important mechanism that influences the distribution, activity, function and abundance of proteins in an organism (Overall et al., 2004). Proteases play a role in diverse physiological processes such as digestion, protein maturation, peptide hormone processing, tissue remodeling, removal of misfolded or damaged proteins and regulation of signaling pathways. Proteolytic processing generates small proteins and peptides and peptidomics technologies have been created as a comprehensive platform to detect and investigate and discover those dynamic alterations (Baggerman et al., 2004; Schrader and Schulz-Knappe, 2001).

The drug discovery process begins with target identification. With the completion of the Human Genome Project, genomics and transcriptomics have flooded the pharmaceutical industry with numerous novel targets that have been difficult to validate in a disease context. A major concern is that message abundance (mRNA) does not necessarily correlate with protein activity or abundance. How can peptidomics technologies be applied in target discovery? Important aspects include:

- the characterization of protease substrates, cleavage products and specificity of inhibitors;
- the discovery of novel peptide hormones as therapeutic targets or leads;
- the discovery of novel peptides as a response to transcriptional activators.

An estimated number of 500 genes encoding proteases have been identified in the human genome and many are of medical importance (Puente et al., 2003). A comprehensive information system of proteases and inhibitors is provided by the MEROPS database (Rawlings et al., 2004). Proteases are attractive drug targets for the treatment of thrombosis (thrombin and factor X), hypertension [angiotensin converting enzyme (ACE) and neutral endopeptidase (NEP)], cancer [matrix metalloproteinases, (MMP) and cathepsins], HIV infection and diabetes [dipeptidylpeptidase IV (DPPIV)]. A major challenge is to understand the physiological functions of novel proteinases in tissues and organs and the pathophysiological effects associated with altered proteolytic activity. To evaluate a protease as a potential drug target, it is important to identify its substrates and bioactive cleavage products. Detailed knowledge about protease substrates has a big impact on the drug development process since specificity of inhibitors, and thus limited side effects can be predicted more accurately.

Traditionally, the specificity of proteases is determined using standard or optimized peptide libraries containing small recognition sites or, for *de novo* screening, phage display libraries (Richardson, 2002). These techniques may prove to be useful as a first screening approach. However, they are insufficient to predict the outcome of inhibiting a proteolytic activity in an *in vivo* situation.

In the following, emerging new treatments for type 2 diabetes mellitus (T2DM) are presented that are based on inhibiting the peptide-processing peptidase DPPIV (Drucker, 2003). Several pharmaceutical companies (Novartis, Merck, GlaxoSmithKline, Prosidion, Bristol-Myers Squibb) have DPPIV inhibitors for the treatment of diabetes in advanced stages of clinical evaluation. DPPIV, which is identical to the lymphocyte glycoprotein CD26, is a multifunctional cell surface serine protease with broad tissue expression (de Meester et al., 2003). A soluble form of DPPIV is shed into the circulation and selectively removes the N-terminal dipeptide from a variety of bioactive peptides with proline or alanine at the second sequence position. DPPIV plays a role in glucose homeostasis through proteolytic inactivation of peptide hormones, termed incretins, released from the gut in response to food ingestion. The incretin hormones GLP-1 (glucagon-like peptide-1) and GIP (glucose-dependent insulinotropic polypeptide or gastric inhibitory peptide) stimulate insulin secretion in a glucose-dependent manner and are rapidly inactivated by DPPIV (Mentlein et al., 1993). Inhibiting DPPIV activity interrupts this pathway of peptide hormone degradation and enhances insulin secretion. Likewise, mice deficient in DPPIV/CD26 are protected against diabetes and obesity (Conarello et al., 2003; Marguet et al., 2000). The positive effect on glucose tolerance in type 2 diabetes rodent models and patients led to the development of specific DPPIV inhibitors that enhance incretin concentration and action (Ahren et al., 2004; Reimer et al., 2002). Apart from its proteolytic activity, DPPIV/CD26 modulates chemotaxis and acts as a costimulatory T-cell molecule. It also interacts with the extracellular matrix proteins fibronectin and collagen, adenosine deaminase, HIV gp120 protein, the chemokine receptor CXCR4, and the tyrosine phosphatase CD45. Furthermore, DPPIV plays an important role in tumor progression. Inhibiting DPPIV activity may therefore be a novel approach to treating cancer (Adams et al., 2004). Susceptibility to DPPIV cleavage has been shown for a wide range of substrates in *in vitro* assays. Truncation of several cytokines and chemokines having a Xaa-Pro motif alters their activity and receptor-binding profile (de Meester et al., 2003). Although of great relevance for predicting potential side effects of DPPIV inhibitors on the immune system, much less is known about endogenous DPPIV substrates. This is in part due to the lack of specific ELISAs for truncated peptides. To overcome these technical limitations, we are currently pursuing a peptidomics study to identify DPPIV substrates in tissues and body fluids, which are relevant *in vivo*. Owing to its broad tissue expression, DPPIV influences several systems by blunting the signaling capacity of peptide mediators. Thus, DPPIV inhibition can also be exploited as a discovery tool to systematically screen for endogenously processed peptides that may subsequently be evaluated as therapeutic peptides. This approach can also be applied to the search for new modes of action of established drugs, for instance to extend the use of DPPIV inhibitors in antitumor therapy.

6.4.2
Peptidomics Applied to in vivo Models

To acquire information about endogenous protease substrates, a discovery method is required to be sensitive enough to detect qualitative differences in the presence of highly abundant proteins. To reveal the plausibility of a peptidomics approach, we chose the obese *fat/fat* mouse that bears a point mutation in the gene encoding carboxypeptidase E (Cpe) (Leiter, 1997; Naggert et al., 1995). Cpe is an ectopeptidase that trims off basic amino acids from peptides remaining after cleavage of pro-hormone processing enzymes. The existence of several well-defined substrates and the potential to screen for novel substrates make the Cpe$^{fat/fat}$ mouse an attractive model system (Bures et al., 2001; Fricker et al., 1996). We performed a peptidomics analysis with the aim of identifying differentially processed peptide hormones in pancreatic tissue that are implicated in type 2 diabetes. Cpe$^{fat/fat}$ have an obese phenotype and elevated blood glucose levels, but treatment of these mice with insulin can normalize blood glucose levels. Abnormalities in the insulin secretory capacity of pancreatic beta-cells in response to glucose uptake is a common feature of type 2 diabetes. In humans, rare mutations in the gene encoding pro-hormoneconvertase-1/3 (PC1/3), which acts proximally to Cpe, have been reported (Jackson et al., 1997). The phenotype in patients bears similarities to the *fat/fat* mouse model. To display alterations in the cleavage products associated with impaired Cpe activity, we compared pancreatic peptide extracts from Cpe$^{fat/fat}$ mice with lean control mice. Figure 6.1 demonstrates how the decreased processing activity of Cpe can be visualized in peptide maps generated by applying peptidomics technologies. Peptide maps of both groups were combined and qualitative and quantitative differences displayed as follows: peptides not differing in abundance in both groups appear in black and shades of gray, whereas peptides increased in Cpe-deficient mice are depicted in red and decreased peptides in blue. The changes observed in Cpe$^{fat/fat}$ mice reveal a characteristic pattern that corresponds to a mass shift of one to two arginine or lysine residues and a corresponding chromatographic shift due to increased hydrophilicity of peptides. Several changes in Cpe$^{fat/fat}$ mice were detected using this technique. As a well established substrate of Cpe in pancreatic beta-cells, the processing of the two insulin forms in Cpe$^{fat/fat}$ mice is strikingly impaired. The reduced Cpe activity causes increased levels of proinsulin in beta-cells which correlates well with the observed phenotype of reduced levels of mature insulin in plasma. In summary, the example highlights how peptidomics technologies can be applied to characterize prohormone processing, proteolytic activity and to identify substrates at the same time.

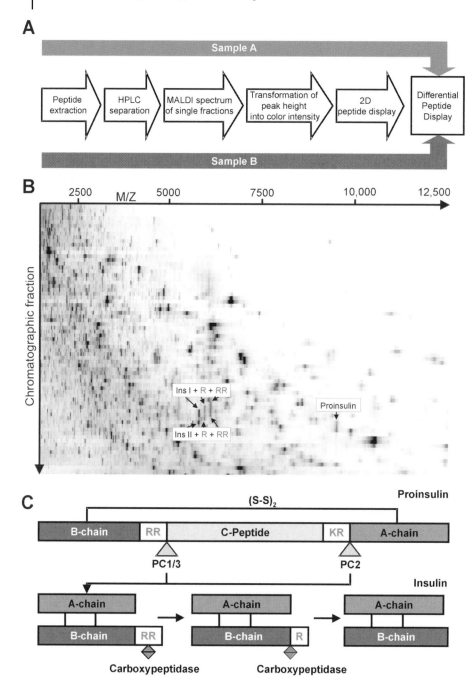

Figure 6.1 Characterization of impaired prohormone processing in mice lacking functional active carboxypeptidase E (Cpe$^{fat/fat}$ mice) by peptidomics.

6.5
Outlook

One century of peptide research has resulted in an enormous source of applications in drug development and clinical diagnostics. The old paradigm of one biological function or disease related to a single gene measured by a single marker or treated by a single drug, is already no longer valid (Miklos and Maleszka, 2001). With an estimate of 20 000–25 000 protein-coding genes in the human genome (Southan, 2004), a multiplicity of three to ten times the number of proteins is expected owing to variations during transcription and translation (Galas, 2001; Kettman et al., 2002). Further posttranslational and proteolytic processing leads to the generation of myriad peptides in a peptidome. At present, the peptidome of most relevant biological systems has not been explored. Peptidomics certainly encompasses a highly dynamic field of research to explore this new molecular *terra incognita*.

In drug development, novel bioactive peptides will be applied as blueprints for new therapeutic principles as is already the case for the incretin hormone GLP-1 and derivatives thereof in the treatment of type 2 diabetes (Holst and Deacon, 2004). Peptidomics technologies today have matured to a state where their application in clinical trials in large patient cohorts (up to 1000 patients) is possible, and the design and success of clinical studies might greatly benefit from these technologies. We hereby coin the term "clinical peptidomics" to refer to this kind of application. Clinical peptidomics will provide support to distinguish between

---

Figure 6.1

A Brief overview of the differential peptide display process: peptides extracted from individual samples are separated by reversed-phase (RP)–HPLC into 96 fractions (first dimension). Each fraction is analyzed by MALDI–TOF–MS to give the mass-to-charge ratio (m/z, second dimension). Signals are extracted and converted to a two-dimensional gel-like peptide display using the Spectromania software package, which transforms the peak height into a color intensity code. Absolute and relative differences between control and Cpe^{fat/fat} mice are calculated applying nonparametric statistics to give rise to a list of significantly altered peptides.
B Differential peptide display of control and Cpe^{fat/fat} mice. Peptide maps of both groups were combined and qualitative and quantitative differences displayed as follows: peptides not differing in abundance in the two groups appear in *black*, whereas peptides

increased in Cpe-deficient mice are depicted in *red* and decreased peptides in *blue*. All of these complex data processing steps were done using our proprietary Spectromania software package. As a prominent example, the two insulin isoforms expressed in mice are highlighted (insulin I, m/z 5801 and insulin II, m/z 5796). Reduced CpE activity causes the accumulation of incompletely processed C-terminally extended insulin I and insulin II peptides with a characteristic mass shift of one or two arginine residues.
C Insulin is released from its precursor by sequential proteolytic processing steps. Based on a current model, the processing of proinsulin is initiated by prohormone convertase 1/3 (PC1/3) cleavage at the B-chain to C-peptide junction followed by PC2 cleavage at the C-peptide to A-chain junction. PC1/3 and PC2 cleave prohormones C-terminally to basic amino acid residues, that are further trimmed off by Cpe to eventually release mature insulin into the circulation.

different types, stages, potential outcomes, reaction to distinct medical treatment strategies and patient-specific individual variations of a disease and response to treatment. Single peptide biomarkers or panels of such will pilot the way to an improved state of clinical development. Peptides are also likely to improve the efficiency of drug development programs. In summary, research on particular peptidomes will complement the human proteome initiatives, and will thus become an important tool in fostering pharmaceutical research.

Acknowledgements

We wish to thank our former colleagues at the Institut für Peptidforschung (now IPF Pharmaceuticals), Prof. Dr. Dr. Wolf-Georg Forssmann for his pioneering work in comprehensive peptide analysis from blood and Dr. Manfred Raida (present address Cellzome AG) for his invaluable work in analytical peptide chemistry. This work was supported in part by grants from the German Ministry for Education and Research, BMBF (FKZ 0312815), and the Lower Saxony Ministry for Economy and Technology (203.19-32329-5-354).

BioVisioN, Peptidomics, Differential Peptide Display and Spectromania are trademarks of BioVisioN AG, Hannover, Germany.

References

ACHER, R., CHAUVET, J. (1953). The structure of bovine vasopressin. *Biochim. Biophys. Acta* 12, 487–488.

ACHER, R., LIGHT, A., DU VIGNEAUD, V. (1958). Purification of oxytocin and vasopressin by way of a protein complex. *J. Biol. Chem.* 233, 116–120.

ADAMS, S., MILLER, G. T., JESSON, M. I., WATANABE, T., JONES, B., WALLNER, B. P. (2004). PT-100, a small molecule dipeptidyl peptidase inhibitor, has potent antitumor effects and augments antibody-mediated cytotoxicity via a novel immune mechanism. *Cancer Res.* 64, 5471–5480.

AEBERSOLD, R., MANN, M. (2003). Mass spectrometry-based proteomics. *Nature* 422, 198–207.

AHREN, B., LANDIN-OLSSON, M., JANSSON, P. A., SVENSSON, M., HOLMES, D., SCHWEIZER, A. (2004). Inhibition of dipeptidyl peptidase-4 reduces glycemia, sustains insulin levels, and reduces glucagon levels in type 2 diabetes. *J. Clin. Endocrinol. Metab.* 89, 2078–2084.

ANDERSON, L., SEILHAMER, J. (1997). A comparison of selected mRNA and protein abundances in human liver. *Electrophoresis* 18, 533–537.

ANDERSON, N. L., ANDERSON, N. G. (2002). The human plasma proteome: history, character, and diagnostic prospects. *Mol. Cell Proteomics* 1, 845–867.

ANDERSSON, L., BLOMBERG, L., FLEGEL, M., LEPSA, L., NILSSON, B., VERLANDER, M. (2000). Large-scale synthesis of peptides. *Biopolymers* 55, 227–250.

ANTELMANN, H., TJALSMA, H., VOIGT, B., OHLMEIER, S., BRON, S., VAN DIJL, J. M., HECKER, M. (2001). A proteomic view on genome-based signal peptide predictions. *Genome Res.* 11, 1484–1502.

BAGGERMAN, G., VERLEYEN, P., CLYNEN, E., HUYBRECHTS, J., DE LOOF, A., SCHOOFS, L. (2004). Peptidomics. *J. Chromatogr. B Analyt. Technol. Biomed. Life Sci.* 803, 3–16.

BANTING, F. G., BEST, C. H., COLLIP, J. B., CAMPBELL, W. R., FLETCHER, A. A. (1991). Pancreatic extracts in the treatment of diabetes mellitus: preliminary report 1922. *CMAJ* **145**, 1281–1286.

BERGQUIST, J., EKMAN, R. (2001). Future aspects of psychoneuroimmunology – lymphocyte peptides reflecting psychiatric disorders studied by mass spectrometry. *Arch. Physiol. Biochem.* **109**, 369–371.

BLENNOW, K., HAMPEL, H. (2003). CSF markers for incipient Alzheimer's disease. *Lancet Neurol.* **2**, 605–613.

BUCKNALL, M., FUNG, K. Y., DUNCAN, M. W. (2002). Practical quantitative biomedical applications of MALDI–TOF mass spectrometry. *J. Am. Soc. Mass Spectrom.* **13**, 1015–1027.

BURES, E. J., COURCHESNE, P. L., DOUGLASS, J., CHEN, K., DAVIS, M. T., JONES, M. D., McGINLEY, M. D., ROBINSON, J. H., SPAHR, C. S., SUN, J., WAHL, R. C., PATTERSON, S. D. (2001). Identification of incompletely processed potential carboxypeptidase E substrates from CpEfat/CpEfat mice. *Proteomics* **1**, 79–92.

BURGUS, R., DUNN, T. F., DESIDERIO, D., WARD, D. N., VALE, W., GUILLEMIN, R. (1970). Characterization of ovine hypothalamic hypophysiotropic TSH-releasing factor. *Nature* **226**, 321–325.

CARRETTE, O., DEMALTE, I., SCHERL, A., YALKINOGLU, O., CORTHALS, G., BURKHARD, P., HOCHSTRASSER, D. F., SANCHEZ, J. C. (2003). A panel of cerebrospinal fluid potential biomarkers for the diagnosis of Alzheimer's disease. *Proteomics* **3**, 1486–1494.

CHEN, Y., YU, P., LUO, J., JIANG, Y. (2003). Secreted protein prediction system combining CJ-SPHMM, TMHMM, PSORT. *Mamm. Genome* **14**, 859–865.

CHEN, Y. D., ZHENG, S., YU, J. K., HU, X. (2004). Application of serum protein pattern model in diagnosis of colorectal cancer. *Zhonghua Zhong Liu Za Zhi* **26**, 417–420.

CLYNEN, E., BAGGERMAN, G., VEELAERT, D., CERSTIAENS, A., VAN DER HORST, D., HARTHOORN, L., DERUA, R., WAELKENS, E., DE LOOF, A., SCHOOFS, L. (2001). Peptidomics of the pars inter-cerebralis–corpus cardiacum complex of the migratory locust, *Locusta migratoria*. *Eur. J. Biochem.* **268**, 1929–1939.

CONARELLO, S. L., LI, Z., RONAN, J., ROY, R. S., ZHU, L., JIANG, G., LIU, F., WOODS, J., ZYCBAND, E., MOLLER, D. E., THORNBERRY, N. A., ZHANG, B. B. (2003). Mice lacking dipeptidyl peptidase IV are protected against obesity and insulin resistance. *Proc. Natl. Acad. Sci. USA* **100**, 6825–6830.

DAVIDSSON, P., BRINKMALM, A., KARLSSON, G., PERSSON, R., LINDBJER, M., PUCHADES, M., FOLKESSON, S., PAULSON, L., DAHL, A., RYMO, L., SILBERRING, J., EKMAN, R., BLENNOW, K. (2003). Clinical mass spectrometry in neuroscience *Proteomics* and peptidomics. *Cell Mol. Biol.* (*Noisy-le-grand*) **49**, 681–688.

DE MEESTER, I., LAMBEIR, A. M., PROOST, P., SCHARPE, S. (2003). Dipeptidyl peptidase IV substrates An update on *in vitro* peptide hydrolysis by human DPPIV. *Adv. Exp. Med. Biol.* **524**, 3–17.

DIMASI, J. A. (2002). The value of improving the productivity of the drug development process: faster times and better decisions. *Pharmacoeconomics* **20** (Suppl. 3), 1–10.

DIMASI, J. A., HANSEN, R. W., GRABOWSKI, H. G. (2003). The price of innovation: new estimates of drug development costs. *J. Health Econ.* **22**, 151–185.

DRUCKER, D. J. (2003). Enhancing incretin action for the treatment of type 2 diabetes. *Diabetes Care* **26**, 2929–2940.

FENN, J. B., MANN, M., MENG, C. K., WONG, S. F., WHITEHOUSE, C. M. (1989). Electrospray ionization for mass spectrometry of large biomolecules. *Science* **246**, 64–71.

FORSSMANN, W. G., SCHULZ-KNAPPE, P., SCHRADER, M., OPITZ, H. G. (1996). Process for determining the status of an organism by peptide measurement [WO 98/07036] 13-8-1996. Published 19-2-1998.

FRANK, R., HARGREAVES, R. (2003). Clinical biomarkers in drug discovery and development. *Nat. Rev. Drug Discov.* **2**, 566–580.

FRICKER, L. D., BERMAN, Y. L., LEITER, E. H., DEVI, L. A. (1996). Carboxypeptidase E activity is deficient in mice with the fat mutation Effect on peptide processing. *J. Biol. Chem.* **271**, 30619–30624.

Gänshirt, D., Harms, F., Schulz-Knappe, P. (2005). *Peptidomics in Drug Development*. Editio Cantor Verlag, Aulendorf.

Galas, D. J. (2001). Sequence interpretation. Making sense of the sequence. *Science* 291, 1257–1260.

Ganz, T. (2004). Defensins: antimicrobial peptides of vertebrates. *C. R. Biol.* 327, 539–549.

Gygi, S. P., Rochon, Y., Franza, B. R., Aebersold, R. (1999). Correlation between protein and mRNA abundance in yeast. *Mol. Cell Biol.* 19, 1720–1730.

Hancock, R. E., Lehrer, R. (1998). Cationic peptides: a new source of antibiotics. *Trends Biotechnol.* 16, 82–88.

Harris, R. B. (1989). Processing of pro–hormone precursor proteins. *Arch. Biochem. Biophys.* 275, 315–333.

Hobbs, R. E. (2003). Using BNP to diagnose, manage, treat heart failure. *Cleve Clin. J. Med.* 70, 333–336.

Holst, J. J., Deacon, C. F. (2004). Glucagon-like peptide 1 and inhibitors of dipeptidyl peptidase IV in the treatment of type 2 diabetes mellitus. *Curr. Opin. Pharmacol.* 4, 589–596.

Ilyin, S. E., Belkowski, S. M., Plata-Salaman, C. R. (2004). Biomarker discovery and validation: technologies and integrative approaches. *Trends Biotechnol.* 22, 411–416.

Jackson, R. S., Creemers, J. W., Ohagi, S., Raffin-Sanson, M. L., Sanders, L., Montague, C. T., Hutton, J. C., O'Rahilly, S. (1997). Obesity and impaired prohormone processing associated with mutations in the human prohormone convertase 1 gene. *Nat. Genet.* 16, 303–306.

Jimenez, C. R., Li, K. W., Dreisewerd, K., Mansvelder, H. D., Brussaard, A. B., Reinhold, B. B., Van der Schors, R. C., Karas, M., Hillenkamp, F., Burbach, J. P., Costello, C. E., Geraerts, W. P. (1997). Pattern changes of pituitary peptides in rat after salt-loading as detected by means of direct, semiquantitative mass spectrometric profiling. *Proc. Natl. Acad. Sci. USA* 94, 9481–9486.

John, H., Standker, L. (2004). Peptide separation and analysis. *J. Chromatogr. B Analyt. Technol. Biomed. Life Sci.* 803, 1–2.

Karas, M., Bachmann, D., Bahr, U., Hillenkamp, F. (1987). Matrix-assisted ultraviolet laser desorption of non-volatile compounds. *Int. J. Mass Spectrom. Ion Processes* 78, 53–68.

Karas, M. (1996). Matrix-assisted laser desorption ionization MS: a progress report. *Biochem. Soc. Trans.* 24, 897–900.

Karl, J., Borgya, A., Gallusser, A., Huber, E., Krueger, K., Rollinger, W., Schenk, J. (1999). Development of a novel, N-terminal-proBNP (NT-proBNP) assay with a low detection limit. *Scand. J. Clin. Lab. Invest. Suppl.* 230, 177–181.

Kelleher, N. L. (2004). Top-down proteomics. *Anal. Chem.* 76, 197A–203A.

Kettman, J. R., Coleclough, C., Frey, J. R., Lefkovits, I. (2002). Clonal proteomics: one gene – family of proteins. *Proteomics* 2, 624–631.

Kochendoerfer, G. (2003). Protein & Peptide Drug Delivery – Third International Conference. Minimally invasive delivery methods 22–23 September 2003, Philadelphia, PA, USA. *IDrugs* 6, 1043–1045.

Lamers, J., Selle, H., Scapozza, L., Crameri, R., Schulz-Knappe, P., Mohring, T., Kellmann, M., Khamenia, V., Zucht, H.-D. (2005). Correlation-associated peptide networks of human cerebrospinal fluid. *Proteomics* 5, 2789–2798.

Leiter, E. H. (1997). Carboxypeptidase E and obesity in the mouse. *J. Endocrinol.* 155, 211–214.

Lesko, L. J., Atkinson, A. J. (2001). Use of biomarkers and surrogate endpoints in drug development and regulatory decision making: criteria, validation, strategies. *Annu. Rev. Pharmacol. Toxicol.* 41, 347–366.

Li, L., Garden, R. W., Sweedler, J. V. (2000). Single-cell MALDI: a new tool for direct peptide profiling. *Trends Biotechnol.* 18, 151–160.

Marguet, D., Baggio, L., Kobayashi, T., Bernard, A. M., Pierres, M., Nielsen, P. F., Ribel, U., Watanabe, T., Drucker, D. J., Wagtmann, N. (2000). Enhanced insulin secretion and improved glucose tolerance in mice lacking CD26. *Proc. Natl. Acad. Sci. USA* 97, 6874–6879.

Mentlein, R., Gallwitz, B., Schmidt, W. E. (1993). Dipeptidyl-peptidase IV hydrolyzes gastric inhibitory polypeptide, glucagon-like peptide-1(7-36)amide, peptide histidine methionine and is responsible for their degradation in human serum. *Eur. J. Biochem.* **214**, 829–835.

MIKLOS, G. L., MALESZKA, R. (**2001**). Protein functions and biological contexts. *Proteomics* **1**, 169–178.

MÖHRING, T., KELLMANN, M., JÜRGENS, M., SCHRADER, M. (**2005**). Top-down identification of endogenous peptides up to 9 kDa in cerebrospinal fluid and brain tissue by nanoelectrospray quadrupole time-of-flight tandem mass spectrometry. *J. Mass Spectrom.* **40**, 214–226.

MUTT, V. (**1992**). On the necessity of isolating peptides. In SCHNEIDER, C. H., EBERLE, A. N. (Eds.), *Peptides.* ESCOM, Leiden, pp. 3–20.

NAGGERT, J. K., FRICKER, L. D., VARLAMOV, O., NISHINA, P. M., ROUILLE, Y., STEINER, D. F., CARROLL, R. J., PAIGEN, B. J., LEITER, E. H. (**1995**). Hyperproinsulinaemia in obese fat/fat mice associated with a carboxypeptidase E mutation which reduces enzyme activity. *Nat. Genet.* **10**, 135–142.

NIWA, T., TAKEUCHI, H., HINO, T., KAWASHIMA, Y. (**1995**). Aerosolization of lactide/glycolide copolymer (PLGA) nanospheres for pulmonary delivery of peptide-drugs. *Yakugaku Zasshi* **115**, 732–741.

OVERALL, C. M., TAM, E. M., KAPPELHOFF, R., CONNOR, A., EWART, T., MORRISON, C. J., PUENTE, X., LOPEZ-OTIN, C., SETH, A. (**2004**). Protease degradomics: mass spectrometry discovery of protease substrates and the CLIP-CHIP, a dedicated DNA microarray of all human proteases and inhibitors. *Biol. Chem.* **385**, 493–504.

OWENS, D. R. (**2002**). New horizons – alternative routes for insulin therapy. *Nat. Rev. Drug Discov.* **1**, 529–540.

PANDEY, A., MANN, M. (**2000**). *Proteomics* to study genes and genomes. *Nature* **405**, 837–846.

PEARSON, J. D., LIN, N. T., REGNIER, F. E. (**1982**). The importance of silica type for reverse-phase protein separations. *Anal. Biochem.* **124**, 217–230.

PETRICOIN, E. F., LIOTTA, L. A. (**2004**). SELDI-TOF-based serum proteomic pattern diagnostics for early detection of cancer. *Curr. Opin. Biotechnol.* **15**, 24–30.

PFÜTZNER, A., KUNT, T., HOHBERG, C., MONDOK, A., PAHLER, S., KONRAD, T., LUBBEN, G., FORST, T. (**2004**). Fasting intact proinsulin is a highly specific predictor of insulin resistance in type 2 diabetes. *Diabetes Care* **27**, 682–687.

PUENTE, X. S., SANCHEZ, L. M., OVERALL, C. M., LOPEZ-OTIN, C. (**2003**). Human and mouse proteases: a comparative genomic approach. *Nat. Rev. Genet.* **4**, 544–558.

PUSCH, W., FLOCCO. M. T., LEUNG, S. M., THIELE, H., KOSTRZEWA, M. (**2003**). Mass spectrometry-based clinical proteomics. *Pharmacogenomics* **4**, 463–476.

RAIDA, M., SCHULZ-KNAPPE, P., HEINE, G., FORSSMANN, W. G. (**1999**). Liquid chromatography and electrospray mass spectrometric mapping of peptides from human plasma filtrate. *J. Am. Soc. Mass Spectrom.* **10**, 45–54.

RAMSTROM, M., BERGQUIST, J. (**2004**). Miniaturized proteomics and peptidomics using capillary liquid separation and high resolution mass spectrometry, *FEBS Lett.* **567**, 92–95.

RAWLINGS, N. D., TOLLE, D. P., BARRETT, A. J. (**2004**). MEROPS: the peptidase database. *Nucleic Acids Res.* **32**, Database issue, D160–D164.

REHFELD, J. F. (**1998**). The new biology of gastrointestinal hormones. *Physiol. Rev.* **78**, 1087–1108.

REHFELD, J. F., GOETZE, J. P. (**2003**). The post-translational phase of gene expression: new possibilities in molecular diagnosis. *Curr. Mol. Med.* **3**, 25–38.

REIMER, M. K., HOLST, J. J., AHREN, B. (**2002**). Long-term inhibition of dipeptidyl peptidase IV improves glucose tolerance and preserves islet function in mice. *Eur J. Endocrinol.* **146**, 717–727.

RICHARDSON, P. L. (**2002**). The determination and use of optimized protease substrates in drug discovery and development. *Curr. Pharm. Des.* **8**, 2559–2581.

RICHTER, R., SCHULZ-KNAPPE, P., SCHRADER, M., STANDKER, L.,

Jürgens, M., Tammen, H., Forssmann, W. G. (**1999**). Composition of the peptide fraction in human blood plasma: database of circulating human peptides. *J. Chromatogr. B Biomed. Sci. Appl.* **726**, 25–35.

Roepstorff, P. (**1997**). Mass spectrometry in protein studies from genome to function. *Curr. Opin. Biotechnol.* **8**, 6–13.

Rose, K., Bougueleret, L., Baussant, T., Bohm, G., Botti, P., Colinge, J., Cusin, I., Gaertner, H., Gleizes, A., Heller, M., Jimenez, S., Johnson, A., Kussmann, M., Menin, L., Menzel, C., Ranno, F., Rodriguez-Tome, P., Rogers, J., Saudrais, C., Villain, M., Wetmore, D., Bairoch, A., Hochstrasser, D. (**2004**). Industrial-scale proteomics: from liters of plasma to chemically synthesized proteins. *Proteomics* **4**, 2125–2150.

Rosenfeld, L. (**2002**). Insulin: discovery and controversy. *Clin. Chem.* **48**, 2270–2288.

Sanger, F. (**1959**). Chemistry of insulin; determination of the structure of insulin opens the way to greater understanding of life processes. *Science* **129**, 1340–1344.

Sasaki, K., Sato, K., Akiyama, Y., Yanagihara, K., Oka, M., Yamaguchi, K. (**2002**). Peptidomics-based approach reveals the secretion of the 29–residue COOH-terminal fragment of the putative tumor suppressor protein DMBT1 from pancreatic adenocarcinoma cell lines. *Cancer Res.* **62**, 4894–4898.

Sato, K., Sasaki, K., Tsao, M. S., Yamaguchi, K. (**2002**). Peptide differential display of serum-free conditioned medium from cancer cell lines. *Cancer Lett.* **176**, 199–203.

Schrader, M., Jürgens, M., Hess, R., Schulz-Knappe, P., Raida, M., Forssmann, W. G. (**1997**). Matrix-assisted laser desorption/ionisation mass spectrometry guided purification of human guanylin from blood ultra-filtrate. *J. Chromatogr. A* **776**, 139–145.

Schrader, M., Schulz-Knappe, P. (**2001**). Peptidomics technologies for human body fluids. *Trends Biotechnol.* **19**, S55–S60.

Schulz-Knappe, P., Schrader, M., Standker, L., Richter, R., Hess, R., Jürgens, M., Forssmann, W. G. (**1997**). Peptide bank generated by large-scale preparation of circulating human peptides. *J. Chromatogr. A* **776**, 125–132.

Schulz-Knappe, P., Zucht, H. D., Heine, G., Jürgens, M., Hess, R., Schrader, M. (**2001**). Peptidomics: the comprehensive analysis of peptides in complex biological mixtures. *Comb. Chem. High Throughput Screen* **4**, 207–217.

Scrivener, E., Barry, R., Platt, A., Calvert, R., Masih, G., Hextall, P., Soloviev, M., Terrett, J. (**2003**). Peptidomics: A new approach to affinity protein microarrays. *Proteomics* **3**, 122–128.

Sechi, S., Oda, Y. (**2003**). Quantitative proteomics using mass spectrometry. *Curr. Opin. Chem. Biol.* **7**, 70–77.

Seidah, N. G., Chretien, M. (**1999**). Proprotein and prohormone convertases: a family of subtilases generating diverse bioactive polypeptides. *Brain Res.* **848**, 45–62.

Sewald, N., Jakubke H-D (**2002**). *Peptides: Chemistry and Biology.* Wiley-VCH, Weinheim.

Sillard, R., Jornvall, H., Carlquist, M., Mutt, V. (**1993**). Chemical assay for cyst(e)ine-rich peptides detects a novel intestinal peptide ZF-1, homologous to a single zinc-finger motif. *Eur. J. Biochem.* **211**, 377–380.

Southan, C. (**2004**). Has the yo-yo stopped? An assessment of human protein-coding gene number. *Proteomics* **4**, 1712–1726.

Steiner, D. F. (**2002**). The prohormone convertases and precursor processing in protein biosynthesis. In Dalbey, R. E., Sigman, D. S. (Eds.), *The enzymes: co- and post-translational proteolysis of proteins,* Vol. XXII. Academic, San Diego.

Stoeckli, M., Chaurand, P., Hallahan, D. E., Caprioli, R. M. (**2001**). Imaging mass spectrometry: a new technology for the analysis of protein expression in mammalian tissues. *Nat. Med.* **7**, 493–496.

Tammen, H., Kreipe, H., Hess, R., Kellmann, M., Lehmann, U., Pich, A., Lamping, N., Schulz-Knappe, P., Zucht, H. D., Lilischkis, R. (**2003**). Expression profiling of breast cancer cells by differential peptide display. *Breast Cancer Res. Treat* **79**, 83–93.

TAMMEN, H., SCHULTE, I., HESS, R., MENCEL, C., KELLMANN, M., MOHRING, T., SCHULZ-KNAPPE, P. (2005). Peptidomic analysis of human blood specimens: comparison between plasma specimens and serum by differential peptide display. *Proteomics* **5**, 414–422.

UTTENWEILER-JOSEPH, S., MONIATTE, M., LAGUEUX, M., VAN DORSSELAER, A., HOFFMANN, J. A., BULET, P. (1998). Differential display of peptides induced during the immune response of Drosophila: A matrix-assisted laser desorption ionization time-of-flight mass spectrometry study. *Proc. Natl. Acad. Sci. USA* **95**, 11342–11347.

VERHAERT, P., UTTENWEILER-JOSEPH, S., DE VRIES, M., LOBODA, A., ENS, W., STANDING, K. G. (2001). Matrix-assisted laser desorption/ionization quadrupole time-of-flight mass spectrometry: an elegant tool for peptidomics. *Proteomics* **1**, 118–131.

VILLANUEVA, J., PHILIP, J., ENTENBERG, D., CHAPARRO, C. A., TANWAR, M. K., HOLLAND, E. C., TEMPST, P. (2004). Serum peptide profiling by magnetic particle-assisted, automated sample processing and MALDI-TOF mass spectrometry. *Anal. Chem.* **76**, 1560–1570.

WALGREN, J. L., THOMPSON, D. C. (2004). Application of proteomic technologies in the drug development process. *Toxicol. Lett.* **149**, 377–385.

WANG, W., ZHOU, H., LIN, H., ROY, S., SHALER, T. A., HILL, L. R., NORTON, S., KUMAR, P., ANDERLE, M., BECKER, C. H. (2003). Quantification of proteins and metabolites by mass spectrometry without isotopic labeling or spiked standards. *Anal. Chem.* **75**, 4818–4826.

WEI, C. M., HEUBLEIN, D. M., PERRELLA, M. A., LERMAN, A., RODEHEFFER, R. J., McGREGOR, C. G., EDWARDS, W. D., SCHAFF, H. V., BURNETT, J. C. JR. (1993). Natriuretic peptide system in human heart failure. *Circulation* **88**, 1004–1009.

WIELAND, T., BODANSZKY, M. (1991). *The World of Peptides. A Brief History of Peptide Chemistry.* Springer-Verlag, Berlin.

YAMAMOTO, K., BURNETT, J. C. JR., JOUGASAKI, M., NISHIMURA, R. A., BAILEY, K. R., SAITO, Y., NAKAO, K., REDFIELD, M. M. (1996). Superiority of brain natriuretic peptide as a hormonal marker of ventricular systolic and diastolic dysfunction and ventricular hypertrophy. *Hypertension* **28**, 988–994.

ZASLOFF, M. (1987). Magainins, a class of antimicrobial peptides from Xenopus skin: isolation, characterization of two active forms, and partial cDNA sequence of a precursor. *Proc. Natl. Acad. Sci. USA* **84**, 5449–5453.

ZOLG, J. W., LANGEN, H. (2004). How industry is approaching the search for new diagnostic markers and biomarkers. *Mol. Cell Proteomics* **3**, 345–354.

7
Protein Biochips in the Proteomic Field

Angelika Lücking and Dolores J. Cahill

Abstract

High density DNA microarray technology has played a key role in the analysis of whole genomes and their gene expression patterns. The ability to study many thousands of individual genes using oligonucleotide or cDNA arrays is now very widespread, with its uses ranging from the profiling of gene expression patterns in whole organisms or tissues to the comparison of healthy and pathological samples. However, despite the success of DNA microarrays, it is obvious that the biological function is executed by biomolecules such as proteins. Protein biochips are therefore emerging to follow DNA microarrays as a possible screening tool. We will present different types of biochips including protein and antibody arrays, as well as carbohydrate, peptide and living cell arrays. Recent progress and current bottlenecks in high-throughput generation of chip content, surface chemistry, molecule attachment, detection methods, and applications in the proteomics field and in drug discovery will be discussed.

7.1
Introduction

The human genome is sequenced and the challenges of understanding the function of the newly discovered genes are being addressed. For this purpose, high-throughput technologies have been developed that allow the monitoring of gene activity at the transcriptional level by analysis of complex expression patterns of a specific tissue. Differential gene expression can be most efficiently monitored using oligonucleotide or cDNA microarrays, where a multitude of cDNAs or oligonucleotides representing individual genes, are immobilized in a small field as capture molecules (Epstein and Butow, 2000; Lockhart and Winzeler, 2000). To study variation in gene expression, complex probes, generated by reverse transcription of mRNA from different tissues and cell lines, are hybridized onto these microarrays. However, despite the success of DNA microarrays in expression

Proteomics in Drug Research
Edited by M. Hamacher, K. Marcus, K. Stühler, A. van Hall, B. Warscheid, H. E. Meyer
Copyright © 2006 Wiley-VCH Verlag GmbH & Co. KGaA, Weinheim
ISBN: 3-527-31226-9

profiling studies, it is obvious that the biomolecules executing a biological function are the proteins. Protein biochips are therefore emerging to follow DNA microarrays as a possible tool to screen for proteins interacting with other biomolecules (e.g., DNA, RNA, proteins, antibodies, peptides, aptamers or other binders) or catalyzing enzymatic reactions (e.g., phosphorylation) (reviewed in: Cahill and Nordhoff, 2003).

Apart from mRNA, which is uniformly formed by four nucleotides, proteins are highly heterogeneous molecules based on the assembly of 22 different amino acids. Owing to their composition, some proteins can be highly water soluble, for example serum proteins, or very insoluble in any solvent, such as keratin. Additionally, processes like alternative splicing, protein and peptide cleavage, multiprotein complex formation and posttranslational modifications such as glycosylation, acetylation, or phosphorylation result in five to ten times more diverse protein species than genes present in an organism or tissue (Anderson and Seilhamer, 1997; Wilkins et al., 1996). Such diversity and complexity of proteins enables adaptation of organisms to different biological systems, to infections and diseases as well as to different environmental requirements. However, this diversity poses a real challenge for the development of protein biochips.

In proteomics, the proteins of an organism or a specific tissue type are studied. The classical method of protein profiling studies in proteomics is two-dimensional gel electrophoresis (2-DE) (Klose, 1975; O'Farrell, 1975) often combined with subsequent protein identification by mass spectrometry (MS) of excised protein spots, e.g., matrix-assisted laser desorption ionization–time-of-flight analysis (MALDI–TOF) (Giavalisco et al., 2003; Gobom et al., 2001). In 2-DE, proteins are separated according to their charge and mass. This technology is widely applied to the comparative study of protein expression patterns, e.g., of drug-treated versus nontreated tissues or diseased versus normal tissues (Marcus et al., 2003; Ostergaard et al., 1997), and has also been used for the comparison of related pathogenic versus nonpathogenic organisms (Mahairas et al., 1996). These experiments can be referred to as *unbiased* or *discovery-oriented* proteomics, generating large data collections. This can be distinguished from *system-oriented* proteomics, where only a defined subset of proteins are analyzed, such as a family of proteins related by function or sequence or proteins belonging to a common pathway. Predominantly, protein biochips will play a role in discovery-oriented proteomics, but it is also expected that they will be applied in system-oriented proteomics where a defined set of proteins or binders will be arrayed and studied.

The aim of system-oriented proteomics is to observe changes in concentration, localization or modification of the subset of proteins of interest. This quantitative view requires the development of analytical microarrays containing antibodies or antibody mimics to capture the antigens of interest (Kusnezow et al., 2003a). A second task of system-oriented proteomics is to elucidate the biological role of a specific protein, which may involve the identification of its interacting proteins or specific substrates where enzymes are being studied. This requires the immobilization of large sets of purified proteins, while optimally maintaining their function and activity during purification and immobilization.

In this chapter, we will present different types of biochips, including protein and antibody arrays, as well as carbohydrate, peptide and living cell arrays. We will discuss recent progress and current bottlenecks in high-throughput generation of chip content, surface chemistry, molecule attachment, detection methods, and also applications in the proteomic field and in drug discovery.

7.2
Technological Aspects

7.2.1
Protein Immobilization and Surface Chemistry

The performance of protein or antibody microarrays is dependent on various factors. One of these is the use of an appropriate microarray surface for the immobilization of the protein or antibody samples. Most conventional microarray surfaces have been adapted from DNA chip technology. DNA can easily be immobilized by electrostatic interactions of the phosphate backbone onto a positively charged surface. In contrast to DNA, as already mentioned, proteins are chemically and structurally much more complex and show variable charges, which may influence the efficiency of protein attachment. Additionally, proteins lose their structure and biochemical activity easily. For example, globular proteins consist of a hydrophilic exterior and a hydrophobic interior. When immobilized on a hydrophobic surface, the inside of the protein turns out, which may destabilize the structure and, simultaneously, the activity of the protein. These considerations demonstrate the complex requirements for protein immobilization.

In principal, microarray supports can be divided into three types, consisting of one-, two and three-dimensional surfaces, respectively (Table 7.1). Protein immobilization on a flat, one-dimensional surface can be performed by covalent attachment via chemical linkers such as polyamine (Madoz-Gurpide et al., 2001; Sreekumar et al., 2001), aldehyde (MacBeath and Schreiber, 2000; Zhu et al., 2000) or epoxy groups (Angenendt et al., 2002; Kusnezow et al., 2003b) or by noncovalent immobilization using poly-L-lysine surfaces (Haab et al., 2001). Several drawbacks, such as smearing effects or a nonhomogeneous spot morphology, have been addressed using this type of surface. The functional groups used for one-dimensional matrixes have also been used in two-dimensional coatings where larger spacers such as polyethylene glycol copolymer (PEG), dendrimers or self-assembled layers (SAM) were positioned between the support matrix and captured protein (Benters et al., 2002; Benters et al., 2001; Jo and Park, 2000; Piehler et al., 1996; Schaeferling et al., 2002). Such layers act as spacers preventing direct protein–surface contact which may avoid protein denaturation and, in addition, eliminate the need for blocking reagents to reduce background binding. It has been shown that using spacers improve signal intensities, as well as the binding efficiency of analytes to the immobilized molecule (Angenendt et al., 2003b; Kusnezow et al., 2003b).

Table 7.1 Surface.

Attachment	Surfaces	Binding mechanism	Advantages/disadvantages
Adsorption Absorption	Poly-L-lysine, Polystyrene, Aminosilane Nitrocellulose PVDF	Electrostatic and hydrophobic forces	No protein modifications required, cheap/random orientation and surface interference
	Polyacrylamide gel Agarose gel	Diffusion	High binding capacity, no protein modifications required, good spot morphology/higher background, random orientation High binding capacity, no protein modifications required, good spot morphology/random orientation
Covalent, random	Epoxy, aldehyde, carbodiimide groups	React with primary amines of lysine and arginine	High densities are available and strong protein attachment/random orientation and surface interference
Covalent, random, coupled to larger layers	SAM, dendrimers or PEG layers with epoxy, aldehyde, carbodiimide groups	React with primary amines of lysine and arginine	High densities are available, strong protein attachment, low surface interference/random orientation
Covalent, orientated	Amino groups, hydrazine	Binds to carbohydrate residues of FC region of antibodies	High densities are available, strong protein attachment, oriented immobilization/surface interference
Covalent, orientated, coupled to larger layers	Amino groups, hydrazine	Binds to carbohydrate residues of FC region of antibodies	High densities are available, strong protein attachment, oriented immobilization, low surface interference/oxidation of carbohydrates is required
Other interactions, oriented	Protein A or G Nickel, avidin	Affinity binding to Fc region of antibodies Affinity binding to His- or Strep-tagged fusion proteins or to biotinylated proteins	Strong and specific interaction between the functional groups, oriented immobilization Strong and specific interaction between the functional groups, oriented immobilization/recombinant fusion proteins have to be available

One further development of this approach has been to link chelating imino-diacetic acid groups to PEG, which in turn can be bound by Cu^{2+} ions and so provide a highly specific binding site for His_6-tagged proteins (Cha et al., 2004). Another variation is gold-coated microarrays, which have the advantage that they can be combined with surface plasma resonance (SPR) and MS for further detection and analysis of the captured molecules (Bieri et al., 1999; Houseman et al., 2002a; Rich and Myszka, 2000). The three-dimensional surface can be a thin gel layer consisting of polyacrylamide or agarose (Afanassiev et al., 2000; Arenkov et al., 2000; Guschin et al., 1997; Lueking et al., 2003b), or have a nitrocellulose or PVDF coating (Kersten et al., 2003; Kukar et al., 2002; Lueking et al., 1999). Owing to the highly porous and hydrophilic matrix, proteins can easily diffuse into the pores and there be immobilized based on hydrophobic interactions. When compared to two-dimensional slides, the three-dimensional structure shows a higher protein binding capacity and detection sensitivity (Angenendt et al., 2002, 2003b). The polyacrylamide or agarose layer provides a native, aqueous environment that minimizes protein denaturation and maintains the enzymatic activity of several enzymes such as horseradish peroxidase, alkaline phosphatase and β-D-glucuronidase in these hydrogel pads (Arenkov et al., 2000). Prestructured surfaces consisting of hydrophilic spots on hydrophobic surfaces have also been reported for protein arraying (Schuerenberg et al., 2000).

Since the main challenge is to preserve proteins in their biologically active state, an oriented attachment using special compounds has several advantages. In contrast to the commonly used random immobilization, oriented attachment leaves the active sites of the protein free and available for interactions. By the presence of chlorine on a glass surface, alcohol groups, which were introduced into small molecules or proteins, have been attached in an oriented manner. In initial screens, a specific protein–protein interaction has been identified (Hergenrother et al., 2000). In a more extensive example, 5800 histidine (His)-tag fusion yeast proteins were immobilized onto nickel coated slides and screened for their ability to interact with proteins and phospholipids. New calmodulin- and phospholipid-interacting proteins were identified (Zhu et al., 2001a). This approach has been extended to other affinity tags such as the glutathione-S-transferase (GST) tag combined with glutathione coated slides, or strep tag or biotinylated proteins combined with streptavidine- or avidine-coated slides (Lesaicherre et al., 2002; Mooney et al., 1996; Ruiz-Taylor et al., 2001; Uttamchandani et al., 2004). Another strategy has been applied for the directed attachment of antibodies. This includes binding via their thiol groups located either in the hinge region or between the light and heavy chains, or binding to maleimide derivates (Horejsi et al., 1997; Zhang and Czupryn, 2002) and immobilization via the carbohydrate residues of the Fc regions either to hydrazide derivates or to protein A- or G-coated slides (Anderson et al., 1997; Arenkov et al., 2000; Johnson et al., 2003; Vijayendran and Leckband, 2001). Compared to random immobilization, higher signal intensities were obtained in most cases (Arenkov et al., 2000; Kusnezow et al., 2003b; Peluso et al., 2003).

However, at this time, there is no one preferred method for the immobilization of proteins or antibodies.

7.2.2
Transfer and Detection of Proteins

Protein arrays are commonly gridded and imaged using the same commercially available automated robot arrayers and scanners as are used to generate DNA arrays. Most of the arraying devices are pin-based systems which transfer nanoliter amounts of the sample using either solid pins, split pins or ring-shaped reservoirs. It has been demonstrated that solid pins are much less sensitive to different sample viscosities than split pins or alternative microdispensing systems such as ink-jet printing or electrospray deposition (Avseenko et al., 2001, 2002; Roda et al., 2000). Using a standard arrayer equipped with solid pins, a highly multiplexed immuno-assay in microarray format has been developed where as little as 400 zeptomoles (zmol) of analyte has been quantified (Angenendt et al., 2003a, 2004b). Imaging systems for protein biochips are mostly fluorescence-based, using either direct labeling of ligands or indirect detection via labeled secondary antibodies that lead to signal amplification. The sensitivity has been further increased by the use of radioisotopes and by rolling circle amplification (RCA) (Morozov et al., 2002; Schweitzer et al., 2002). Combining antibody chips to capture proteins with MS, the proteins can be identified and their posttranslational modifications analyzed (Austen et al., 2000; Davies et al., 1999; Morozov et al., 2003). A novel label-free readout was recently reported, based on the autofluorescence of the amino acids tryptophan and tyrosine, which enables the direct detection of proteins and their natural interactions without steric interference due to chemical alteration (Striebel et al., 2004). However, the sensitivity obtained by this type of detection is a factor of five lower than that of the detection of fluorescently labeled samples.

7.2.3
Chip Content

For the generation of protein biochips, especially for the discovery-oriented approach, large numbers of antigens or binding molecules are required. Depending on the type of microarray to be generated, different biomolecules have to be produced, which is a major technical challenge calling for highly parallel and preferably automated recombinant expression systems.

High-throughput subcloning of open reading frames (ORF) from the genome of humans, *Saccaromyces cerevisiae* (*S. cerevisiae*), *Arabidopsis. thaliana* and *Caenorhabditis elegans* (*C. elegans*) have been described for the generation of protein biochips (Kersten et al., 2003; Reboul et al., 2003; Wiemann et al., 2003; Zhu et al., 2000, 2001a). Such large-scale subcloning approaches are strongly dependent on progress in the respective genome sequencing projects and the annotation of those sequences (Heyman et al., 1999; Walhout et al., 2000). One disadvantage of this approach is that previously uncharacterized proteins may be absent, thus possibly posing limitations for its use as a discovery tool. Additionally, correct annotation of the expressed sequences remains difficult due to differential splicing. For these reasons, this approach has proved most valuable in the production of chips containing proteins from well-characterized organisms, such as *S. cerevisiae* and *C. elegans* (Reboul et al., 2003; Zhu et al., 2001a). This effort in creating each individual cDNA expression construct can be circumvented by the generation of arrayed cDNA expression libraries leading to thousands of cDNA expression products in parallel (Büssow et al., 1998; Gutjahr et al., submitted; Holz et al.,

2001; Lueking et al., 2000). The use of cDNA expression libraries eliminates the need to construct individual expression clones for every protein of interest. By introducing a sequence coding for an affinity tag (e.g., His tag, GST tag) to the 5′ end of the cDNA insert, resulting expression clones can be rapidly identified by the presence of the His- or GST-tagged fusion protein. *Escherichia. coli* (*E. coli*) is most commonly used in automated recombinant expression and purification as it is a robust and convenient host organism (Braun et al., 2002; Brizuela et al., 2002; Büssow et al., 2000). However, many eukaryotic proteins end up in cytoplasmic inclusion bodies when they are expressed in *E. coli* and can only be recovered in the denatured state, yet even in the denatured state, the retained display of linear epitopes is sufficient for antigen–antibody screening purposes, although not for answering questions concerning the functionality or biological role of these proteins. Therefore, eukaryotic expression systems as *S. cerevisiae* (Holz et al., 2002, 2003; Zhu et al., 2000, 2001a) or *Pichia pastoris* have been adapted to high-throughput expression and purification (Boettner et al., 2002; Lueking et al., 2000, 2003a). These systems are also able to perform posttranslational modifications. Albala and coworkers have previously applied a baculovirus expression system to the 96-well format and expressed 72 unique human cDNA clones in high-throughput, of which 42% gave a soluble product (Albala et al., 2000). Alternatively, cell-free expression systems lead very often to soluble and functional proteins. He and Taussig have created a protein *in situ* array (PISA) by combining *in vitro* expression of functional antibody fragments and their immobilization in one step (He and Taussig, 2001). In a similar approach, plasmid DNA containing the sequence coding for the GST tag was immobilized together with an antibody against the GST tag on a microarray. Following *in vitro* expression based on a mammalian reticulocyte lysate, the subsequent binding of the expressed GST-tagged protein to the anti-GST antibody has been shown. Such a protein biochip was used successfully for protein–protein interaction studies and for analysis of binding epitopes (Ramachandran et al., 2004). In a proof of principle experiment, direct cell-free expression in nanowells was performed, followed by monitoring enzymatic activity of these expressed proteins (Angenendt et al., 2004a). Although all these experiments were preliminary, they have the potential to be automated and could be applied to a larger number of proteins.

Direct cloning of ORFs has also been used for living cell arrays where the cDNAs, cloned in mammalian expression vectors, were spotted onto microscope slides. Mammalian cells growing onto these printed areas took up the cDNA and expressed the corresponding gene product (Ziauddin and Sabatini, 2001).

Similar to protein biochips, the generation of antibody arrays requires a large number of high-affinity, high-specificity protein binding ligands, ideally one for each protein of the proteome of interest. Taking all the different posttranslational modifications into account, this means the generation of more than 100 000 protein binders for the human proteome. Additionally, those binders should have a known binding behavior (ideally high k_{on}, low k_{off}), affinity and specificity. This is a major task and would ideally be performed in a united and concerted effort within the scientific community. Some excellent initiatives have already started to address this, such as the Swedish Human Proteome Resource (HPR); www.biotech.Kth.se/molbio/hpr/. In the HPR, polyclonal antibodies are generated which will be affinity-purified and screened against tissue arrays. Another alternative is to generate thousands of poly- or monoclonal antibodies by mouse immunization, which is expensive and raises ethical and intellectual property issues. The use of recombinant antibodies as an alternative has therefore been explored (Holt et al., 2000). Antigen-binding fragments such as Fab or ScFvs provide simple antibody formats which can be affinity-selected *in vitro* by display technologies such as phage or ribosome display (Schaffitzel et al., 1999; Walter et al., 2001). Alternatively, single-stranded oligonucleotides called aptamers have been developed by systematic evolution of ligands via an exponential enrichment (SELEX) process to bind proteins (Jayasena, 1999). They appear to be promising new array probes as they can be photo-crosslinked to the recognized proteins with very low background from other proteins in the sample. Since no other proteins are immobilized onto those arrays, nonspecific protein stains can be used to detect the ligands (Brody and Gold, 2000; Brody et al., 1999).

7.3
Applications of Protein Biochips

In recent years, different formats of biochips such as protein arrays, peptide arrays, antibody arrays, living cell arrays, tissue arrays, and carbohydrates arrays have been developed (Figure 7.1). They are associated with different applications and show specific advantages as well as definite bottlenecks.

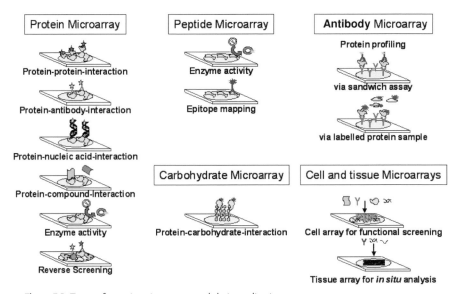

Figure 7.1 Types of protein microarrays and their applications.

Protein arrays comprised of immobilized proteins are an emerging biochip format. One bottleneck in manufacturing such high-throughput, high-content protein arrays is the efficient production of large numbers of proteins. This is a major technical challenge, calling for highly parallel, preferably automated recombinant expression systems as discussed in the previous section. Protein biochips have been used for protein and (auto-) antibody profiling, the study of protein–ligand interactions where the ligand can be either other proteins, peptides, DNA or RNA, and for the determination of enzymatic activity and substrate specificity of classes of enzymes.

The specificity and crossreactivity of antibodies has been successfully determined using high-content protein biochips (Lueking et al., 1999, 2003b; Michaud et al., 2003). When antibodies are used extensively as diagnostic tools, the characterization of their binding specificity is of prime importance. Additionally, these characterized antibodies would also be very interesting for the generation of highly specific antibody arrays. Antibody–antigen interactions can be exploited in medically relevant fields such as autoimmune diseases. Screening protein arrays with sera or plasma from autoimmune patients would not only allow the identification of potentially new autoantigens, but also the diagnosis and subtyping of the autoimmune disease based on the presence of specific autoantibodies, hence profiling the antibody repertoire of patients with autoimmune disease (Lueking et al., 2003b; Robinson et al., 2002). Approaches using λgt11 phage library plaque-lift filters or tissue extracts separated by 1-DE or 2-DE (Latif et al., 1993; Pohlner et al., 1997) may be problematic as the further analysis of identified putative autoantigens is very labor-intensive. Although mass spectrometric identification of separated proteins has improved greatly, characterization often requires expensive protein sequencing, which in turn needs high quantities of purified protein. Using microarray technology for the profiling of such an autoantibody repertoire would greatly simplify this process, since protein arrays contain large numbers of proteins, the sequence of which can be determined since they come from an ordered recombinant source. Moreover, using protein arrays for diagnostic purposes would minimize variations occurring in natural tissue extracts, thereby increasing reproducibility. Once disease-specific antigens are known, it is possible to create small, easy-to-evaluate diagnostic protein arrays. By combining the cDNA expression library approach with protein microarrays, the humoral autoimmune repertoire of dilated cardiomyopathy (DCM) patients has been profiled to develop a disease-associated protein chip (Horn et al., submitted; Figure 7.2). In a first step, plasma samples from patients were analyzed via protein arrays consisting of approximately 37 200 redundant human proteins. This data set obtained was compared to the data set derived from incubations with plasma of age- and sex-matched control persons. From these results, a smaller protein subset containing 50–200 different, purified proteins has been generated and subjected to further qualitative and quantitative analysis of a larger patient cohort using protein microarrays. Following bioinformatical analysis, several protein antigens have been determined as being associated with heart failure (Horn et al., 2005). A protein array consisting of 196 distinct biomolecules representing major

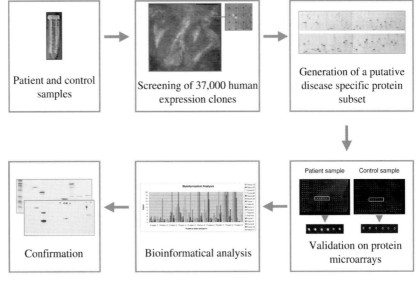

Figure 7.2 Profiling the antibody repertoire of autoimmune patients by combining protein microarray technology with large cDNA expression libraries. In a first step, serum samples from patients were analysed onto protein arrays consisting of approximately 37,200 redundant human proteins. The data sets obtained were compared to the data sets deriving from incubations with serum of age and sex matched control persons. From these results, a smaller protein subset containing approximately 50–200 different, purified proteins will be generated and subjected to further qualitative and quantitative on protein biochips. Following bioinformatical analysis, protein antigens will be determined which may be associated with the respective autoimmune disease following confirmation by a second method such as western immunoblotting.

autoantigens were probed with serum from patients with different autoimmune diseases, such as systemic lupus erythematosus (SLE), Sjögren syndrome, and rheumatoid arthritis (RA), and distinct autoantibody patterns have been shown indicating their suitability for diagnosis (Robinson et al., 2002). Using fluorescence-labeled secondary antibodies, autoantibody binding was detected in a linear range between 1 and 900 ng/mL, which is four- to eightfold more sensitive than the conventional ELISA method. Recently, it has been shown that DNA vaccines encoding autoantigens induce specific immune tolerance (Garren et al., 2001; Ruiz et al., 1999; Urbanek-Ruiz et al., 2001). To study the autoantibody B-cell response in acute and chronic experimental autoimmune encephalomyelitis (EAE), a mouse model for multiple sclerosis (MS), an antigen microarray consisting of 232 proteins specific for the myelin proteome was generated (Robinson et al., 2003). Putative autoantigens have been determined, followed by an antigen-specific tolerizing therapy which leads to the reduction of epitope spreading of autoreactive B-cell response and a lower relapse rate. Such analysis of immune response can also be applied to other diseases. For example, to analyze the humoral immune response to cancer, solubilized proteins from the LoVo colon adenocarcinoma cell line were separated into 1760 fractions, which were arrayed in microarray format and incubated with sera from newly diagnosed patients with colon cancer. One fraction exhibiting a strong reactivity to colon cancer sera was subjected to MS, leading to the identification of a putative antigen (Nam et al., 2003). Another immediate application is the use of allergen arrays to screen for the presence of particular IgE molecules in a patient sample. The traditional approach involves the use of simple extracts from potential allergens. Such extracts are commonly used in skin-prick tests to determine the possible source of an allergic reaction in the patient. By combining array technology with recombinant allergens (e.g., pollen and fungus proteins) relatively large arrays have been produced for screening purposes (Deinhofer et al., 2004; Hiller et al., 2002; Jahn-Schmid et al., 2003; Wiltshire et al., 2000). These arrays can also readily accommodate nonprotein allergens such as latex, etc.

Protein microarrays are suitable for studying protein–ligand interactions where the ligand can be a protein, peptide, DNA, RNA, an oligosaccharide or a chemical compound. In an initial experiment, Ge employed a low-density protein array for studying interactions with protein, DNA, RNA and small chemical ligand probes (Ge, 2000). In a recent approach the interaction of a restriction enzyme to double-stranded DNA was monitored on a micrometer-sized monolayer surface using atomic force microscopy (O'Brien et al., 2000). In a discovery-oriented approach, a yeast proteome chip containing 5800 recombinant proteins was generated and used to identify known, as well as new, calmodulin binders (Zhu et al., 2001a). In addition, lipid binding specificity was profiled using phosphoinoside-doped liposomes. In a very focused approach, the interaction between the DNA repair protein RAD51B and the histone proteins have been shown on a low-content protein array consisting of chromatin-related proteins (Coleman et al., 2003). As an another example, the particular conserved cytoplasmic motif KVGFFKR from the platelet membrane protein integrin was used for interaction studies in a system-oriented approach. This motif has previously been shown to play a critical role in the regulation of activation of the platelet integrin $\alpha_{IIb}\beta_3$ (Larkin et al., 2004; Stephens et al., 1998). The tagged peptide (biotin-KVGFFKR) was screened against a high-density array of approximately 37 000 *E. coli* clones expressing recombinant human proteins (Büssow et al., 1998; Lueking et al., 1999) and 13 different proteins were identified as binding the labeled peptide. Using peptide pull-down assays and coprecipitation experiments, the interaction between a putative chloride channel, ICln and integrin $\alpha_{IIb}\beta_3$ has been confirmed (Larkin et al., 2004). Such an experiment reveals the enormous potential of a protein array approach, not only in identifying novel protein interactions, but also towards elucidating biological pathways in general.

Interactions between proteins and carbohydrates are essential for various biological processes as the carbohydrates contained in glycoproteins, glycolipids and proteoglycans are involved in recognition processes such as cell adhesion, migration and signalling. To profile such interactions, carbohydrate microarrays containing polysaccharides, natural glycoconjugates, and mono- and oligo-saccharides coupled to carrier molecules have been developed (Feizi et al., 2003;

Galanina et al., 2003; Wang, 2003; Wang et al., 2002; Willats et al., 2002). However, this technology is in its infancy and its impact on lead or therapeutic target discovery is, as yet, unclear.

Different enzyme activities, including phosphatase, peroxidase, galactosidase, restriction enzymes and protein kinases have been detected on protein, peptide and nanowell microarrays (Angenendt et al., 2004a; Arenkov et al., 2000; Bulyk et al., 1999; Burns-Hamuro et al., 2003; Houseman et al., 2002a; Kramer et al., 2004; MacBeath and Schreiber, 2000; Reineke et al., 2001; Uttamchandani et al., 2004; Zhu et al., 2000). For example, phosphorylation of proteins by protein kinases plays a central role in regulating cellular processes and it is suggested that this contributes to many diseases such as diabetes, inflammation or cancer. Therefore, protein kinases have a high potential as drug target candidates, resulting in a strong interest in identifying new kinases and their substrates. Zhu et al. have created protein arrays of *S. cerevisiae* kinases (Zhu et al., 2000). In total, 119 known or predicted proteins kinases were expressed, purified as GST fusion proteins, and arrayed, crosslinked on a protein chip and assayed with 17 different substrates for autophosphorylation by treatment with radiolabeled ATP (Zhu et al., 2000). In a broader experiment, the barley protein kinase CK2alpha was incubated onto a protein microarray with 768 purified proteins deriving from a barley expression library. From these 768 proteins, 21 potential targets including new as well as known substrates of CK2alpha, such as high mobility group proteins or calreticulin have been identified (Kramer et al., 2004). Similar approaches were performed using peptide arrays which contain presynthesized or on-spot synthesized peptides (Houseman et al., 2002b; Reineke et al., 2001). Peptide arrays have also been used for antibody epitope mapping or determination of protease substrate specificity, and they seem to be suitable to profile protein–protein interactions that are based on linear epitopes or domains such as SH2, SH2 PH, PDZ, WW, or kringel (Apweiler et al., 2001; Reineke, 2004; Salisbury et al., 2002). These applications thus demonstrate that peptide arrays will have an important impact on proteome studies, enabling analysis of molecular recognition events and identification of biological active and functionally relevant epitopes.

To profile the status of key proteins in a proteome with minute quantities of cells, reverse phase protein arrays (RPPA) have been developed (reviewed in Charboneau et al., 2002). Paweletz et al. have used such an approach to analyze prosurvival checkpoint proteins in patient matched normal prostate epithelium, prostate intraepithelial neoplasia and invasive prostate cancer. Each type of cell was isolated from ten individuals, lysed, arrayed on slides and probed with different antibodies (Paweletz et al., 2001). It has been shown that cancer progression was associated with increased phosphorylation of Akt and decreased phosphorylation of ERK. In a similar, system-oriented approach, proteins in tissue lysates or subcellular fractions have been chromatographically separated and approximately 600–1200 different protein fractions were used for array analysis with specific antibodies (Madoz-Gurpide et al., 2001). In contrast to the classical 2-DE approach combined with western immunoblotting, the RPPA approach requires much less biological material, such as cell lysates and antibodies, due to the small chip format.

An alternative way to profile proteins is the use of antibody arrays which correspond to an immunoassay in micro format. The capture antibodies are immobilized and subsequently incubated with a biological sample, and bound proteins are detected following washing. Detection is carried out by using labeled samples, or using a second labeled antibody recognizing another epitope in the protein of interest. Different cytokine-specific antibody arrays have been created and used for measuring cytokine secretion after stimulation with lipopolysaccharide or TNF, either alone or together with an anti-inflammatory drug (Moody et al., 2001; Schweitzer et al., 2002). Using such arrays containing a collection of cytokine antibodies, global patterns of cytokine expression can be connected to biological relevant and clinical useful information. However, it is not surprising that many of the commercially available antibody arrays are directed against cytokines since cytokine-specific antibodies have been developed and commercialized for many years for use in diagnostics. Additionally, cytokines are proteins which are secreted by the cell, resulting in relatively simple mixtures in contrast to whole cellular lysates. Experiments done by Haab and coworkers have demonstrated that the use of antibodies directed against intracellular proteins is more difficult. They have spotted 115 well-characterized antibodies and incubated them with mixtures of fluorescence-labeled protein antigens (Haab et al., 2001). Of these antibodies, 60% gave a qualitative result about the presence of their antigen, but only 23% enabled a quantitative analysis. It has also been shown that an increasing amount of other nonantigen proteins in the sample results in higher background. This can be overcome by using a sandwich assay, which involves detection of the antigen in the unlabeled sample by a second specific antibody. However, the resulting higher specificity of such an approach has the disadvantage that this approach requires two specific binding reagents. In a larger study about squamous cell carcinoma, an antibody array consisting of 368 antibodies was incubated with 0.5 µg protein derived from laser capture microdissection (LCM) (Knezevic et al., 2001). Eleven proteins were determined to be differentially expressed in epithelium or stroma near the cancer cells. Sreekumar has used antibody arrays containing 146 antibodies to analyze the changes in protein expression level in a colon carcinoma cell line following ionizing radiation treatment (Sreekumar et al., 2001). Although only 14% of these antibodies performed well, 11 proteins have been identified that show reproducible changes in their abundance or phoshorylation state. Using an antibody array consisting of up to 360 antibodies, posttranslational modifications such as protein tyrosine phosphorylation, ubiquitination, and acetylation have been monitored in mammalian cells cultivated under different conditions (Ivanov et al., 2004). Such discovery-oriented approaches may have a strong impact on identification of new biomarkers which may improve diagnosis, prognosis and customized medicine.

A strong impact on target validation in clinical studies is expected from tissue microarray technology (TMA) (Kononen et al., 1998). Tissue microarrays allow the cost-effective high-throughput *in situ* analysis of specific molecular targets, either at the DNA, RNA or protein level, at once. They consist of miniaturized collections of arrayed tissues deriving from pathologically evident biopsies from, for example, tumor tissue or carcinomas (reviewed in: Kononen et al., 1998). They enable linking of molecular data with various tumor and patient data, such as clinicopathological information, survival and treatment response. Most of the current TMA studies have been done in cancer research.

However, it is evident that target genes identified, for example, by microarray screening have to be functionally validated which is usually done by molecular and cell-based assays on a gene-by-gene basis. In order to screen for drug–target interaction, specific biochemical assays based on the function of the target (such as enzymatic activity) are designed. This is, of course, rate-limiting for the increasing number of targets arising from the different genomic and proteomic studies.

Living cell arrays are based on the addition of (mammalian) cells and a lipid transfection reagent to microscope slides printed with cDNAs cloned in expression vectors (Ziauddin and Sabatini, 2001). Following transfection of these cells, the microarray presents clusters of cells overexpressing the specific gene product. In a proof of principle study, 192 distinct cDNAs cloned in expression vectors were used for living cell arrays to identify proteins involved in tyrosine kinase signalling, apoptosis, and cell adhesion (Ziauddin and Sabatini, 2001). Future applications are foreseen in screening biologically active substances that affect a given cellular phenotype, or in a combination with the approach of RNA interference (RNAi) to study gene underexpression or loss of gene function (Fuchs et al., 2004).

7.4
Contribution to Pharmaceutical Research and Development

Applications of protein array technology such as target identification and characterization, target validation, diagnostic marker identification and validation, preclinical study monitoring and patient typing seem to be feasible. We have recently reviewed the value of recent combinations of efforts in genomics, proteomics and biochip technology and their impact on the overall drug development process (Huels et al., 2002). For the first time, tools are available to study disturbances within biological systems, such as disease or drug treatment, on the gene and protein expression levels.

Proteins as targets dominate pharmaceutical research and development, with ligand–receptor interactions and enzymes as the vast majority, comprising ~45% and ~28%, respectively, of the targets (Drews, 2000). Additionally, many therapeutic proteins, especially humanized antibodies, are in clinical development. The ultimate tool for high-throughput and significant screening would be to test new leads or new targets in a highly parallel manner. Some examples for applications in this direction already exist. Recently, an immunosensor array has been developed that enables the simultaneous detection of clinical analytes (Rowe et al., 1999). Here, capture antibodies and analytes were arrayed on microscope slides using flow chambers in a crosswise fashion. This current format is low-density (6 × 6 pattern), but has high-throughput potential, as it involves automated image analysis and microfluidics; it is already becoming one of the future formats for enzyme activity testing and other assays (Cohen et al., 1999). In another study, small sets of active enzymes were immobilized in a hydrophilic gel matrix. Enzymatic cleavage of the substrate could be detected and inhibitors blocked the reaction (Arenkov et al., 2000). More recently, an enzyme array that is suitable for assays of enzyme inhibition has been reported (Park and Clark, 2002). Initial publications in the area of receptor–ligand interaction studies in a microarray format have shown that the interaction of immobilized compounds and proteins in solutions can be determined (MacBeath and Schreiber, 2000; Mangold et al., 1999; Zhu et al., 2001b). This technology allows high-throughput screening of ligand–receptor interactions with small sample volumes.

The multiparallel possibilities of protein array applications have the potential not only to allow the optimization of preclinical, toxicological and clinical studies through better selection and stratification of individuals, but also to affect how diagnostics are used in drug development.

Acknowledgements

Parts of the authors' work are funded by the German Ministry of Education and Research (BMBF) Grant 0811870 (BioFuture) and from the Health Education Authority and Science Foundation Ireland (SFI), Dublin 2, Ireland.

References

AFANASSIEV, V., HANEMANN, V., WOLFL, S. (2000). Preparation of DNA and protein micro arrays on glass slides coated with an agarose film. *Nucleic Acids Res.* **28**, E66.

ALBALA, J. S., FRANKE, K., MCCONNELL, I. R., PAK, K. L., FOLTA, P. A., RUBINFELD, B., DAVIES, A. H., LENNON, G. G., CLARK, R. (2000). From genes to proteins: high-throughput expression and purification of the human proteome. *J. Cell Biochem.* **80**, 187–191.

ANDERSON, G. P., JACOBY, M. A., LIGLER, F. S., KING, K. D. (1997). Effectiveness of protein A for antibody immobilization for a fiber optic bio-sensor. Biosens Bioelectron **12**, 329–336.

ANDERSON, L., SEILHAMER, J. (1997). A comparison of selected mRNA and protein abundances in human liver. *Electrophoresis* **18**, 533–537.

ANGENENDT, P., GLOKLER, J., KONTHUR, Z., LEHRACH, H., CAHILL, D. J. (2003a). 3D protein microarrays: performing multiplex immunoassays on a single chip. *Anal. Chem.* **75**, 4368–4372.

ANGENENDT, P., GLÖKLER, J., MURPHY, D., LEHRACH, H., CAHILL, D. J. (2002). Towards antibody microarrays: a comparison of current microarray support materials. *Anal. Biochem.* **309**, 253–260.

ANGENENDT, P., GLÖKLER, J., SOBEK, J., LEHRACH, H., CAHILL, D. J. (2003b). Next generation of protein microarray support materials: evaluation for protein and antibody microarray applications. *J. Chromatogr.* **1009**, 97–104.

ANGENENDT, P., NYARSIK, L., SZAFLARSKI, W., GLOKLER, J., NIERHAUS, K. H., LEHRACH, H., CAHILL, D. J., LUEKING, A. (2004a). Cell-free protein expression and functional assay in nanowell chip format. *Anal. Chem.* **76**, 1844–1849.

ANGENENDT, P., WILDE, J., KIJANKA, G., BAARS, S., CAHILL, D. J., KREUTZBERGER, J., LEHRACH, H., KONTHUR, Z., GLOKLER, J. (2004b). Seeing better through a MIST: evaluation of monoclonal recombinant antibody fragments on microarrays. *Anal. Chem.* **76**, 2916–2921.

APWEILER, R., ATTWOOD, T. K., BAIROCH, A., BATEMAN, A., BIRNEY, E., BISWAS, M., BUCHER, P., CERUTTI, L., CORPET, F., CRONING, M. D., et al. (2001). The InterPro database, an integrated documentation resource for protein families, domains and functional sites. *Nucleic Acids Res.* **29**, 37–40.

ARENKOV, P., KUKHTIN, A., GEMMELL, A., VOLOSHCHUK, S., CHUPEEVA, V., MIRZABEKOV, A. (2000). Protein microchips: use for immunoassay and enzymatic reactions. *Anal. Biochem.* **278**, 123–131.

AUSTEN, B. M., FREARS, E. R., DAVIES, H. (2000). The use of seldi proteinchip arrays to monitor production of Alzheimer's betaamyloid in transfected cells. *J. Pept. Sci.* **6**, 459–469.

AVSEENKO, N. V., MOROZOVA, T., ATAULLAKHANOV, F. I., MOROZOV, V. N. (2001). Immobilization of proteins in immunochemical microarrays fabricated by electrospray deposition. *Anal. Chem.* **73**, 6047–6052.

AVSEENKO, N. V., MOROZOVA, T. Y., ATAULLAKHANOV, F. I., MOROZOV, V. N. (2002). Immunoassay with multicomponent protein microarrays fabricated by electrospray deposition. *Anal. Chem.* **74**, 927–933.

BENTERS, R., NIEMEYER, C. M., DRUTSCHMANN, D., BLOHM, D., WOHRLE, D. (2002). DNA microarrays with PAMAM dendritic linker systems. *Nucleic Acids Res.* **30**, E10.

BENTERS, R., NIEMEYER, C. M., WOHRLE, D., DRUTSCHMANN, D., BLOHM, D. (2001). Dendrimer-activated solid supports for nucleic acid and protein microarrays. DNA microarrays with PAMAM dendritic linker systems. *Chembiochem.* **2**, 686–694.

BIERI, C., ERNST, O. P., HEYSE, S., HOFMANN, K. P., VOGEL, H. (1999). Micropatterned immobilization of a G protein-coupled receptor and direct detection of G protein activation, *Nat. Biotechnol.* **17**, 1105–1108.

BOETTNER, M., PRINZ, B., HOLZ, C., STAHL, U., LANG, C. (2002). High-throughput screening for expression of heterologous proteins in the yeast *Pichia pastoris*. *J. Biotechnol.* **99**, 51–62.

BRAUN, P., HU, Y., SHEN, B., HALLECK, A., KOUNDINYA, M., HARLOW, E., LABAER, J. (2002). Proteome-scale purification of human proteins from bacteria. *Proc. Natl. Acad. Sci. USA* **99**, 2654–2659.

BRIZUELA, L., RICHARDSON, A., MARSISCHKY, G., LABAER, J. (2002). The FLEXGene repository: exploiting the fruits of the genome projects by creating a needed resource to face the challenges of the post-genomic era. *Arch. Med. Res.* **33**, 318–324.

BRODY, E. N., GOLD, L. (2000). Aptamers as therapeutic and diagnostic agents. *J. Biotechnol.* **74**, 5–13.

BRODY, E. N., WILLIS, M. C., SMITH, J. D., JAYASENA, S., ZICHI, D., GOLD, L. (1999). The use of aptamers in large arrays for molecular diagnostics. *Mol Diagn.* **4**, 381–388.

BULYK, M. L., GENTALEN, E., LOCKHART, D. J., CHURCH, G. M. (1999). Quantifying DNA-protein interactions by double-stranded DNA arrays. *Nat. Biotechnol.* **17**, 573–577.

BURNS-HAMURO, L. L., MA, Y., KAMMERER, S., REINEKE, U., SELF, C., COOK, C., OLSON, G. L., CANTOR, C. R., BRAUN, A., TAYLOR, S. S. (2003). Designing isoform-specific peptide disruptors of protein kinase A localization. *Proc. Natl. Acad. Sci. USA* **100**, 4072–4077.

BÜSSOW, K., CAHILL, D., NIETFELD, W., BANCROFT, D., SCHERZINGER, E., LEHRACH, H., WALTER, G. (1998). A method for global protein expression and antibody screening on high-density filters of an arrayed cDNA library. *Nucleic Acids Res.* **26**, 5007–5008.

BÜSSOW, K., NORDHOFF, E., LÜBBERT, C., LEHRACH, H., WALTER, G. (2000). A human cDNA library for high-throughput protein expression screening. *Genomics* **65**, 1–8.

CAHILL, D. J., NORDHOFF, E. (2003). Protein arrays and their role in proteomics. *Adv. Biochem. Eng. Biotechnol.* **83**, 177–187.

CHA, T., GUO, A., JUN, Y., PEI, D., ZHU, X. Y. (2004). Immobilization of oriented protein molecules on poly(ethylene glycol)-coated Si(111). *Proteomics* **4**, 1965–1976.

CHARBONEAU, L., SCOTT, H., CHEN, T., WINTERS, M., PETRICOIN 3RD, E. F., LIOTTA, L. A., PAWELETZ, C. P.,

GRUBB, R. L., CALVERT, V. S., WULKUHLE, J. D., et al. (2002). Utility of reverse phase protein arrays: applications to signalling pathways and human body arrays. *Brief Funct. Genomic Proteomic* **1**, 305–315.

COHEN, C. B., CHIN-DIXON, E., JEONG, S., NIKIFOROV, T. T. (1999). A microchip-based enzyme assay for protein kinase A. *Anal. Biochem.* **273**, 89–97.

COLEMAN, M. A., MILLER, K. A., BEERNINK, P. T., YOSHIKAWA, D. M., ALBALA, J. S., WISEMAN, S. B., SINGER, T. D., KURDISTANI, S. K., GRUNSTEIN, M. (2003). Identification of chromatin-related protein interactions using protein microarrays. *Proteomics* **3**, 2101–2107.

DAVIES, H., LOMAS, L., AUSTEN, B. (1999). Profiling of amyloid beta peptide variants using SELDI protein chip arrays. *BioTechniques* **27**, 1258–1261.

DEINHOFER, K., SEVCIK, H., BALIC, N., HARWANEGG, C., HILLER, R., RUMPOLD, H., MUELLER, M. W., SPITZAUER, S. (2004). Microarrayed allergens for IgE profiling. *Methods* **32**, 249–254.

DREWS, J. (2000). Drug discovery: a historical perspective. *Science* **287**, 1960–1964.

EPSTEIN, C. B., BUTOW, R. A. (2000). Microarray technology – enhanced versatility, persistent challenge. *Curr. Opin. Biotechnol.* **11**, 36–41.

FEIZI, T., FAZIO, F., CHAI, W., WONG, C. H. (2003). Carbohydrate microarrays – a new set of technologies at the frontiers of glycomics. *Curr. Opin. Struct. Biol.* **13**, 637–645.

FUCHS, U., DAMM-WELK, C., BORKHARDT, A. (2004). Silencing of disease-related genes by small interfering RNAs. *Curr. Mol. Med.* **4**, 507–517.

GALANINA, O. E., MECKLENBURG, M., NIFANTIEV, N. E., PAZYNINA, G. V., BOVIN, N. V. (2003). GlycoChip: multiarray for the study of carbohydrate-binding proteins. *Lab Chip* **3**, 260–265.

GARREN, H., RUIZ, P. J., WATKINS, T. A., FONTOURA, P., NGUYEN, L. T., ESTLINE, E. R., HIRSCHBERG, D. L., STEINMAN, L. (2001). Combination of gene delivery and DNA vaccination to protect from and reverse Th1 auto-immune disease via deviation to the Th2 pathway. *Immunity* **15**, 15–22.

GE, H. (2000). UPA, a universal protein array system for quantitative detection of protein–protein, protein–DNA, protein–RNA and protein–ligand interactions. *Nucleic Acids Res.* **28**, e3.

GIAVALISCO, P., NORDHOFF, E., LEHRACH, H., GOBOM, J., KLOSE, J., MUELLER, M., EGELHOFER, V., THEISS, D., SEITZ, H. (2003). Extraction of proteins from plant tissues for two-dimensional electro-phoresis analysis: A calibration method that simplifies and improves accurate determination of peptide molecular masses by MALDI-TOF MS. *Electrophoresis* **24**, 207–216.

GOBOM, J., SCHUERENBERG, M., MUELLER, M., THEISS, D., LEHRACH, H., NORDHOFF, E. (2001). α-Cyano-4–hydroxycinnamic acid affinity sample preparation. A protocol for MALDI-MS peptide analysis in proteomics. *Anal. Chem.* **73**, 434–438.

GUSCHIN, D., YERSHOV, G., ZASLAVSKY, A., GEMMELL, A., SHICK, V., PROUDNIKOV, D., ARENKOV, P., MIRZABEKOV, A. (1997). Manual manufacturing of oligonucleo-tide, DNA, and protein microchips. *Anal. Biochem.* **250**, 203–211.

GUTJAHR, C., MURPHY, D., LUEKING, A., KOENIG, A., JANITZ, M., O'BRIEN, J., KORN, B., HORN, S., LEHRACH, H., CAHILL, D. J. (2005). Mouse protein arrays from a TH1 cell cDNA library for antibody screening and serum profiling. *Genomics* **85**, 285–296.

HAAB, B. B., DUNHAM, M. J., BROWN, P. O. (2001). Protein microarrays for highly parallel detection and quantitation of specific proteins and antibodies in complex solutions. *Genome Biol.* **2**, research00040001–00040013.

HE, M., TAUSSIG, M. J. (2001). Single step generation of protein arrays from DNA by cell-free expression and *in situ* immobilisation (PISA method). *Nucleic Acids Res.* **29**, E73–73.

HERGENROTHER, P. J., DEPEW, K. M., SCHREIBER, S. L. (2000). Small-molecule microarrays: covalent attachment and screening of alcohol-containing small molecules on glass slides. *J. Am. Chem. Soc.* **122**, 7849–7850.

HEYMAN, J. A., CORNTHWAITE, J., FONCERRADA, L., GILMORE, J. R., GONTANG, E., HARTMAN, K. J., HERNANDEZ, C. L., HOOD, R., HULL, H. M., LEE, W. Y., et al. (1999). Genome-scale cloning and expression of individual open reading frames using topoisomerase I-mediated ligation. *Genome Res.* **9**, 383–392.

HILLER, R., LAFFER, S., HARWANEGG, C., HUBER, M., SCHMIDT, W. M., TWARDOSZ, A., BARLETTA, B., BECKER, W. M., BLASER, K., BREITENEDER, H., et al. (2002). Microarrayed allergen molecules: diagnostic gatekeepers for allergy treatment. *FASEB J.* **16**, 414–416.

HOLT, L. J., ENEVER, C., DE WILDT, R. M., TOMLINSON, I. M. (2000). The use of recombinant antibodies in proteomics. *Curr. Opin. Biotechnol.* **11**, 445–449.

HOLZ, C., HESSE, O., BOLOTINA, N., STAHL, U., LANG, C. (2002). A micro-scale process for high-throughput expression of cDNAs in the yeast *Saccharomyces cerevisiae*. *Protein Expr. Purif.* **25**, 372–378.

HOLZ, C., LUEKING, A., BOVEKAMP, L., GUTJAHR, C., BOLOTINA, N., LEHRACH, H., CAHILL, D. J. (2001). A human cDNA expression library in yeast enriched for open reading frames. *Genome Res.* **11**, 1730–1735.

HOLZ, C., PRINZ, B., BOLOTINA, N., SIEVERT, V., BUSSOW, K., SIMON, B., STAHL, U., LANG, C. (2003). Establishing the yeast *Saccharomyces cerevisiae* as a system for expression of human proteins on a proteome-scale. *J. Struct. Funct. Genomics* **4**, 97–108.

HOREJSI, R., KOLLENZ, G., DACHS, F., TILLIAN, H. M., SCHAUENSTEIN, K., SCHAUENSTEIN, E., STEINSCHIFTER, W. (1997). Interheavy disulfide bridge in immunoglobulin G (IgG) reacting with dithionitrobenzoate. A unique feature in serum proteins. *J. Biochem. Biophys. Methods* **34**, 227–236.

HORN, S., LUEKING, A., MURPHY, D., STAUDT, A., GUTJAHR, C., SCHULTE, K., KOENIG, A., LANDSBERGER, M., LEHRACH, H., FELIX, S. B., CAHILL, D. (2005). Profiling humoral auto-immune repertoire of dilated cardiomyopathy (DCM) patients and development of a disease-associated protein chip. *Proteomics*, in press.

HOUSEMAN, B. T., HUH, J. H., KRON, S. J., MRKSICH, M. (2002a). Peptide chips for the quantitative evaluation of protein kinase activity. *Nat. Biotechnol.* **20**, 270–274.

HOUSEMAN, B. T., MRKSICH, M., HUH, J. H., KRON, S. J. (2002b). Towards quantitative assays with peptide chips: a surface engineering approach. *Trends Biotechnol.* **20**, 279–281.

HUELS, C., MUELLNER, S., MEYER, H. E., CAHILL, D. J. (2002). The impact of protein biochips and microarrays on the drug development process. *Drug Discov. Today* **7**, S119–124.

IVANOV, S. S., CHUNG, A. S., YUAN, Z. L., GUAN, Y. J., SACHS, K. V., REICHNER, J. S., CHIN, Y. E. (2004). Antibodies Immobilized as arrays to profile protein post-translational modifications in mammalian cells. *Mol. Cell Proteomics* **3**, 788–795.

JAHN-SCHMID, B., HARWANEGG, C., HILLER, R., BOHLE, B., EBNER, C., SCHEINER, O., MUELLER, M. W. (2003). Allergen microarray: comparison of microarray using recombinant allergens with conventional diagnostic methods to detect allergen-specific serum immunoglobulin E. *Clin. Exp. Allergy* **33**, 1443–1449.

JAYASENA, S. D. (1999). Aptamers: an emerging class of molecules that rival antibodies in diagnostics. *Clin. Chem.* **45**, 1628–1650.

JO, S., PARK, K. (2000). Surface modification using silanated poly(ethylene glycol)s. *Biomaterials* **21**, 605–616.

JOHNSON, C. P., JENSEN, I. E., PRAKASAM, A., VIJAYENDRAN, R., LECKBAND, D., VIJAYENDRAN, R. A., LECKBAND, D. E. (2003). Engineered protein a for the orientational control of immobilized proteins. A quantitative assessment of heterogeneity for surface-immobilized proteins. *Bioconjug. Chem.* **14**, 974–978.

KERSTEN, B., FEILNER, T., KRAMER, A., WEHRMEYER, S., POSSLING, A., WITT, I., ZANOR, M. I., STRACKE, R., LUEKING, A., KREUTZBERGER, J., et al. (2003). Generation of Arabidopsis protein chip for antibody and serum screening. *Plant Mol. Biology* **52**, 999–1010.

KLOSE, J. (**1975**). Protein mapping by combined isoelectric focusing and electrophoresis of mouse tissues A novel approach to testing for induced point mutations. *Humangenetik* **26**, 231–143.

KNEZEVIC, V., LEETHANAKUL, C., BICHSEL, V. E., WORTH, J. M., PRABHU, V. V., GUTKIND, J. S., LIOTTA, L. A., MUNSON, P. J., PETRICOIN 3RD, E. F., KRIZMAN, D. B. (**2001**). Proteomic profiling of the cancer microenvironment by antibody arrays. *Proteomics* **1**, 1271–1278.

KONONEN, J., BUBENDORF, L., KALLIONIEMI, A., BARLUND, M., SCHRAML, P., LEIGHTON, S., TORHORST, J., MIHATSCH, M. J., SAUTER, G., KALLIONIEMI, O. P. (**1998**). Tissue microarrays for high-throughput molecular profiling of tumor specimens. *Nat. Med.* **4**, 844–847.

KRAMER, A., FEILNER, T., POSSLING, A., RADCHUK, V., WESCHKE, W., BURKLE, L., KERSTEN, B. (**2004**). Identification of barley CK2alpha targets by using the protein microarray technology. *Phytochemistry* **65**, 1777–1784.

KUKAR, T., ECKENRODE, S., GU, Y., LIAN, W., MEGGINSON, M., SHE, J. X., WU, D. (**2002**). Protein microarrays to detect protein–protein interactions using red and green fluorescent proteins. *Anal. Biochem.* **306**, 50–54.

KUSNEZOW, W., HOHEISEL, J. D., JACOB, A., WALIJEW, A., DIEHL, F. (**2003a**). Solid supports for microarray immunoassays. Antibody microarrays: an evaluation of production parameters. *J. Mol. Recognit.* **16**, 165–176.

KUSNEZOW, W., JACOB, A., WALIJEW, A., DIEHL, F., HOHEISEL, J. D. (**2003b**). Antibody microarrays: an evaluation of production parameters. *Proteomics* **3**, 254–264.

LARKIN, D., MURPHY, D., REILLY, D. F., CAHILL, M., SATTLER, E., HARRIOTT, P., CAHILL, D. J., MORAN, N. (**2004**). ICln, a novel integrin alphaIIbbeta3-associated protein functionally regulates platelet activation. *J. Biol. Chem.* **279**, 27286–27293.

LATIF, N., BAKER, C. S., DUNN, M. J., ROSE, M. L., BRADY, P., YACOUB, M. H. (**1993**). Frequency and specificity of antiheart antibodies in patients with dilated cardiomyopathy detected using SDS-PAGE and western blotting. *J. Am. Coll. Cardiol.* **22**, 1378–1384.

LESAICHERRE, M. L., LUE, R. Y., CHEN, G. Y., ZHU, Q., YAO, S. Q. (**2002**). Intein-mediated biotinylation of proteins and its application in a protein microarray. *J. Am. Chem. Soc.* **124**, 8768–8769.

LOCKHART, D. J., WINZELER, E. A. (**2000**). Genomics, gene expression and DNA arrays. Nature **405**, 827–836.

LUEKING, A., HOLZ, C., GOTTHOLD, C., LEHRACH, H., CAHILL, D. (**2000**). A system for dual protein expression in *Pichia pastoris* and *Escherichia coli*. *Protein Expr. Purif.* **20**, 372–378.

LUEKING, A., HORN, M., EICKHOFF, H., BÜSSOW, K., LEHRACH, H., WALTER, G. (**1999**). Protein microarrays for gene expression and antibody screening. *Anal. Biochem.* **270**, 103–111.

LUEKING, A., HORN, S., LEHRACH, H. J. C. D. (**2003a**). A dual-expression vector allowing expression in *E. coli* and *P pastoris*, including new modifications. *Methods Mol. Biol.* **205**, 31–42.

LUEKING, A., POSSLING, A., HUBER, O., BEVERIDGE, A., HORN, M., EICKHOFF, H., SCHUCHARDT, J., LEHRACH, H., CAHILL, D. J. (**2003b**). A nonredundant human protein chip for antibody screening and serum profiling. *Mol. Cell Proteomics* **2**, 1342–1349.

MACBEATH, G., SCHREIBER, S. L. (**2000**). Printing proteins as microarrays for high-throughput function determination. *Science* **289**, 1760–1763.

MADOZ-GURPIDE, J., WANG, H., MISEK, D. E., BRICHORY, F., HANASH, S. M. (**2001**). Protein based microarrays: a tool for probing the proteome of cancer cells and tissues. *Proteomics* **1**, 1279–1287.

MAHAIRAS, G. G., SABO, P. J., HICKEY, M. J., SINGH, D. C., STOVER, C. K. (**1996**). Molecular analysis of genetic differences between *Mycobacterium bovis* BCG and virulent *M. bovis*. *J. Bacteriol.* **178**, 1274–1282.

MANGOLD, U., DAX, C. I., SAAR, K., SCHWAB, W., KIRSCHBAUM, B., MULLNER, S. (**1999**). Identification and characterization of potential new therapeutic targets in inflammatory and autoimmune diseases. *Eur. J. Biochem.* **266**, 1184–1191.

MARCUS, K., MOEBIUS, J., MEYER, H. E. (2003). Differential analysis of phosphorylated proteins in resting and thrombin-stimulated human platelets. *Anal. Bioanal. Chem.* **376**, 973–993.

MICHAUD, G. A., SALCIUS, M., ZHOU, F., BANGHAM, R., BONIN, J., GUO, H., SNYDER, M., PREDKI, P. F., SCHWEITZER, B. I. (2003). Analyzing antibody specificity with whole proteome microarrays. *Nat. Biotechnol.* **21**, 1509–1512.

MOODY, M. D., VAN ARSDELL, S. W., MURPHY, K. P., ORENCOLE, S. F., BURNS, C. (2001). Array-based ELISAs for high-throughput analysis of human cytokines. *Biotechniques* **31**, 186–190, 192–184.

MOONEY, J. F., HUNT, A. J., MCINTOSH, J. R., LIBERKO, C. A., WALBA, D. M., ROGERS, C. T. (1996). Patterning of functional antibodies and other proteins by photolithography of silane monolayers. *Proc. Natl. Acad. Sci. USA* **93**, 12287–12291.

MOROZOV, V. N., GAVRYUSHKIN, A. V., DEEV, A. A. (2002). Direct detection of isotopically labeled metabolites bound to a protein microarray using a charge-coupled device. *J. Biochem. Biophys. Methods* **51**, 57–67.

MOROZOV, V. N., MOROZOVA, T. Y., JOHNSON, K. L., NAYLOR, S. (2003). Parallel determination of multiple protein metabolite interactions using cell extract, protein microarrays and mass spectrometric detection. *Rapid Commun. Mass Spectrom.* **17**, 2430–2438.

NAM, M. J., MADOZ-GURPIDE, J., WANG, H., LESCURE, P., SCHMALBACH, C. E., ZHAO, R., MISEK, D. E., KUICK, R., BRENNER, D. E., HANASH, S. M. (2003). Molecular profiling of the immune response in colon cancer using protein microarrays: occurrence of autoantibodies to ubiquitin C-terminal hydrolase L3. *Proteomics* **3**, 2108–2115.

O'BRIEN, J., STICKNEY, J. T., PORTER, M. D. (2000). Preparation and characterisation of self-assembled double-stranded DNA (dsDNA) microarrays for protein:

dsDNA screening using atomic force microscopy. *Langmuir* **16**, 9559–9567.

O'FARRELL, P. H. (1975). High resolution two-dimensional electrophoresis of proteins. *J. Biol. Chem.* **250**, 4007–4021.

OSTERGAARD, M., RASMUSSEN, H. H., NIELSEN, H. V., VORUM, H., ORNTOFT, T. F., WOLF, H., CELIS, J. E. (1997). Proteome profiling of bladder squamous cell carcinomas: identification of markers that define their degree of differentiation. *Cancer Res.* **57**, 4111–4117.

PARK, C. B., CLARK, D. S. (2002). Sol-gel encapsulated enzyme arrays for high-throughput screening of biocatalytic activity. *Biotechnol. Bioeng.* **78**, 229–235.

PAWELETZ, C. P., CHARBONEAU, L., BICHSEL, V. E., SIMONE, N. L., CHEN, T., GILLESPIE, J. W., EMMERT-BUCK, M. R., ROTH, M. J., PETRICOIN, I. E., LIOTTA, L. A. (2001). Reverse phase protein microarrays which capture disease progression show activation of pro-survival pathways at the cancer invasion front. *Oncogene* **20**, 1981–1989.

PELUSO, P., WILSON, D. S., DO, D., TRAN, H., VENKATASUBBAIAH, M., QUINCY, D., HEIDECKER, B., POINDEXTER, K., TOLANI, N., PHELAN, M., et al. (2003). Optimizing antibody immobilization strategies for the construction of protein microarrays. *Anal. Biochem.* **312**, 113–124.

PIEHLER, J., BRECHT, A., GECKELER, K. E., GAUGLITZ, G. (1996). Surface modification for direct immunoprobes. *Biosens. Bioelectron.* **11**, 579–590.

POHLNER, K., PORTIG, I., PANKUWEIT, S., LOTTSPEICH, F., MAISCH, B. (1997). Identification of mitochondrial antigens recognized by antibodies in sera of patients with idiopathic dilated cardiomyopathy by two-dimensional gel electrophoresis and protein sequencing. *Am. J. Cardiol.* **80**, 1040–1045.

RAMACHANDRAN, N., HAINSWORTH, E., BHULLAR, B., EISENSTEIN, S., ROSEN, B., LAU, A. Y., WALTER, J. C., LABAER, J. (2004). Self-assembling protein microarrays. *Science* **305**, 86–90.

REBOUL, J., VAGLIO, P., RUAL, J. F., LAMESCH, P., MARTINEZ, M., ARMSTRONG, C. M., LI, S., JACOTOT, L., BERTIN, N., JANKY, R., et al. (2003).

C. elegans ORFeome version 11: experimental verification of the genome annotation and resource for proteome-scale protein expression. *Nat. Genet.* **34**, 35–41.

REINEKE, U. (**2004**). Antibody epitope mapping using arrays of synthetic peptides. *Methods Mol. Biol.* **248**, 443–463.

REINEKE, U., VOLKMER-ENGERT, R., SCHNEIDER-MERGENER, J. (**2001**). Applications of peptide arrays prepared by the SPOT-technology. *Curr. Opin. Biotechnol.* **12**, 59–64.

RICH, R. L., MYSZKA, D. G. (**2000**). Advances in surface plasmon resonance biosensor analysis. *Curr. Opin. Biotechnol.* **11**, 54–61.

ROBINSON, W. H., DIGENNARO, C., HUEBER, W., HAAB, B. B., KAMACHI, M., DEAN, E. J., FOURNEL, S., FONG, D., GENOVESE, M. C., DE VEGVAR, H. E., et al. (**2002**). Autoantigen microarrays for multiplex characterization of autoantibody responses. *Nat. Med.* **8**, 295–301.

ROBINSON, W. H., FONTOURA, P., LEE, B. J., NEUMAN DE VEGVAR, H. E., TOM, J., PEDOTTI, R., DIGENNARO, C. D., MITCHELL, D. J., FONG, D., HO, P., et al. (**2003**). Protein microarrays guide tolerizing DNA vaccine treatment of autoimmune encephalomyelitis. *Nat. Biotechnol.* **21**, 1033–1039.

RODA, A., GUARDIGLI, M., RUSSO, C., PASINI, P., BARALDINI, M. (**2000**). Protein microdeposition using a conventional ink-jet printer. *Biotechniques* **28**, 492–496.

ROWE, C. A., SCRUGGS, S. B., FELDSTEIN, M. J., GOLDEN, J. P., LIGLER, F. S. (**1999**). An array immunosensor for simultaneous detection of clinical analytes. *Anal. Chem.* **71**, 433–439.

RUIZ, P. J., GARREN, H., RUIZ, I. U., HIRSCHBERG, D. L., NGUYEN, L. V., KARPUJ, M. V., COOPER, M. T., MITCHELL, D. J., FATHMAN, C. G., STEINMAN, L. (**1999**). Suppressive immunization with DNA encoding a self-peptide prevents autoimmune disease: modulation of T cell costimulation. *J. Immunol.* **162**, 3336–3341.

RUIZ-TAYLOR, L. A., MARTIN, T. L., ZAUGG, F. G., WITTE, K., INDERMUHLE, P., NOCK, S., WAGNER, P. (**2001**). Monolayers of derivatized poly(L-lysine)-grafted poly(ethylene glycol) on metal oxides as a class of biomolecular interfaces. *Proc. Natl. Acad. Sci. USA* **98**, 852–857.

SALISBURY, C. M., MALY, D. J., ELLMAN, J. A. (**2002**). Peptide microarrays for the determination of protease substrate specificity. *J. Am. Chem. Soc.* **124**, 14868–14870.

SCHAEFERLING, M., SCHILLER, S., PAUL, H., KRUSCHINA, M., PAVLICKOVA, P., MEERKAMP, M., GIAMMASI, C., KAMBHAMPATI, D. (**2002**). Application of self-assembly techniques in the design of biocompatible protein microarray surfaces. *Electrophoresis* **23**, 3097–3105.

SCHAFFITZEL, C., HANES, J., JERMUTUS, L., PLUCKTHUN, A. (**1999**). Ribosome display: an *in vitro* method for selection and evolution of antibodies from libraries. *J. Immunol. Methods* **231**, 119–135.

SCHUERENBERG, M., LUEBBERT, C., EICKHOFF, H., KALKUM, M., LEHRACH, H., NORDHOFF, E. (**2000**). Prestructured MALDI-MS sample supports. *Anal. Chem.* **72**, 3436–3442.

SCHWEITZER, B., ROBERTS, S., GRIMWADE, B., SHAO, W., WANG, M., FU, Q., SHU, Q., LAROCHE, I., ZHOU, Z., TCHERNEV, V. T., et al. (**2002**). Multiplexed protein profiling on microarrays by rolling-circle amplification. *Nat. Biotechnol.* **20**, 359–365.

SREEKUMAR, A., NYATI, M. K., VARAMBALLY, S., BARRETTE, T. R., GHOSH, D., LAWRENCE, T. S., CHINNAIYAN, A. M. (**2001**). Profiling of cancer cells using protein microarrays: discovery of novel radiation-regulated proteins. *Cancer Res.* **61**, 7585–7593.

STEPHENS, G., O'LUANAIGH, N., REILLY, D., HARRIOTT, P., WALKER, B., FITZGERALD, D., MORAN, N. (**1998**). A sequence within the cytoplasmic tail of GpIIb independently activates platelet aggregation and thromboxane synthesis. *J. Biol. Chem.* **273**, 20317–20322.

STRIEBEL, H. M., SCHELLENBERG, P., GRIGARAVICIUS, P., GREULICH, K. O. (**2004**). Readout of protein microarrays using intrinsic time resolved UV fluorescence for label-free detection. *Proteomics* **4**, 1703–1711.

URBANEK-RUIZ, I., RUIZ, P. J., PARAGAS, V., GARREN, H., STEINMAN, L.,

FATHMAN, C. G. (**2001**). Immunization with DNA encoding an immunodominant peptide of insulin prevents diabetes in NOD mice. *Clin. Immunol.* **100**, 164–171.

UTTAMCHANDANI, M., CHEN, G. Y., LESAICHERRE, M. L., YAO, S. Q., LUE, R. Y., ZHU, Q., LI, D. (**2004**). Site-specific peptide immobilization strategies for the rapid detection of kinase activity on microarrays. Site-specific immobilization of biotinylated proteins for protein microarray analysis. Enzymatic profiling system in a small-molecule microarray. Intein-mediated biotinylation of proteins and its application in a protein microarray. Developing site-specific immobilization strategies of peptides in a microarray. *Methods Mol. Biol.* **278**, 191–204.

VIJAYENDRAN, R. A., LECKBAND, D. E. (**2001**). A quantitative assessment of hetero-geneity for surface-immobilized proteins. *Anal. Chem.* **73**, 471–480.

WALHOUT, A. J., TEMPLE, G. F., BRASCH, M. A., HARTLEY, J. L., LORSON, M. A., VAN DEN HEUVEL, S., VIDAL, M. (**2000**). GATEWAY recom-binational cloning: application to the cloning of large numbers of open reading frames or ORFeomes. *Methods Enzymol.* **328**, 575–592.

WALTER, G., KONTHUR, Z., LEHRACH, H. (**2001**). High-throughput screening of surface displayed gene products. *Comb. Chem. High Throughput Screen* **4**, 193–205.

WANG, D. (**2003**). Carbohydrate microarrays. *Proteomics* **3**, 2167–2175.

WANG, D., LIU, S., TRUMMER, B. J., DENG, C., WANG, A. (**2002**). Carbohydrate micro-arrays for the recognition of cross-reactive molecular markers of microbes and host cells. *Nat. Biotechnol.* **20**, 275–281.

WIEMANN, S., MEHRLE, A., BECHTEL, S., WELLENREUTHER, R., PEPPERKOK, R., POUSTKA, A. (**2003**). CDNAs for functional genomics and proteomics:

the German Consortium. *C. R. Biol.* **326**, 1003–1009.

WILKINS, M. R., PASQUALI, C., APPEL, R. D., OU, K., GOLAZ, O., SANCHEZ, J. C., YAN, J. X., GOOLEY, A. A., HUGHES, G., HUMPHERY-SMITH, I., et al. (**1996**). From proteins to proteomes – large-scale protein identification by 2-dimensional electrophoresis and amino-acid-analysis. *Bio-Technology* **14**, 61–65.

WILLATS, W. G., RASMUSSEN, S. E., KRISTENSEN, T., MIKKELSEN, J. D., KNOX, J. P. (**2002**). Sugar-coated microarrays: a novel slide surface for the high-throughput analysis of glycans. *Proteomics* **2**, 1666–1671.

WILTSHIRE, S., O'MALLEY, S., LAMBERT, J., KUKANSKIS, K., EDGAR, D., KINGSMORE, S. F., SCHWEITZER, B. (**2000**). Detection of multiple allergen-specific IgEs on microarrays by immunoassay with rolling circle ampli-fication. *Clin. Chem.* **46**, 1990–1993.

ZHANG, W., CZUPRYN, M. J. (**2002**). Free sulfhydryl in recombinant monoclonal antibodies. *Biotechnol. Prog.* **18**, 509–513.

ZHU, H., BILGIN, M., BANGHAM, R., HALL, D., CASAMAYOR, A., BERTONE, P., LAN, N., JANSEN, R., BIDLINGMAIER, S., HOUFEK, T., et al. (**2001a**). Global analysis of protein activities using proteome chips. *Science* **293**, 2101–2105.

ZHU, H., BILGIN, M., BANGHAM, R., HALL, D., CASAMAYOR, A., BERTONE, P., LAN, N., JANSEN, R., BIDLINGMAIER, S., HOUFEK, T., et al. (**2001b**). Global analysis of protein activities using proteome chips. *Science* **293**, 2101–2105.

ZHU, H., KLEMIC, J. F., CHANG, S., BERTONE, P., CASAMAYOR, A., KLEMIC, K. G., SMITH, D., GERSTEIN, M., REED, M. A., SNYDER, M. (**2000**). Analysis of yeast protein kinases using protein chips. *Nat. Genet.* **26**, 283–289.

ZIAUDDIN, J., SABATINI, D. M. (**2001**). Microarrays of cells expressing defined cDNAs. *Nature* **411**, 107–110.

8
Current Developments for the *In Vitro* Characterization of Protein Interactions

Daniela Moll, Bastian Zimmermann, Frank Gesellchen and Friedrich W. Herberg

Abstract

In pharmaceutical research it has become increasingly important to speed up the discovery and validation of new drug candidates. Besides drug design based on structural information, high-throughput screening (HTS) (Ohtsuka et al., 1999) of potential drugs against accessible targets has become standard practice. An inherent disadvantage of HTS is the occurrence of false positives, complicating and prolonging the validation process. HTS must therefore be complemented by a rational approach and independent methods to validate putative interactions. Furthermore, to aid in rational drug design, precise knowledge of an entire protein interaction network should be generated. Biomolecular interaction analysis (BIA) focuses on the quantification of binding events between small molecules, proteins, peptides and other biomolecules thus providing a valuable tool for generating highly accurate and physiologically relevant binding data. Several BIA methods based on different physical principles are available and are compared here. Biosensors utilizing the optical principle of surface plasmon resonance (SPR) are commonly used in a chip-based format and allow the determination of distinct rate and equilibrium binding constants. Several BIA methods are based on fluorescence labeling and can be applied in a variety of *in vivo* and *in vitro* assays. Fluorescence polarization (FP) provides a fast and simple assay setup commonly used for characterization of inhibitor action. Amplified luminescence proximity homogeneous assay (AlphaScreen), represents a homogeneous, bead-based assay easily adaptable to HTS with a robust luminescence readout. Isothermal titration calorimetry (ITC) is a truly label-free method in a homogeneous assay format, additionally yielding thermodynamic parameters. Examples for the application of each method are given and advantages and drawbacks are discussed.

Proteomics in Drug Research
Edited by M. Hamacher, K. Marcus, K. Stühler, A. van Hall, B. Warscheid, H. E. Meyer
Copyright © 2006 Wiley-VCH Verlag GmbH & Co. KGaA, Weinheim
ISBN: 3-527-31226-9

8.1
Introduction

Characterization of biological interactions is an integral component in the drug discovery process. Rational drug design requires an in-depth knowledge of the binding properties of relevant components in order to develop substances which bind to their target with improved affinity. This may result in lower dosage requirement, greater specificity, better drug efficacy, reduced side effects, and less drug resistance. Biomolecular interaction analysis (BIA) has proved to be a helpful tool in determining binding affinity and can nowadays be performed using a wide variety of methods. Depending on the experimental setup and the detection principle, several methods based on different physical parameters have been developed for BIA.

Each method has advantages and drawbacks and can in principle be performed in solid phase or solution-based/homogeneous assay format (Figure 8.1). Biosensors employing surface plasmon resonance (SPR) are generally used in a solid phase format requiring the coupling of one interaction partner to a matrix; the same is true for the recently widely used protein array technology (Figure 8.1A). On the other hand solution-based methods often require the labeling of one interaction partner, commonly performed with fluorescent dyes or radioactive isotopes (Figure 8.1B). Applications here include stopped flow (SF), fluorescence polarization (FP), or bead-based assays like AlphaScreen (Figure 8.1C). The

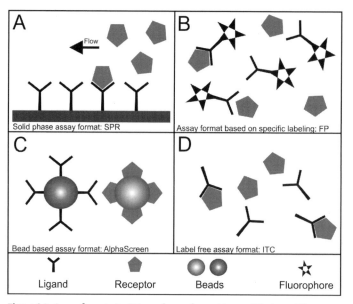

Figure 8.1 Assay formats in BIA can be performed in a solid phase (**A**) or in solution-based (**B–D**) assay format. Applications are given in the figure. For details see text.

AlphaScreen system is well suited for screening specific interactions without getting detailed binding data. Calorimetric methods, for example isothermal titration calorimetry (ITC) (Figure 8.1D), do not require immobilization or labeling; however, a considerable amount of receptor and ligand is generally needed. In addition to basic kinetic parameters ITC characterizes the thermodynamics and stoichiometry of the interaction. Apart from the equilibrium binding constant K_A, values for free energy (ΔG), enthalpy (ΔH) and entropy (ΔS) are generated.

The readout in a BIA experiment can be an equilibrium binding constant (K_A or K_D) or, preferably, distinct association and dissociation rate constants, k_{ass} and k_{diss}, respectively. This requires an experimental setup which allows monitoring of association separately from dissociation in real-time.

8.2
The Model System: cAMP-Dependent Protein Kinase

To validate interaction assays, a suitable biological model system is helpful. cAMP-dependent protein kinase (PKA) is used as such a model system in the assays described below. PKA is the main target for the second messenger cAMP in eukaryotic cells and, since it has been implicated in various human diseases, the development of cAMP analogs modulating PKA activity has provoked increasing interest.

In its inactive state, PKA is a tetrameric holoenzyme consisting of two regulatory (R, with four isoforms in humans R_α^I, R_β^I, R_α^{II}, R_β^{II}) and two catalytic (C, with several isoforms in humans) subunits. The binding of four molecules of cAMP to the dimeric R subunit leads to the dissociation of the holoenzyme releasing the active C subunits (Taylor et al., 2004).

8.3
Real-time Monitoring of Interactions Using SPR Biosensors

Modern biosensors based on SPR have become the standard for real-time analysis of biomolecular interactions over the last decade. The main advantages of SPR technology are the potential to separately detect association and dissociation kinetics combined with high sensitivity and reproducibility, as well as minimal sample consumption. The ability to measure several flow cells simultaneously permits online subtraction of control surfaces. The accurate data generated by SPR biosensors allow researchers to assign function to biomolecular interactions of interest such as protein–protein, protein–DNA, DNA–DNA, protein–lipid, protein–carbohydrate and small molecule drug–receptor interactions. SPR biosensor technology is therefore increasingly applied in modern pharmacological research at almost all stages of the drug development process.

The detection system utilized in commercially available Biacore biosensors (Biacore, Sweden) is also based on SPR, an optical phenomenon that arises when light is totally internally reflected from a metal-coated interface between two media of differing refractive index, in the case of the Biacore system a gold-coated glass surface and the buffer solution in the flow cell. A resonant coupling between the incident light energy and surface plasmons in the gold surface occurs at a specific angle of the incident light, absorbing a part of the light energy and causing a drop in the reflected light intensity at this specific angle. The resonance angle is dependent on the refractive index of the medium in the flow cell. Binding events at the sensor surface cause characteristic changes in the refractive index and can therefore be measured directly.

Before starting an experiment, one interactant, termed the ligand, has to be immobilized on a functionalized sensor surface and the other interaction partner, termed the analyte, is applied using a microfluidics system. A variety of coupling strategies can be applied, either covalent via chemical modification or noncovalent via specific capturing, employing anchoring molecules or antibodies (for an overview of immobilization strategies see Hahnefeld et al., 2004; Herberg and Zimmermann, 1999). Careful experimental design is indispensable in order to prevent measuring artifacts. Experimental parameters such as ligand density, flow rate, buffer composition and concentration range of the analyte have to be considered. Adequate control surfaces with biochemical and biophysical parameters comparable to the specific surface must be selected. The analysis of a whole set of different analyte concentrations results in a complete data set of binding curves with different shapes. This data set allows the calculation of association and dissociation rate constants (k_{ass} and k_{diss}, respectively). For data evaluation a number of software packages are available that employ automated fitting algorithms to simulate biological interactions with varying complexity.

As an example, the binding of two different isoforms of the PKA regulatory subunit to the immobilized catalytic subunit was monitored in the presence or absence of MgATP (Figure 8.2). The different shapes of the binding curves describe the kinetic properties of the interacting proteins and demonstrate the varying interaction patterns of target proteins under different physiological conditions (for details see Figure 8.2 legend). In the presence of MgATP as a low molecular weight cofactor, the type Iα R subunit displayed a rather rapid association which was significantly diminished in the absence of cofactor. Under this condition the type Iβ isoform did not bind at all.

Recent refinements of the technology now also allow investigation of the binding of low molecular weight substances and thus its integration into nearly all stages of the drug development process. These include target validation, secondary screening, hit validation, lead optimization, assay development, serum protein binding, for example absorption, distribution, metabolism, excretion (ADME) studies, and immunoreactivity (Zimmermann et al., 2002). Further areas of application are the characterization of antibody–antigen interactions, quality control of recombinant proteins and analysis of nucleic acid interactions. The flexibility of SPR-based systems and their broad applicability for the analysis of almost every biomolecular binding event allows a comprehensive analysis of protein interactions.

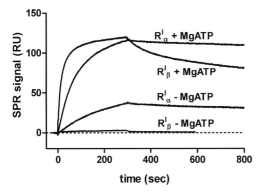

Figure 8.2 Binding analysis using a SPR biosensor. Interaction of a covalently immobilized PKA catalytic subunit with two different regulatory isoforms (R'_α and R'_β, respectively) in the absence and presence of MgATP as a low molecular weight cofactor using a Biacore 3000 system. The association phase was monitored for 300 s and the dissociation was initiated by buffer injection and followed for 500 s.

8.4
ITC in Drug Design

ITC directly determines the heat released (or absorbed) by the binding of a ligand to a receptor, thereby providing a direct route to a thermodynamic characterization of noncovalent equilibrium interactions between drugs and their therapeutic targets *in vitro*. The number of articles published on ITC has doubled every two years since 1994. Interaction analyses of proteins with small molecules (physiological ligands), drugs, inhibitors, nucleotides, polysaccharides or lipids have been described (Cliff and Ladbury, 2003).

The popularity of ITC in drug development is based on its ability to characterize the complete thermodynamics of equilibrium binding by determining the stoichiometry of the interaction (n), the equilibrium binding constants K_A and K_D, the free energy (ΔG), the enthalpy (ΔH) and the entropy (ΔS). This provides important insights into the interaction mechanism. ΔH reflects the strength of the interaction between drug and target relative to those with the solvent, primarily due to hydrogen bond formation and van der Waals interactions. When ΔH is negative, binding is enthalpically favored. Favorable enthalpy requires correct placement of hydrogen bond acceptor and donor groups at the binding interface.

A positive $T\Delta S$ (see also Gibbs–Helmholtz equation below) results in entropically favored binding. Favorable entropy changes are primarily due to hydrophobic interactions, due to an increase in solvent entropy from burial of hydrophobic groups and release of water upon binding, as well as minimal loss of conformational degrees of freedom. A drug with a high affinity to its target has a favorable negative ΔH, a characteristic of hydrogen bond formation, but an unfavorable negative ΔS. Typically, these drugs have a large degree of flexibility, as well as high polarity, potentially causing problems with membrane permeability *in vivo*.

On the other hand, a drug with a favorable $T\Delta S$, indicating that binding is driven by hydrophobic interactions, but with an unfavorable ΔH, may be poorly soluble and conformational restraints may lead to a lack of adaptability (Luque and Freire, 2002). Substances displaying a negative ΔH and a positive $T\Delta S$ in ITC measurements most likely qualify as a drug. These thermodynamic parameters are directly connected with the free energy (ΔG) and the K_D via the Gibbs-Helmholtz equation (Perrin, 1926):

$$\Delta G = \Delta H - T\,\Delta S = -RT\ln\frac{1}{K_D} \tag{8.1}$$

where R is the universal gas constant and T the absolute temperature.

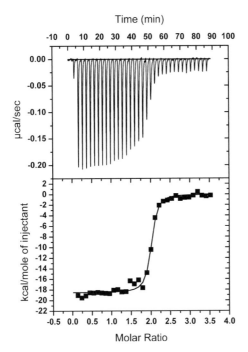

Figure 8.3 Analysis of ligand binding using ITC.

A Titration of 1.9 µM monomeric regulatory subunit with 54 µM cAMP (5 µL injection steps) in 150 mM NaCl, 20 mM MOPS, 1 mM β-mercaptoethanol, pH 7 at 20.1 °C (VP-ITC MicroCalorimeter; MicroCal LLC., Northampton MA), raw data. The rate of heat released is plotted as a function of time.

B Plot of heat exchange per mole of injectant (integrated areas under the respective peaks in **A**) versus molar ratio of protein after subtraction of buffer control (not shown). The best least squares fit applying a model where two cAMP molecules per regulatory subunit monomer are bound (one binding site model) was performed with the software MicroCal Origin. The turning point of the binding isotherm (molar ratio of 2) indicates that two cAMP molecules bind to one regulatory subunit.

A typical ITC experiment (for experimental setup and instrumentation see Pierce et al., 1999) consists of sequential injections of a ligand with extremely accurately adjusted volume into the receptor solution. Both the ligand and the receptor have to be solved in exactly the same buffer. It is crucial that the heat of dilution is subtracted by titration of the same ligand into buffer only.

As an example of ITC, the binding of a ligand, here cAMP, to one of its receptors, a regulatory subunit of PKA, was tested as shown in Figure 8.3. This process is an exothermic binding process. Panel A shows the raw data obtained when 5 μL volumes of cAMP (54 μM) in the syringe were injected into the solution containing the regulatory subunit (1.9 μM) in the sample cell at 20 °C. Panel B shows the binding isotherm (after subtraction of the buffer control) that has been analyzed using the software MicroCal Origin (Microcal LLC., Northampton, MA, USA).

From the binding isotherm a binding stoichiometry of $n = 2$ was determined assuming a single set of identical binding sites. The calculated equilibrium dissociation constant (K_D) was 6.4 nM, the enthalpy (ΔH) was -18.6 kcal mol^{-1} and the entropy (ΔS) was -0.025 kcal mol^{-1} K^{-1}. The calculated free enthalpy (ΔG) via the Gibbs-Helmholtz equation was -11.0 kcal mol^{-1}. These results correlate very well with the role of cAMP as a second messenger in the cell. In this biological system cAMP has a high affinity to the regulatory subunit indicated by a negative ΔH and a poor membrane permeability indicated by an unfavorable negative ΔS. This information is a prerequisite for the development of cAMP analogs capable of permeating the cell membrane while maintaining their ability to bind to the regulatory subunit with high affinity.

8.5
Fluorescence Polarization, a Tool for High-Throughput Screening

Fluorescence polarization (FP) is also generally performed in a homogeneous assay format and provides a rapid and easy-to-use method complementary to other BIA technologies. The first reported observation of the phenomenon fluorescence was made in 1565 by a Spanish physician, Nicolás Monardes, who observed light emission by an extract of *Lignum nephriticum* in water. About 300 years later Stokes published the outstanding paper "On the Change of Refrangibility of Light" (Stokes, 1852). In a footnote he named the detected phenomenon "fluorescence" for the first time. The basic principle of fluorescence is that a fluorophore absorbs a high-energy photon (shorter wavelength), and re-emits it as a lower energy photon (longer wavelength). Characteristic fluorophores are quinine (present in tonic water), rhodamin B (often used as a cell stain), acridin orange (a DNA stain) and fluorescein. It took another half a century to understand another physical attribute of fluorophores, termed polarization (synonymic: anisotropy). In a FP assay only fluorophores oriented parallel to the axis of incident light absorb the photon and enter the excited state. The excited state has a limited lifetime before light emission occurs. During this time the fluorophore rotates and the light is emitted in an altered direction resulting in a depolarization. Since the excitation plane has rotated as well, re-excitation of the fluorophore by the incident light is prevented. In a watery, nonviscous solution, the fluorophores are oriented randomly and therefore display a polarization signal close to zero. If the fluorophore interacts with a large macromolecule the rotation of the fluorophore is slowed down upon binding. In contrast to the free fluorophore, the bound fluorophore does not rotate out of the plane of excitation during the lifetime of the excited state meaning the emitted light remains polarized. The intensity of emitted polarized light correlates with the amount of fluorophore bound to the macromolecule. Francis Perrin described the relationship between observed polarization and physical parameters quantitatively in the equation that bears his name:

$$p = p_0 \frac{1}{1 + \left(1 - \frac{1}{3}p_0\right)\frac{RT}{V\eta}\tau} \tag{8.2}$$

where p is the polarization, p_0 is the polarization in the absence of rotation (limiting polarization), R is the universal gas constant and τ is the lifetime of the excited state. Therefore, like any other molecule, the rotational speed of a fluorophore is dependent on the effective molar volume (V) of the rotating unit, the absolute temperature (T) and the viscosity (η) of the surrounding solution. With viscosity and temperature held constant, only the effective molar volume determines the polarization signal and therefore gives a measure for complex formation.

As a guideline, with a fluorescently labeled ligand smaller than 5 kDa and a lifetime of the fluorophore in the nanosecond range, depolarization can be measured. Note that the FP signal is not influenced by the concentration of the fluorophore, because the ability of a fluorophore to depolarize light is not a function of its concentration, but rather a function of its ability to rotate while it is still in the excited state. According to this, the interaction between a fluorescently labeled small ligand and a macromolecule (50–100 kDa) can be detected in a homogeneous assay suitable for HTS. The standard format today is a 384-well microtiter plate with a typical total assay volume of 40 µL, which can be potentially extended to the 1536-well plate format. Possible applications include protein binding studies, nucleic acid hybridization or enzymatic assays. Not surprisingly, FP is used increasingly in the drug discovery process on therapeutic targets such as kinases, phosphatases, proteases, G-protein-coupled receptors and nuclear receptors. For an overview describing FP assays in drug discovery see the special issue of "Combinatorial Chemistry and High Throughput Screening" 2003, volume 6, issue 3.

A competitive dose-response FP assay is commonly used for screening of pharmaceuticals. In this assay the concentration of receptor and fluorescently labeled ligand A remain fixed. To achieve maximum sensitivity the concentrations of both interacting partners should be around two to five times the K_D. The concentration of the competing ligand B is varied over at least six orders of magnitude. The resulting polarization signal is then plotted against the logarithm of the concentration of ligand B, resulting in a sigmoidal dose–response curve.

From this, IC_{50} values (inhibitory concentration 50%) can be extracted, i.e., where 50% of the receptor binding sites at a given concentration are occupied by ligand B under equilibrium conditions. With a known binding constant (K_D) of ligand A to the receptor an inhibitory equilibrium binding constant (K_i) can be estimated from the IC_{50} values (Burlingham and Widlanski, 2003; Cheng and Prusoff, 1973), describing the affinity of ligand B to the receptor.

8.6
AlphaScreen as a Pharmaceutical Screening Tool

Amplified luminescence proximity homogeneous assay, or AlphaScreen, is a homogeneous bead-based assay for measuring biomolecular interactions in a microplate format suitable for HTS. It is based on physical proximity between donor and acceptor beads. The proximity between these beads is mediated by the interaction of biomolecules which are bound to the respective beads. Signal generation is dependent on the release of singlet oxygen from the donor beads, by means of an embedded photosensitizer. The photosensitizer converts ambient molecular oxygen to the excited singlet state upon illumination with laser light. Singlet oxygen has a half-life of about 4 µs before returning to the ground state, a time span that allows for diffusion of up to 200 nm from the donor bead (Ullman et al., 1994). If an acceptor bead is in the immediate vicinity of a donor bead – due to a specific interaction of the bead-bound biomolecules – energy of the singlet oxygen is transferred to thioxene derivatives embedded in the acceptor beads which in turn elicits a luminescence readout (Figure 8.4A). It should be noted that the luminescence is emitted at a lower wavelength than the laser light used for excitation (520–620 vs. 680 nm), resulting in a very low background that is further reduced by a time-resolved measurement, i.e. signal detection is temporally separated from excitation.

An AlphaScreen assay is generally conducted in a 384-well plate in a total volume of 25 µL, but can be scaled down to the 1536-well format with assay volumes of 5 µL. Since the donor beads are light sensitive, the assay has to be performed under dimmed light. After all components have been added, the luminescence signal is measured, usually after an incubation of 1 h.

The first consideration when setting up an AlphaScreen assay is the choice of an assay format. In most cases, the decision will depend on the interaction partners under investigation and on the biological tools available for their detection. The interaction partners can be coupled to the beads directly via reductive amination of reactive aldehyde groups, similar to the immobilization on a Biacore sensor chip (see above). The usefulness of this approach is limited by the reaction conditions, which may not be appropriate for maintaining the biologically active conformation of the biomolecule. Therefore the biomolecule of interest is usually not coupled to the beads directly, but instead captured via an antibody, also preventing steric hindrance. While not strictly necessary, it is often convenient to use a biotinylated molecule which can be captured by streptavidin-coated donor beads.

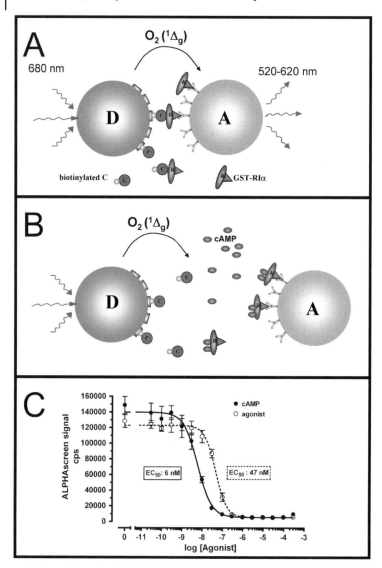

Figure 8.4 AlphaScreen principle and readout.

A The generation of a luminescence signal in AlphaScreen is initiated by the release of singlet oxygen [$O_2(^1\Delta_g)$] from donor beads (*D*) upon illumination with laser light ($\lambda = 680$ nm). If an acceptor bead (*A*) is in immediate proximity (~ 200 nm) the chemical energy of the singlet oxygen is converted into a luminescence signal ($\lambda = 520$–620 nm). Interaction between the beads is mediated by a specific biomolecular interaction, in this example the association between catalytic (*C*) and regulatory subunits (R_α^I) of PKA.
B Upon addition of cAMP (or a cAMP agonist), C and R subunits dissociate, resulting in a decrease in luminescence as the singlet oxygen decays without an acceptor bead in its vicinity.
C Luminescence signal obtained (counts per second, cps) is plotted against the logarithm of agonist concentration and fitted to a sigmoidal dose–response model.

In the so-called direct assay format, a biotinylated component binds to an antibody directly coupled to acceptor beads or to a protein captured by this antibody (Figure 8.4A). In the indirect assay format, the antibody used for capturing the biomolecule is in turn bound by a secondary antibody or by Protein A conjugated to the acceptor beads. In principle, any method capable of capturing the interaction partners to donor and acceptor beads, respectively, is suitable for setting up an AlphaScreen assay.

The actual readout of AlphaScreen is the luminescence signal generated by the acceptor beads. Depending on the assay setup, a signal increase or a signal decrease can be generated. In the simplest case, if the biomolecule under investigation can be detected directly, for example by an antibody, an increase in concentration of this molecule will correspond with a luminescence increase. In other cases, for example with small molecule ligands, it might not be possible to monitor the molecule of interest directly. In this case a competition type of assay can be devised, which generates a signal decrease. An example is given below describing an assay to detect cAMP agonist action (Figure 8.4).

The components involved are streptavidin-coated donor beads and anti-GST acceptor beads (Perkin Elmer LAS, Boston, USA) and – as the biological system – the PKA subunits as cAMP sensors. Upon binding of cAMP to each regulatory (R) subunit, the PKA holoenzyme dissociates releasing the catalytic (C) subunits. In the absence of cAMP the holoenzyme complex is intact and C and R subunits, captured to AlphaScreen donor and acceptor beads, are brought into proximity generating a luminescence signal. If cAMP or a cAMP agonist is added, the signal decreases as the subunits and thus the beads, dissociate (Figure 8.4B).

During assay development, it became evident that a direct coupling of the PKA subunits to the beads was unsuitable, probably due to loss of biological activity during the coupling procedure. Instead, a more indirect approach was selected employing biotinylated C- and glutathione-S-transferase (GST)-tagged R subunits which are captured to streptavidin-coated donor beads and anti-GST antibody-coated acceptor beads, respectively. This strategy resulted in the generation of a robust AlphaScreen signal that was decreased in a dose-dependent manner by cAMP or cAMP agonists (Figure 8.4C). In order to compare agonist potency, EC_{50} values were determined from dose–response curves.

The assay displayed a very good signal-to-background ratio and was highly reproducible. The EC_{50} values obtained were different from those reported previously in ligand binding assays (Schaap et al., 1993), probably due to inherent differences in assay design. However, the relative potency of cAMP and the agonist was reflected correctly with a difference of approximately one order of magnitude.

8.7
Conclusions

BIA has been proven to serve as an important tool in several stages of the drug development process. In each step of the traditional drug discovery process (identification of a drug target, validation and optimization of hit, lead and candidate substances) an interaction between the biological component (drug target) and the potential drug has to be described quantitatively. Since the number of accessible drug targets is limited and, on the other hand, companies are being forced to speed up the entire drug development process, technologies for protein interaction analyses have to be applied at different points of the development process. The technologies have to be customized to fit the needs and requirements of each step. These requirements can differ in throughput, physical parameters, accuracy or biological relevance.

The entire drug discovery process takes on average 12 years, and almost a decade of this time is used in the preclinical part. It is therefore more and more important to speed up the design and development of new drug candidates. To meet this requirement researchers have to rely increasingly on HTS technologies. Since drugs are usually either inhibitors or activators of their target proteins, a precise characterization of the interaction between drug and target is very important.

Table 8.1 Comparison of different methods used in BIA. *LMW*: Low molecular weight.

	SPR	FP	ITC	AlphaScreen
Modification of interaction partner	Coupling	Labeling	None	Coupling
Assay format	Solid phase (chip-based)	Homogeneous	Homogeneous	Homogeneous (bead-based)
Amount of receptor/ligand	Low	Low	Medium	High
Readout	SPR	Fluorescence	Heat	Luminescence
Potential for miniaturization and automation	Medium	High	Low	Very high
Advantages	Distinct rate constants, real time measurement	Fast, easy assay setup	Thermo-dynamic parameters	Robust assay design, HTS compatible
Limitations	Mass transfer limitations, analysis speed	Need of labeled LMW compound	High protein consumption	Relative binding data only
Resulting physical parameters	k_{ass}, k_{diss}, K_D, EC_{50}	K_D, EC_{50}	ΔH, ΔG, ΔS, n, K_D	EC_{50}

As an example, a novel drug to treat chronic myelogenous leukemia (CML), Gleevec (Imatinib), was designed as an inhibitor of one specific receptor. Owing to the extremely high specificity of the drug, Gleevec treatment offers significant advantages compared to the standard therapy, avoiding the typical side effects of classical cytostatic or radiation therapy. Gleevec interferes with the specific molecular mechanism responsible for uncontrolled growth by inhibition of the constitutively active BCR-ABL protein (Smith et al., 2004).

The example of Gleevec also illustrates that the intracellular protein network has to be described on a molecular level, identifying as many interaction partners as possible and thereby understanding entire protein interaction networks. This is especially important in order to validate the drug target and to ensure specificity of the drug candidate. It allows elimination of false positives early on in the drug development process, thus saving time and money. For this purpose several methods can be applied some of which are reviewed in this article. Each method has its own advantages and drawbacks, summarized in Table 8.1.

Depending on the biological compounds, the number of screenings required, and the accuracy of the physical parameter needed, one or a combination of several BIA methods should be chosen. The price per assay depends not only on the costs of chemicals and hardware for the respective assay but also on the availability of receptor and ligand.

SPR offers direct determination of the association and dissociation rate constants in real time; however, it requires the immobilization of one interaction partner. FP offers a fast and easy-to-use assay setup, but is limited by physical parameters. No labeling is needed using ITC, but depending on the biological system, high amounts of receptor and ligand are required. Any kind of molecular interaction can be measured with AlphaScreen; however, no kinetic constants can be obtained.

Acknowledgements

We gratefully acknowledge Dr. Christian Hammann (University of Kassel, Germany) for helpful discussions regarding the ITC measurements. This work was supported by the BMBF 031U202F, EU CRAFT QLK2-CT-2002-72419 and the DFG He1818/4-1.

References

BURLINGHAM, B. T., WIDLANSKI, T. S. (2003). An intuitive look at the relationship of K_i and IC_{50}. A more general use for the Dixon Plot. *J. Chem. Educ.* **80**, 214–218.

CHENG, Y., PRUSOFF, W. H. (1973). Relationship between the inhibition constant (K1) and the concentration of inhibitor which causes 50 per cent inhibition (I50) of an enzymatic reaction. *Biochem. Pharmacol.* **22**, 3099–3108.

CLIFF, M. J., LADBURY, J. E. (2003). A survey of the year 2002 literature on applications of isothermal titration calorimetry. *J. Mol. Recognit.* **16**, 383–391.

HAHNEFELD, C., DREWIANKA, S., HERBERG, F. W. (2004). Determination of kinetic data using surface plasmon resonance biosensors. *Methods Mol. Med.* **94**, 299–320.

HERBERG, F. W., ZIMMERMANN, B. (1999). Analysis of protein kinase interactions using biomolecular interaction analysis. In HARDIE, D. G. (Ed.), *Protein Phosphorylation – A Practical Approach.* Oxford University Press, pp. 335–371.

LUQUE, I., FREIRE, E. (2002). Structural parameterization of the binding enthalpy of small ligands. *Proteins* **49**, 181–190.

OHTSUKA, T., HATA, Y., IDE, N., YASUDA, T., INOUE, E., INOUE, T., MIZOGUCHI, A., TAKAI, Y. (1999). nRap GEP: a novel neural GDP/GTP exchange protein for rap1 small G protein that interacts with synaptic scaffolding molecule (S-SCAM). *Biochem. Biophys. Res. Commun.* **265**, 38–44.

PERRIN, F. (1926). Polarisation de la lumière de fluorescence Vie moyenne des molécules dans l'état excité. *J. Phys.* **7**, 390–401.

PIERCE, M. M., RAMAN, C. S., NALL, B. T. (1999). Isothermal titration calorimetry of protein–protein interactions. *Methods* **19**, 213–221.

SCHAAP, P., VAN MENTS-COHEN, M., SOEDE, R. D., BRANDT, R., FIRTEL, R. A.,

DOSTMANN, W., GENIESER, H. G., JASTORFF, B., VAN HAASTERT, P. J. (1993). Cell-permeable non-hydrolyzable cAMP derivatives as tools for analysis of signaling pathways controlling gene regulation in Dictyostelium. *J. Biol. Chem.* **268**, 6323–6331.

SMITH, J. K., MAMOON, N. M., DUHE, R. J. (2004). Emerging roles of targeted small molecule protein–tyrosine kinase inhibitors in cancer therapy. *Oncol. Res.* **14**, 175–225.

STOKES, G. G. (1852). On the change of refrangibility of light. *Philos. Trans. R Soc. London*, 463–562.

TAYLOR, S. S., YANG, J., WU, J., HASTE, N. M., RADZIO-ANDZELM, E., ANAND, G. (2004). PKA: a portrait of protein kinase dynamics. *Biochim. Biophys. Acta* **1697**, 259–269.

ULLMAN, E. F., KIRAKOSSIAN, H., SINGH, S., WU, Z. P., IRVIN, B. R., PEASE, J. S., SWITCHENKO, A. C., IRVINE, J. D., DAFFORN, A., SKOLD, C. N. (1994). Luminescent oxygen channeling immunoassay: measurement of particle binding kinetics by chemiluminescence. *Proc. Natl. Acad. Sci. USA* **91**, 5426–5430.

ZIMMERMANN, B., HAHNEFELD, C., HERBERG, F. W. (2002). Applications of biomolecular interaction analysis in drug development. *Targets* **1**, 66–73.

9

Molecular Networks in Morphologically Intact Cells and Tissue– Challenge for Biology and Drug Development

Walter Schubert, Manuela Friedenberger and Marcus Bode

Abstract

The hierarchy of cell function comprises at least four distinct functional levels: genome, transcriptome, proteome, and toponome. The toponome is the entirety of all protein networks traced out as patterns directly on the single cell level in the natural environment of cells *in situ* (e.g., tissues). We have developed a photonic microscopic robot technology (MELK) capable of tagging and imaging hundreds (and possibly thousands) of different molecular components (e.g., proteins) in morphologically intact cells and tissue. MELK addresses the fact that each protein must be at the right time at the right place in the right concentration in the cell to interact with other proteins assembled in a spatially organized network. The resulting toponome maps are subcellular compositions of geometric objects representing combinatorial molecular patterns (CMPs) which are images of the molecular networks enciphering the myriad different functionalities. The MELK toponome technology can be interlocked with the industrialized drug discovery process leading to the selection of valid targets and drug leads on the basis of high-throughput/high-content screening.

9.1
Introduction

The costs for developing new drugs are steadily increasing. According to the "Start Centre for the Study of Drug Development" (2001), the average developing costs per drug added up to US$ 802 million in 2000. Yet the clinical success rate remains unsatisfactory. An improvement from the current 1 : 5 to 1 : 3 for each approved drug would be expected to decrease the costs by approximately US$ 235 million (Tufts, 2002).

According to Lehmann Brothers & McKinsey (2001) the low attrition rate in the pharmaceutical industry is essentially due to poor target selection, suggesting that the major problem is entrenched at the basis of any drug development. This

Proteomics in Drug Research
Edited by M. Hamacher, K. Marcus, K. Stühler, A. van Hall, B. Warscheid, H. E. Meyer
Copyright © 2006 Wiley-VCH Verlag GmbH & Co. KGaA, Weinheim
ISBN: 3-527-31226-9

conclusion is strongly supported by the fact that the mean net present value (NPV) per target program and new chemical entity (NCE) per average drug company was US$ 264 million per year before 1995 (classical pharmaceutical models). When new models emerged (genomics-based) the NPV decreased to US$ 34 million in 2001, although the average number of target candidates increased during the time period (1995–2001) by a factor of 4. Over these years, large scale expression profiling methods were established, enabling researchers to screen very effectively for proteins, which are up-regulated in diseases. However, as suggested by Lehman Brothers and McKinsey's analysis, the resulting flood of information does not reveal the relevant biological functions necessary for an efficient target selection process. Consequently – and clearly, in part, caused by the latter target-finding problem – current drug discovery programs are still beset by overall low productivity, as documented by Pharmaceutical Manufactures of America (PhRMH) over recent years. Although the pharmaceutical industry is pursuing several strategies to tackle this problem, it is undeniable that better target selection is one of the problems to be solved by innovative science (Jessen, 2005).

The present paper summarizes a new approach of target selection by addressing molecular networks directly in morphologically intact cells and tissue.

9.2
A Metaphor of the Cell

To illustrate the contextual nature and functional coding principle of proteins in a cell, the written language may be taken as a metaphor: Not the letters, but only the correct *topological* context of letters (their assembly as words, sentences, texts, etc.) reveals meaning. As illustrated in Figure 9.1, counting the numbers of letters in two sentences with a totally different meaning results in almost the same number of different letters. Hence, one cannot distinguish these sentences simply on the basis of the profiles of letters. Only the letter "p" exists in only one of these sentences. However, comparing the sentences on the level of words, they differ much among themselves. Three words specifically designate sentence 1, and three words designate sentence 2. Consequently, only the level of the syntax gives the distinction.

How is meaning codified in biological systems? Assuming that the language metaphor applies, and if, according to this metaphor, proteins can be metaphorically interpreted as letters (signs) composed to specific syntactic symbols (clusters, complexes, etc.) which codify cell functions, a separation of cellular proteins by extracting them from their biological context, will consequently lead to a loss of functional information. Moreover, it must be considered that a known molecular function of a protein can not simply be translated into its cellular function: a protein with a known molecular function "x" exerts a certain *cellular* function by a spatially and temporally determined interplay with other specific proteins within the cell (network 1). The interaction of the same protein x with other proteins (network 2) is likely to generate another cellular function for this protein. Therefore,

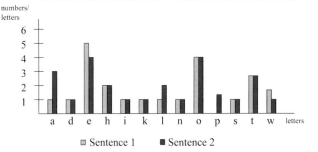

distinguishing letters: only letter ‚p' for Sentence 2

distinguishing words:

	Sentence 1	Sentence 2
	trees	pool
	woods	sea
	in	and

Figure 9.1 Language metaphor illustrating the relationship between letters and syntax, symbolizing the relationship between proteins and the toponome (entire assembly of protein network motifs). Only the level of the syntax, but not the profile of letters, permits a distinction of the two sentences.

the molecular function(s) of proteins must be distinguished from their cellular function(s), which are given by the different spatial arrangements of any given protein(s). Given these considerations, the next important step is to map and decipher the cellular functions of proteins by thorough analysis of their spatial arrangements in morphologically intact cells and tissues on a proteome-wide scale.

In this regard the language metaphor is interesting, because it implies that there may be rules for protein arrangements in cells encompassing a "grammar" of syntactic and semantic structures: the higher level order within a cell's proteome. A similar image of the cell was drawn by Bruce Alberts suggesting that the entire cell can be viewed as an assembly of interlocked protein machines (Alberts, 1998). Whatever metaphor is used to develop an idea or vision of the cell, understanding how protein networks interact in time and space to generate function is a major challenge of the postgenomic era, and clearly, this information must be extracted from the intact cell. Existing methods for analyzing protein–protein interactions (Alberts 1998; Dandekar et al., 1998; Enright et al., 1999; Gavin et al., 2002; Giot et al., 2003; Ho et al., 2002; Hughes et al., 2000; Ito et al., 2001; Marcotte et al., 1999; Overbeek et al., 1999; Pellegrini et al., 1999; Roberts et al., 2000; Uetz et al., 2000) have limitations (Huynen and Bork, 1998; von Mering et al., 2002) and generally cannot reveal when and where interactions occur *in vivo*. In particular we need methods that indicate not only when and where proteins are assembled in a cell, but that also include proteins which interact only weakly and/or transiently, rather than being restricted to those proteins which interact so strongly that the resulting protein complex will withstand purification.

9.3
Mapping Molecular Networks as Patterns: Theoretical Considerations

Theoretical concepts for mapping molecular networks in the cell must consider the principles that cells have established during their evolution to generate the myriad cell functionalities: one may assume that in every cell many proteins are not stochastically distributed but are spatially and temporally highly organized. In every cell every protein that follows these rules must be present at the right time, in the right concentration and at the right location to interact with other proteins. The resulting spatially determined network of proteins generates the different functionalities. Consequently every protein network in a cell is characterized by a spatially determined contextual pattern of proteins. Potentially, these patterns can be imaged by using appropriate techniques able to localize every protein of a pattern independently in one cell, finally generating a "mosaic": images of proteins as spatial contexts. Given an appropriate number of data points per cell resolving the different subcellular compartments with sufficient optical resolution, practically all protein networks can be found as unique patterns. The gain of biological information would consist of maps providing the subcellular protein concentration *and* protein arrangements in one and the same data set. The latter serves as a basis for a systems biology understanding of proteins in the physiological context of cell structures and tissues (Schubert, 2003). A photonic imaging technology termed *m*ultiepitope *l*igand *k*artographie (MELK) was developed (Schubert, 2003) that principally allows us to localize random numbers of proteins in one sample by iterative rounds of tagging proteins (Schubert, 1990) (see below). Preliminary analyses using this technology in combination with a new marker protein (Schubert et al., 1989) indicated that the regeneration of skeletal muscle fibers is based on a reprogramming of vascular endothelial cells to form myogenic cells (Schubert, 1992). This observation was made possible because the simultaneous localization of nine cell surface proteins *in situ* directly uncovered the transition of an endothelial phenotype to a muscle cell phenotype within the endomysial tube of regenerating muscle fibers. The significance of vascular endothelial cells in muscle regeneration was subsequently confirmed by others (De Angelis et al., 1999), providing a first proof of principle of the MELK technology.

Together these data suggested that complex cellular mechanisms can be detected directly in tissue when a sufficient number of cellular differentiation proteins are simultaneously localized as a projection on cells and tissue structures. Referring to the language metaphor: only the visualization of letters in their correct context on pages reveals the syntactic structures, i.e., the assembly of letters to words and sentences. In general, the analysis of every topologically organized system, such as the protein system in the cell, requires the independent localization of its parts to detect its semantic structure.

9.4
Imaging Cycler Robots

To systematically analyze the organization of proteins in cells and tissue, MELK (Schubert, 2003) was established in our lab as a system of robotic workstations. These robots, consisting of a multipipette handling unit, a fluorescence microscope, and a CCD camera, apply large tag-libraries (antibodies, lectins or any other kinds of ligands) on fixed cells or tissues to localize their molecular components. Each tag is conjugated to the same specific dye, for example FITC, and applied sequentially on the sample. Each labeling step is followed by an imaging and a bleaching step, resulting in iterative rounds of labeling, imaging and bleaching (repetitive incubation–imaging–bleaching cycles). The result of these imaging procedures is a protein fingerprint (or molecular component fingerprint) of every subcellular data point in the corresponding sample. By comparative analysis of these fingerprints (diseases vs. controls, etc.), specific patterns of certain cell states are found and thereby functionally mapped. An important advantage of this technology is that the spatial co-compartmentalization of the cellular proteins is preserved and can be analyzed cell by cell in the context of tissue. The resulting context-dependent protein information, which would be lost in cell homogenates or lists of proteins, supports the view that protein networks are spatially determined functional units in every single cell (different functions = different cellular protein networks = unique cellular protein patterns). The scheme in Figure 9.2 illustrates that single normal and abnormal cells are specifically characterized by the relative order of proteins detected as cellular fingerprints. However, when the cells are destroyed by homogenization, the quantitative profile of the corresponding proteins shows no difference in this example. The latter profiles would lead to the conclusion that the identified proteins are not relevant for this disease. However, the mapping of the protein pattern networks (Figure 9.2, right part) leads to the opposite conclusion. The MELK robotic readout in combination with appropriate computer models leads to the identification of cells, which are specifically characterized by protein fingerprints. For example, the application of MELK has confirmed by the analysis of invasive cells (immune cells, tumor cells), that the simultaneous mapping of 20–100 proteins of the cell surface results in highly specific protein patterns (or fingerprints), uniquely characterizing disease processes (to be published). These recent experiences support the view that every cell obeys a quasi infinite "data space" to generate different patterns enciphering different functionalities. However, the patterns which are actually used by cells *in situ* (*in vivo*), are highly restricted and nonrandom, as expected. Consequently the detection of these patterns is important to understanding the organization and functional coding properties within the cellular proteome. MELK supports the search for relevant patterns by providing large data spaces or "reading frames" to map functional clusters of proteins, i.e., in diseases. From a purely mathematical point of view the gain of information provided by MELK compared to methods extracting proteins from cell homogenates is enormous: given 20 proteins, which are isolated from a cell- or tissue homogenate (classical proteome analysis), and given that every protein can be detected at 250 different concentrations (0–250), the resulting protein profile would contain a maximum number of 25 020 different possibilities of protein combinations at different concentrations. However, every cell has the potential to combine the proteins in every single subcellular compartment in a different way. Given, that in every cell out of 1 million cells, 2000 subcellular data points can be measured (MELK), the resulting maximum of possible combinations would be 1 million × 2000 × 25 020. Hence, the biological information increases exponentially compared to the classical procedures of protein analysis using MELK. MELK provides such reading frames and enables us to trace out this information by comparing cells in diseases or in different functional states (walking through large proteome fractions

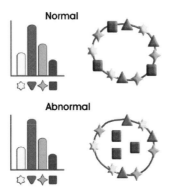

Figure 9.2 Comparison of cellular protein patterns (representing a spatially determined protein network) and protein profiles.

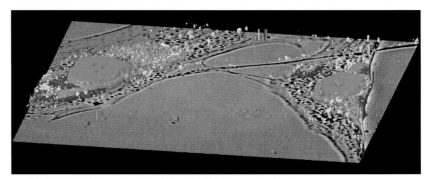

Figure 9.3 Subcellular toponome fingerprint of two muscle cells.
Cell on the *right*: fusion-competence. Cell on the *left*: migration status.
Note the different *colors* representing different vesicle types,
each of which is characterized by a unique protein assembly.

with subcellular resolution). On the basis of reference maps for specific cell types, disease-specific cellular fingerprints can be traced out surprisingly fast, sometimes within hours or few days (Nattkemper et al., 2001). Figure 9.3 illustrates a subcellular protein reference map, referred to as partial toponome map, of two muscle cells. The densely packed vesicles are easily recognized. Each vesicle contains a specific protein assembly directly designating different vesicle types. As shown in Figure 9.3, the two cells show a clearly different subcellular order of the different vesicle types. The latter signify different functional states of these cells: cell on the right (fusion competence), cell on the left (migration status).

9.5
Formalization of Network Motifs as Geometric Objects

A formal description of protein networks as unique architectures, or patterns in the cell, by mathematical methods will be essential for a quantitative analysis of the cellular proteome. Indeed a complete mathematization of the cell on the basis of MELK toponome data sets is an important next step towards the "grammar" of the complete network of proteins encompassing the cell's functionalities. We define a topological assembly of proteins exerting a given cell function as a protein-network-motif. This is illustrated in the scheme shown in Figure 9.4. A molecular network of the cell surface membrane of two cells is shown illustrating different proteins forming complexes. The single proteins are detected by MELK and imaged as photonic signals. Primarily, every protein signal is detected as a grey value (resulting from a given fluorescence intensity), which gives a relative measure for the concentration of each given protein. It has turned out to be efficient in routine procedures, when this primary data set is binarized, i.e., every protein is expressed as present or absent related to a certain grey value threshold level (present [1], absent [0] = 1 bit). Hence, the primary image information consisting in grey value distributions for every protein can be replaced by relatively simple geometric descriptions of combinatorial binary vectors in $x/y/z$ coordinates. As shown in Figure 9.4, one can readily recognize which combinatorial molecular pattern (CMP) is present or absent within the biological structure. In the present example it is shown that CMPs change along the cell surface membrane. Biologically, this difference can be classified as different supramolecular domains of the cell surface membrane. It is of particular functional interest to search for common features among these CMPs by comparative analyses. As illustrated in Figure 9.4 the protein 1 in cell 1 is the only protein, which is present in all domains (CMPs) of the same cell surface membrane. This protein coexists with other proteins in different combinations in the different domains (in Figure 9.4 such variable linkages are indicated by wildcards [*]). The protein 7, however, is inversely correlated with protein 1 (Figure 9.4 [0] = protein 7). By contrast, the same protein 7 in cell 2 is common to all domains. We could show by high-throughput MELK analysis that the detection of such common features and differences among CMPs – together termed toponome motifs (or protein network-motifs) – are highly characteristic

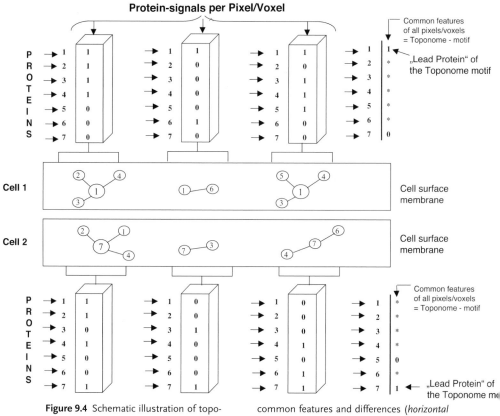

Figure 9.4 Schematic illustration of topo-nome mapping of molecular networks on the cell surface of two different cells. The proteins 1 to 7 are associated as complexes in different combinations. These patterns can be traced out by MELK robotic workstations as binary combinatorial codes [combinatorial molecular patterns (CMPs)]. The resulting CMPs show common features and differences (*horizontal comparison in the scheme*). All CMPs assemble to form a toponome motif (protein-network-motif) illustrated on the *right*. [1] Lead protein, [*] proteins that are variably associated with the lead protein; [0] proteins that are inversely correlated with the lead protein.

for specific cell functions or cell types (details will be published elsewhere). Proteins, which are present in all domains (Figure 9.4: protein 1 in cell 1, and protein 7 in cell 2) are of particular functional interest. We term such proteins lead proteins (Schubert, 2003). Selective inhibition of such proteins, but not the inhibition of the variably associated proteins (wildcards), leads to significant alterations of given cell functions as indicated by experiments involving appropriate biological model systems (e.g., cells *in vitro* or mouse models) (to be published). We can conclude from these experiments that lead proteins within a network of protein complexes play a dominating hierarchical role. Such experiments, based on geometric binary code data uncovering the lead proteins, are the basis for systematic analyses of protein networks within the cell. Figure 9.5 illustrates systematically the results of experiments performed with migratory cells: one

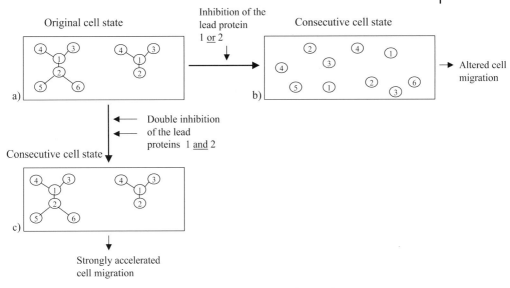

Figure 9.5 Schematic illustration of the experimental analysis of molecular networks of the cell surface membrane (for details see text).

migrating cell is characterized by two different lead proteins on the cell surface membrane (Figure 9.5a). The experimental inhibition of either one of these proteins (protein 1 *or* protein 2) leads to the same functional results: first, a destruction of the protein pattern occurs (Figure 9.5b: the proteins of the primary network still reside in the membrane, but are topologically disassembled); second, a severe alteration of cell migration results as a consequence of the topological destruction of the molecular networks of the cell surface membrane. Based on these results, one would expect that a functional inhibition of both lead proteins (Figure 9.5: double inhibition of proteins 1 and 2 simultaneously) may lead to even stronger effects: a more severe restriction of the ability to migrate. However, experimentally, the opposite was observed: the CMP pattern was largely preserved (Figure 9.5c), and the cell showed a multifold increase in the migratory speed. This permits the conclusion that it is not possible to predict subcellular functions of proteins on the basis of a simple molecular logic, because the architecture and internal topological relationships among the proteins place constraints on the cellular functions of each single protein of the network.

Overall, it appears to be essential to combine MELK toponome measurements with experiments allowing us to (1) detect protein networks as unique geometric patterns, and (2) decipher the precise (hidden) cellular function of each protein as part of the network architecture. Furthermore, lead proteins are obviously important, because they appear to function as central elements controlling the correct topology of the whole interrelated protein network. Finally, lead proteins found in disease-specific networks are likely to be first order target candidates for drug-lead-finding procedures.

9.6
Gain of Functional Information: Perspectives for Drug Development

We have established a system of toponome analysis. It consists of the following categories: measurement (MELK), filtering (comparison of cellular toponome patterns in multidimensional data spaces), predictions (of disease specific targets = lead proteins) and experimental examination (biological models, by which target candidates can be validated). With the aid of this deciphering apparatus it has been possible to identify targets and drug leads in neurological diseases and cancer (Schubert, 1999; Schubert, 2001). A critical target molecule in amyotrophic lateral sclerosis (ALS) found by MELK (Schubert, 1999) was recently confirmed by an independent knock out mouse model (Mohamed et al., 2002) providing further proof of principle of the predictive power of MELK. This suggests that MELK toponome analysis is an efficient way of finding proteins which are highly linked to the pathogenesis of diseases.

MELK can be combined in different ways with other proteome analyses and high-throughput procedures in the field of drug discovery. It may be particularly interesting to perform MELK analyses of drug candidates which have been found by other methods (e.g. proteomics profiling), in order to prioritize drug candidates on the basis of hierarchical toponome protein data. On the other hand MELK provides a platform for the topological validation of target candidates identified by classical genomics and proteomics approaches.

References

ALBERTS, B. (1998). The cell as a collection of protein machines: preparing the next generation of molecular biologists. *Cell* **92**, 291–294.
DE ANGELIS, L., BERGHELLA, L., COLETTA, M., LATTANZI, L., ZANCHI, M., CUSELLA-DE ANGELIS, M. G., PONZETTO, C., COSSU, G. (1999). Skeletal myogenic progenitors originating from embryonic dorsal aorta coexpress endothelial and myogenic markers and contribute to postnatal muscle growth and regeneration. *J. Cell Biol.* **147**, 869–878.
DANDEKAR, T., SNEL, B., HUYNEN, M., BORK, P. (1998). Conservation of gene order: a fingerprint of proteins that physically interact. *Trends Biochem. Sci.* **23**, 324–328.
ENRIGHT, A. J., ILIOPOULOS, I., KYRPIDES, N. C., OUZOUNIS, C. A. (1999). Protein interaction maps for complete genomes based on gene fusion events. *Nature* **402**, 86–90.

GAVIN, A. C., BOSCHE, M., KRAUSE, R., GRANDI, P., MARZIOCH, M., BAUER, A., SCHULTZ, J., RICK, J. M., MICHON, A. M., CRUCIAT, C. M., et al. (2002). Functional organization of the yeast proteome by systematic analysis of protein complexes. *Nature* **415**, 141–147.
GIOT, L., BADER, J. S., BROUWER, C., CHAUDHURI, A., KUANG, B., LI, Y., HAO, Y. L., OOI, C. E., GODWIN, B., VITOLS, E., et al. (2003). A protein interaction map of *Drosophila melanogaster*. *Science* **302**, 1727–1736.
HO, Y., GRUHLER, A., HEILBUT, A., BADER, G. D., MOORE, L., ADAMS, S. L., MILLAR, A., TAYLOR, P., BENNETT, K., BOUTILIER, K., et al. (2002). Systematic identification of protein complexes in *Saccharomyces cerevisiae* by mass spectrometry. *Nature* **415**, 180–183.
HUGHES, T. R., MARTON, M. J., JONES, A. R., ROBERTS, C. J., STOUGHTON, R., ARMOUR, C. D., BENNETT, H. A.,

COFFEY, E., DAI, H., HE, Y. D., et al. (**2000**). Functional discovery via a compendium of expression profiles. *Cell* **102**, 109–126.

HUYNEN, M. A., BORK, P. (**1998**). Measuring genome evolution. *Proc. Natl. Acad. Sci. USA* **95**, 5849–5856.

ITO, T., CHIBA, T., OZAWA, R., YOSHIDA, M., HATTORI, M., SAKAKI, Y. (**2001**). A comprehensive two-hybrid analysis to explore the yeast protein interactome. *Proc. Natl. Acad. Sci. USA* **98**, 4569–4574.

JESSEN, T. (**2005**). *Future drug discovery*. Business Briefing, p. 10.

LEHMAN BROTHERS, MC KINSEY & CO. (**2001**). *The Fruits of Genomics*.

MARCOTTE, E. M., PELLEGRINI, M., NG, H. L., RICE, D. W., YEATES, T. O., EISENBERG, D. (**1999**). Detecting protein function and protein–protein interactions from genome sequences. *Science* **285**, 751–753.

MOHAMED, H. A., MOSIER, D. R., ZOU, L. L., SIKLOS, L., ALEXIANU, M. E., ENGELHARDT, J. I., BEERS, D. R., LE, W. D., APPEL, S. H. (**2002**). Immunoglobulin Fc gamma receptor promotes immunoglobulin uptake, immunoglobulin-mediated calcium increase, and neurotransmitter release in motor neurons. *J. Neurosci. Res.* **69**, 110–116.

NATTKEMPER, T. W., RITTER, H. J., SCHUBERT, W. (**2001**). A neural classifyer enabling high-throughput topological analysis of lymphocytes in tissue sections. *IEEE Trans. Inf. Technol. Biomed.* **5**, 138–149.

OVERBEEK, R., FONSTEIN, M., D'SOUZA, M., PUSCH, G. D., MALTSEV, N. (**1999**). The use of gene clusters to infer functional coupling. *Proc. Natl. Acad. Sci. USA* **96**, 2896–2901.

PELLEGRINI, M., MARCOTTE, E. M., THOMPSON, M. J., EISENBERG, D., YEATES, T. O. (**1999**). Assigning protein functions by comparative genome analysis: protein phylogenetic profiles. *Proc. Natl. Acad. Sci. USA* **96**, 4285–4288.

ROBERTS, C. J., NELSON, B., MARTON, M. J., STOUGHTON, R., MEYER, M. R., BENNETT, H. A., HE, Y. D., DAI, H., WALKER, W. L., HUGHES, T. R., et al. (**2000**). Signaling and circuitry of multiple MAPK pathways revealed by a matrix of global gene expression profiles. *Science* **287**, 873–880.

SCHUBERT, W., ZIMMERMANN, K., CRAMER, M., STARZINSKI-POWITZ, A. (**1989**). Lymphocyte antigen Leu19 as a molecular marker of regeneration in human skeletal muscle. *Proc. Natl. Acad. Sci. USA* **86**, 307–311.

SCHUBERT, W. (**1990**). Multiple antigen – mapping microscopy of human tissue. *Adv. Analytical Cell Pathol.*, 97–98.

SCHUBERT, W. (**1992**). Antigenic determinants of T lymphocyte a/b receptor and other leukocyte surface proteins as differential markers of skeletal muscle regeneration: detection of spatially and timely restricted patterns by MAM microscopy. *Eur J. Cell Biol.* **58(2)**, 395–410.

SCHUBERT, W. (**1999**). US Patents 09/367,011, 09/802,305, 09/801,414.

SCHUBERT, W. (**2001**). US Patent 6,150,173 (2000), EP 1 069 431 (00 114 8543).

SCHUBERT, W. (**2003**). Topological proteomics, toponomics MELK technology. *Adv. Biochem. Eng. Biotechnol.* **83**, 189–211.

TUFTS (**2002**). *CSDD Impact Report*, Vol. 4, No. 5.

UETZ, P., GIOT, L., CAGNEY, G., MANSFIELD, T. A., JUDSON, R. S., KNIGHT, J. R., LOCKSHON, D., NARAYAN, V., SRINIVASAN, M., POCHART, P., et al. (**2000**). A comprehensive analysis of protein–protein interactions in *Saccharomyces cerevisiae*. *Nature* **403**, 623–627.

VON MERING, C., KRAUSE, R., SNEL, B., CORNELL, M., OLIVER, S. G., FIELDS, S., BORK, P. (**2002**). Comparative assessment of large-scale data sets of protein–protein interactions. *Nature* **417**, 399–403.

III
Applications

Proteomics in Drug Research
Edited by M. Hamacher, K. Marcus, K. Stühler, A. van Hall, B. Warscheid, H. E. Meyer
Copyright © 2006 Wiley-VCH Verlag GmbH & Co. KGaA, Weinheim
ISBN: 3-527-31226-9

10
From Target to Lead Synthesis

Stefan Müllner, Holger Stark, Päivi Niskanen, Erich Eigenbrodt, Sybille Mazurek and Hugo Fasold

Abstract

The search for biomarkers and drugs is extremely time-consuming and cost-intensive. Thus, new and reliable strategies for the fast identification of drug targets as well as lead compounds are required. The small molecule adaptation on research targets (SMART) approach described in this chapter represents such a concept. We have focused on rheumatoid arthritis (RA), a human autoimmune disease, as one of the most challenging problems in of modern medicine, and Leflunomide (Arava), an innovative drug with complex mode-of-action features. Application of the SMART approach not only led to the identification and validation of new targets for that drug, but also to the design of a potential lead structure.

10.1
Introduction

Global pharmaceutical industries today spend billions of dollars per year on the development of new drugs and therapies for the major indication areas, e.g., cardiovascular, central nervous system, metabolism, inflammation/immunology and oncology. All companies are under increasing pressure to reduce development costs and at the same time to come up with more specific drugs with fewer side effects based on patentable new structures, yet there remains a very high risk that such a new drug would fail even after approval. Over the last 10 years all pharma firms have invested heavily in new technologies, e.g., genomics, high-throughput screening, robotics, protein crystallography, etc., with a now visible effect on overall development risks and costs.

As Protagen AG is a specialist in applied protein science, we have designed a concept for target discovery and lead identification, the small molecule adaptation on research targets (SMART) approach (Figure 10.1). The concept is based on common knowledge and on the hypothesis that every drug sold in the market

Proteomics in Drug Research
Edited by M. Hamacher, K. Marcus, K. Stühler, A. van Hall, B. Warscheid, H. E. Meyer
Copyright © 2006 Wiley-VCH Verlag GmbH & Co. KGaA, Weinheim
ISBN: 3-527-31226-9

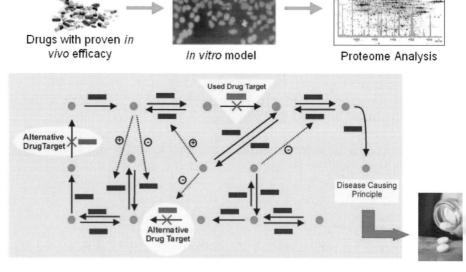

Figure 10.1 The SMART approach is built on drugs with proven *in vitro* efficacy, common pharmacological *in vitro* models, and high resolution proteomics technologies. Alternative drug targets are subsequently identified and evaluated for their potential in drug screening programs.

interacts with more than one single protein (drug target) and that the synergistic combination of high-end technologies in protein sciences, e.g., matrix-assisted laser desorption ionization (MALDI) and electrospray ionization (ESI) mass spectrometry (MS), plasmon resonance, high-resolution 2-dimensional gel electrophoresis, affinity chromatography, etc., allows the in depth elucidation of the mode of action of any of these therapeutic compounds and leads to the identification of a whole set of new "druggable" targets which then serve as a rational basis for the design of new pharma actives.

A prerequisite for successful application of the SMART approach is the careful selection of a drug with an interesting pharmacological profile, proven *in vivo* efficacy in humans and good target indication. Furthermore, all pharmacological models subjected to proteome analysis should be widely known and scientifically accepted. Ideally, the same pharmacological models which were published during development of the drug are employed.

Apart from the effective treatment of all the different types of cancer, the therapy for rheumatoid arthritis (RA), a human autoimmune disease, is one of the most challenging problems in modern medicine. Both the etiology and the mechanism of the disease process (Feldmann et al., 1996; Vyse and Todd, 1996), as well as the mode of action of immunosuppressive and anti-inflammatory drugs, currently under clinical investigation or proven therapeutics, are not completely understood.

Chronic inflammation of the disease-affected joints is one of the main characteristics and anti-inflammatory drugs are therefore commonly used for treatment.

Leflunomide (Arava) is an oral immunomodulatory agent, which is considered effective for the treatment of RA. Leflunomide is a disease-modifying antirheumatic drug that is approved for treatment of RA (Bartlett et al., 1991; Strand et al., 1999; Smolen et al. 1999; Emery et al. 2000; Breedveld and Dayer, 2000). *In vivo*, Leflunomide is rapidly converted into its pharmacologically active metabolite A77 1726 (Herrmann et al., 2000). Although the precise mode of action of Leflunomide *in vivo* remains elusive, A77 1726 has been shown *in vitro* to inhibit reversibly dihydro-orotate dehydrogenase (DHODH), which catalyzes a rate-limiting step in the *de novo* synthesis of pyrimidines (Cherwinski et al., 1995; Williamson et al., 1996). The inhibition of DHODH activity by A77 1726 might explain part of its mechanism of action in suppressing inflammation.

Leflunomide is a potent noncytotoxic inhibitor of the proliferation of stimulated B and T-lymphocytes, which depend on *de novo* pyrimidine synthesis to fulfill their metabolic needs (Breedveld and Dayer, 2000; Herrmann et al., 2000). This antiproliferative effect can be antagonized *in vitro* by the addition of uridine or cytidine to the cell culture medium, underscoring the significance of this mechanism of action (Cao et al., 1995; Williamson et al., 1995; Zielinski et al., 1995; Silva et al., 1996). Furthermore, Leflunomide blocks tumor necrosis factor (TNF)-α-mediated cellular responses in T-cells by inhibiting nuclear factor κB, a mechanism that also depends on pyrimidine biosynthesis (Manna and Aggarwal, 1999; Manna et al., 2000). In addition, A77 1726 exerts a direct inhibitory effect on cyclo-oxygenase (COX)-2 activity, both *in vitro* and *in vivo* (Hamilton et al., 1999; Burger et al., 2003).

Finally, it has been reported that, at higher concentrations, A77 1726 inhibits different types of receptor and nonreceptor tyrosine kinases that are involved in cytokine and growth factor signaling (Mattar et al., 1993; Xu et al., 1996; Elder et al.,1997). However, several effects, such as the inhibition of the switch in immunoglobulin from IgM to IgG (Siemasko et al., 1996) or inhibition of various tyrosine kinases cannot be completely reversed by the addition of uridine (Silva et al., 1997), indicating that Leflunomide may possess additional molecular mechanisms of action.

The diverse and different pharmacological effects on several central biosynthetic pathways and signaling cascades that are reported for Leflunomide recommended this small molecule drug for employment of the SMART approach for the identification of alternative targets and finally, lead structure identification.

10.2
Materials and Methods

10.2.1
Cells and Culture Conditions

The preparation of cellular protein extracts was carried out either at 4 °C or on ice.

RAW 264.7 mouse monocyte/macrophage cells were cultured in Dulbecco's modified Eagle's medium (Sigma) at 37 °C and 10% CO_2. Fetal calf serum (10%) and penicillin/streptomycin (50 U mL^{-1} and 50 µg mL^{-1}, respectively) were added. About 2×10^8 confluently grown cells were harvested and washed with cold phosphate-buffered saline (PBS). Cells were pooled and centrifuged at $200 \times g$ for 10 min at 4 °C. The pellet was homogenized in 10 mL lysis buffer using a glass homogenizer. Lysis buffer was 20 mM Tris, 2 mM $MgCl_2 \times 6\ H_2O$, 1 mM DTT (pH 7.0) containing protease inhibitors (1 µg mL^{-1} Leupeptin, 2 mM benzamidine/HCl, 0.05 mM Pefabloc SC). The cytosolic fraction was obtained by centrifugation at $22\,000 \times g$ for 35 min at 4 °C. The supernatant was diluted with lysis buffer to a volume of 30 mL and directly applied to the affinity column. An aliquot of 2 mL was retained and used as reference.

10.2.2
In Vitro Activity Testing

Novikoff cells were seeded in a plastic flask with a cell count of 0.1×10^6 cells. The compounds were dissolved in DMSO, diluted with the same volume of Hanks buffer and added to the culture medium. Assigned concentration is the end concentration of the compound in the liquid culture medium. Cell cultures were incubated for 4 days, and the vital cells were counted under a light microscope. The IC_{50} value represents the average of $N = 6$. Untreated cell culture served as control.

10.2.3
Affinity Chromatography

For the identification of cytosolic proteins with binding affinity to Leflunomide extracts of RAW 264.7 mouse monocyte/macrophage cells were applied at a flow rate of 1 mL min^{-1} to an A 95 0277-derivatized (Figure 10.2) Fractogel column (1×5 cm, internal diameter×length) equilibrated with water. The flow-through was collected and the program was started with an elution rate of 3 mL min^{-1}. The column was washed with deionized water and with PBS, followed by a high salt elution step using a gradient from 0 to 0.5 M NaCl, and a second washing step with PBS. The specific elution was carried out with an A 77 1726 gradient. Remaining proteins were removed by successive washing steps with water, 6 M urea and with 60% acetonitrile/0.1% TFA. Each fraction except the acetonitrile

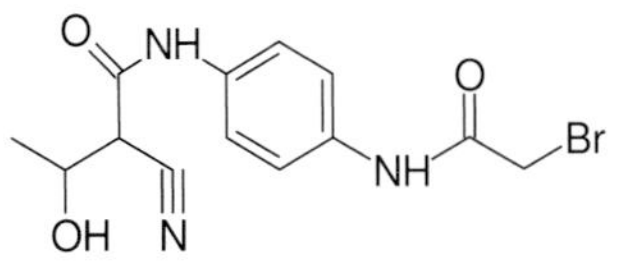

Structure of A 95 0277

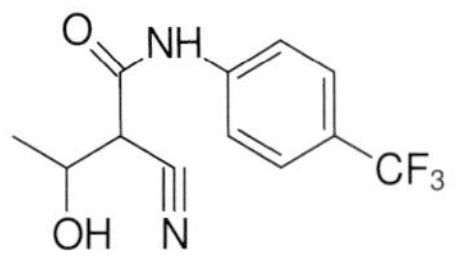

Structure of A 77 1726

Figure 10.2 Molecular structure of the Leflunomide metabolite A 77 1726 and the derivative A95 0277 which was successfully coupled to the activated Fractogel matrix used for affinity chromatography as well as to the BIAcore chip used for binding studies.

fraction and the flow through was dialyzed against water. After dialysis, the fractions were frozen and lyophilized. Lyophilized protein fractions were dissolved in 1 mL water respectively and protein concentration was determined using the bicinchoninic acid method (Pierce). Fractions were lyophilized a second time and protein concentration was adjusted to 1 mg mL^{-1} with the corresponding volume of SDS-PAGE sample buffer.

10.2.4
Electrophoresis and Protein Identification

SDS-PAGE was performed as previously described (Laemmli, 1970). Overnight, 10–17% gradient gels were run with 14 A h^{-1}. Analytical PAA gels were silver-stained according to the method of Heukeshoven and Dernick (1985), and preparative gels were stained with Coomassie brilliant blue.

For the identification of proteins, preparative gels with a total protein load of 1 mg per gel were run and proteins were visualized by Coomassie Blue R staining. Leflunomide binding proteins were analyzed using preparative SDS-PAGE. Coomassie-stained protein bands were identified by MS.

10.2.5
BIAcore Analysis

The surface matrix of a carboxymethylated sensor chip CM5 (Pharmacia Biosensor, Uppsala, Sweden) was activated by injection of 35 µL of 0.05 M N-hydroxy-succinimide (NHS)/0.2 M N-ethyl-N'(dimethylaminopropyl)carbodiimide (EDC). The BIAcore IFC forms four parallel flow cells on the sensor chip. Three of them were derivatized with 40 mM cystamin-dihydrochloride/ethanolamine in the ratios of 4 : 1, 2 : 1 and 1 : 1. The fourth flow cell remained underivatized and was used as a negative control. Excess reactive structures on the surface matrix were saturated with 1 M ethanolamine hydrochloride (pH 8.5). Reactive groups were activated with 0.1 M DTT (dissolved in sodium borate buffer, pH 8,5). Coupling of the Leflunomide derivative A 95 0277 was performed at a flow rate of 5 µL min^{-1}. The substrate concentration was set at 10 mg mL^{-1} in sodium borate buffer (pH 7.5). Potential Leflunomide-binding proteins were tested for their binding specificity using BIAcore analysis. Commercially available enzymes [malic dehydrogenase (MDH), lactic dehydrogenase (LDH), glyceraldehyde 3-P dehydrogenase (GAPDH), pyruvate kinase, DHODH and actin] were diluted in HBS buffer to get concentrations of 5 µM, 1 µM, 0.5 µM, 0.1 µM, 50 nM and 10 nM. Binding data were analyzed using the BIAcore 2000 control software and the BIAcore evaluation software.

10.2.6
Synthesis of Acyl Cyanides

Some of the starting compounds were commercially available only as acids or anhydride. The respective acyl chlorides were obtained according to the thionyl chloride method of Smith et al. (1991). Reactions with cyanotrimethylsilane were completed under water-free conditions and in an argon atmosphere (Hünig and Schaller, 1982). Glassware was dried in an oven directly before use.

To 1.0 equivalent of acyl chloride, 1.0–1.4 equivalents of cyanotrimethylsilane were added by syringe under stirring in an ice bath. After 15 min the ice bath was removed and the stirring was continued at room temperature for 2–4 h.

Vacuum distillation with the microdistillation equipment was used to purify the acyl cyanides.

^{1}H and ^{13}C spectra were recorded on AM 250 and Avance 400; tetramethylsilane (TMS) was used as a reference. Frequencies of proton measurements were 250 MHz, 300 MHz or 400 MHz and frequencies of carbon measurements were 50.3 MHz, 62.9 MHz, 75.5 MHz or 100.6 MHz. MS spectra were measured on a Fisons MALDI–time-of-flight (TOF) spectrometer or a Fisons Platform II ESI spectrometer. CNH elemental analyses were done on a Foss Heraeus CHN-O-Rapid device.

10.2.6.1 Methyl 5-cyano-5-oxopentanoate
Yield: 0.58 g (3.74 mmol, 12%).
^{1}H NMR (CDCl$_3$) δ 1.97 (*m*, 2H, J = 7.0 Hz), 2.34 (*t*, 2H, *J* = 7.0 Hz), 2.81 (*t*, 2H, *J* = 7.1 Hz), 3.63(s, 3H)
^{13}C NMR(CDCl$_3$) δ 17.60 31.84 43.96 51.67 113.02 (CN) 172.52 176.03
IR: cyanide (CN) peak (–2230 cm^{-1}) and carbonyl (C=O) peak (–1725 cm^{-1}) were found from the IR spectrum.
Elemental analysis:
Measured: C, 53.94%; H, 5.87%; N, 8.66%.
Calculated: C, 54.19%; H, 5.85%; N, 9.03%.

10.2.6.2 **Methyl 6-cyano-6-oxohexanoate**
Yield: 2.06 g (12.18 mmol, 43%).
^{1}H NMR (CDCI3) 5, 1.65–1.80 (*m*, 4H), 2.36 (*t*, 2H, *J* = 7.0 Hz), 2.78 (*t*, 2H, *J* = 7.0 Hz),
3.68 (s, 3H)
^{13}C NMR (CDCI3) 5, 21.92, 23.53, 33.18, 44.58, 51.52, 113.02 (CN) 173.08, 176.30
IR: CN peak (–2150 cm^{-1}) and carbonyl (C=O) peak (–1685 cm^{-1}) were found from
the IR spectrum.
Elemental analysis:
Measured: C, 56.88%; H, 6.69%; N, 7.85%.
Calculated: C, 56.80%; H, 6.55%; N, 8.28%.

10.2.6.3 **Methyl-5-cyano-3-methyl-5-oxopentanoate**
Yield: 5.65 g (33.40 mmol, 86% overall yield).
^{1}H NMR (CDCl$_3$) 5, 1.06 (*d*, 3H, *J* = 6.6 Hz), 2.34 (*dd*, 2H, *J* = 6.6 Hz, 1.6 Hz),
2.71 (*d*, 2H, *J* = 7.3Hz), 2.87–2.93 (*m*, 1H), 3.68 (*s*, 3H)
MS: (TofSpec Linear LDI+) M[H$^+$] = 170 peak was found.
Elemental analysis:
Measured: C, 56.66%; H, 6.57%; N, 8.07%.
Calculated: C, 56.80%; H, 6.55%; N, 8.28%.

10.3
Results

Cells involved in the immune response and inflammatory processes include
PMNs, platelets, lymphocytes, plasma cells, and macrophages. In particular,
macrophages play a key role in this network. They are responsible for phagocytosis,
major histocompatibility complex II (MHC II)-mediated antigen presentation and
they are the most active secreting cells of the organism. Through the release of
cytokines, complement factors, proteases and acute-phase proteins, macrophages
are involved in initiation, regulation and suppression of the immune response
and are responsible for effective antigen presentation to T-lymphocytes. Furthermore, they also express disease-relevant receptors for various cytokines (Kirchner
M., 1996) and the release of pro-inflammatory cytokines can be triggered *in vitro*
by stimulation with bacterial lipopolysaccharide (Gessani et al., 1993; Wright et
al., 1990).

In the first differential proteomics approach for the identification and characterization of proteins involved in both the inflammatory cascade and Leflunomide
mode of action (Dax et al., 1998), the murine monocyte/macrophage cell line
RAW 264.7 was employed and the RAW 264.7 proteomes of normal, unstimulated
RAW 264.7 macrophages with the respective LPS-stimulated and/or drug-treated
were compared. This cell line was chosen since RAW 264.7 cells have been used
as a model system to study immune responses (Bahl et al., 1994) successfully
before.

In this study, cofilin and keratinocyte lipid-binding proteins were identified as proteins linked directly to the LPS-stimulatory effect and Leflunomide action. Cofilin belongs to the group of small actin-binding proteins and is expressed in all cell types. It forms complexes with actin monomers and actin filaments, and the reversible polymerization/depolymerization process of actin is in part regulated by the interaction with the phosphorylated/unphosphorylated form of cofilin (Abe et al., 1996; Gunsalus et al., 1995; Edwards et al., 1994; Suzuki et al., 1995; Lappalainen and Drubin, 1997). It has been shown that cofilin is an important component in T-cell activation (Samstag et al., 1994, 1996).

The functional role of keratinocyte lipid-binding protein is still not clear. It belongs to the family of intracellular lipid-binding proteins which are known to be overexpressed in skin tumors (Krieg et al., 1993). The recombinant protein shows a high affinity to long-chain fatty acids, such as oleic acid, arachidonic acid and eicosanoids (Kane et al., 1996), and might therefore regulate the inflammatory cascade by that means.

A different approach to elucidate the Leflunomide mode of action was chosen by Ortlepp et al. (1998). A mouse B cell line, A 20N, was exposed to increasing concentrations of this drug over an extended period of time. This approach was based on the rationale that the cell will respond to the drug treatment through overexpression of major target proteins. The new cell line generated, A 20T, exhibited a 40-fold higher tolerance to Leflunomide than its parental cell line (Kirschbaum et al., 1999). The protein patterns of both cell lines were compared using both 1-dimensional SDS gradient and 2-dimensional gel electrophoresis. Indeed, at least one protein with an apparent molecular weight of 135 kDa, designated p135, appeared with stronger intensity in A 20T cells. It was demonstrated that p135 belongs to the family of DEAH-box RNA helicases. A database search with the murine cDNA sequence was used to identify the human homologue and by using comparative cDNA analysis, the human cDNA of p135 was cloned. The study revealed that p135 shares 41% sequence homology with the splicing factor Prp16p of yeast.

Owing to the fact that screening departments in preclinical drug development have to serve as the bridge between medicinal chemical synthesis and *in vitro* pharmacology, the target-driven drug discovery programs as well as the proteins selected as bases for the design of (high-throughput) screening assays have to meet several prerequisites.

Receptors or enzymes are the preferred drug targets; since binding and enzyme inhibitor assays are standard in a research laboratory, they can be established easily and offer the possibility of automation. Transport proteins, ion channels and transcription factors are less attractive targets for drug screening programs, because automation is costly and not always possible. In addition, the requirement for highly sophisticated equipment and expertise is decisive.

Considering all the needs of the synthetic chemist and assay developer, the ideal target protein for screening should be an enzyme which can be expressed in high yield by recombinant technologies, can be purified in functional form, does not require cofactors, is stable on plastic and glass surfaces and tolerates a broad range of organic solvents. Despite the scientific relevance and attractiveness of the proteins described above and in view of the features necessary, none of these new targets were suitable for screening programs.

In order to find more appropriate target proteins for high-throughput screening, affinity chromatography was chosen to look for Leflunomide-binding proteins.

Membrane fractions of several cell lines derived from immunocompetent cells were previously analyzed by photoaffinity labeling (Williamson et al., 1995) and the key enzyme in pyrimidine biosynthesis, DHODH, was identified as one of the major targets of Leflunomide. For further elucidation of its mode of action, the cytosolic fraction of the monocyte/macrophage cell line RAW 264.7 was subjected to affinity chromatography to identify all cytosolic binding partners of Leflunomide. The cytosolic preparation was applied to an affinity column where the Leflunomide derivative A 95 0277 was coupled as a ligand (Figure 10.2).

To minimize nonspecific protein–drug interactions, several washing steps including a high salt elution step with NaCl were performed before the specific elution with A 77 1726 was carried out. Successive washing steps with water followed by urea and acetonitrile removed tightly bound proteins and lipids (Mangold et al., 1999). SDS-PAGE was used to analyze the protein fractions obtained by affinity chromatography. Figure 10.3 gives a detailed view of the fraction specifically eluted by the drug.

A 77 1726 fraction

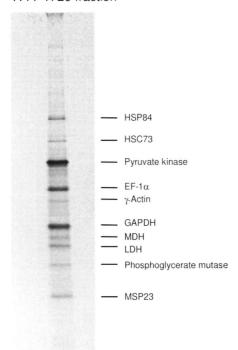

— HSP84

— HSC73

— Pyruvate kinase

— EF-1α
— γ-Actin

— GAPDH
— MDH
— LDH
— Phosphoglycerate mutase

— MSP23

Figure 10.3 Detailed view of the protein fraction specifically eluted by A 77 1726 from the affinity column. Analytical 10–17% gradient SDS-PAGE, silver-stained.

All of the ten marked cytosolic potential binding partners of A 77 1726 were identified by subsequent protein sequence analysis of internal peptide fragments. Four of them, GAPDH, phosphoglycerate mutase, LDH and pyruvate kinase belong to the second part of the glycolytic enzyme complex. Two heat shock proteins (HSPs) HSP84 and HSC73 and MDH, were also specifically eluted with A 77 1726. Furthermore, actin, elongation factor 1α (EF-1α) and the macrophage 23 kDa stress protein were identified as well.

The results obtained by affinity chromatography gave good evidence but are not proof of a direct involvement of these proteins in the Leflunomide mode of action. However, it was very interesting that four enzymes of the glycolytic pathway as well as MDH could be eluted specifically.

Besides the fact that enzymes in general are preferred targets for drug screening programs, it is important that these proteins are members of the glycolytic enzyme complex. The total complex contains the enzymes hexokinase, GAPDH, phosphoglycerate kinase, phospoglyceromutase, enolase, pyruvate kinase, LDH, nucleotide diphosphate kinase, adenylate kinase, and glucose 6-phosphate dehydrogenase. It also contains an AU-rich mRNA molecule for stabilization of the whole complex, which is bound to the cytoskeleton (Eigenbrodt et al., 1994). In addition, affinity interaction between A 77 1726 and particular proteins may result in the coelution of proteins which are not involved in but interacting with this complex.

BIAcore analysis was performed to countervalidate identified A 77 1726 enzymic binding partners and to narrow down the number of possible target proteins. The Leflunomide derivative A 95 0277 was coupled covalently to the carboxy-methylated dextran matrix of a commercially available CM5-chip using the surface thiol method. BIAcore analysis of commercially available proteins, GAPDH, pyruvate kinase, LDH, phosphoglycerate mutase, MDH and actin was used to further investigate binding specificity and protein drug affinity. Binding of these proteins was tested in relation to their protein concentration and compared with one already known and well characterized Leflunomide binding partner, DHODH. All five enzymes were diluted in HBS buffer to yield final concentrations of 10 nM, 50 nM, 0.1 μM, 0.5 μM, 1 μM, and 5 μM respectively. DHODH was kindly provided by Prof. Dr. Loeffler (University of Marburg). Owing to the low protein amount available, the enzyme could not be tested at a concentration of 5 μM. Figure 10.4 shows the binding capacity according to the protein concentration. Interestingly, binding of GAPDH exhibited an exponential time course, whereas LDH showed a saturated binding curve.

In addition to these four enzymes, MDH and actin had also been identified as potential Leflunomide binding partners. Their binding affinity to the derivatized BIAcore chip was also compared with LDH. MDH showed a saturated binding curve similar to LDH (Figure 10.5). Actin did not interact with the BIAcore chip at all.

The results obtained so far suggested that Leflunomide obviously interferes with more than one intracellular target protein and that key enzymes in energy production might be directly or indirectly inhibited.

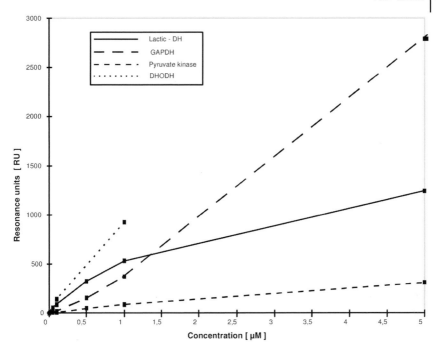

Figure 10.4 Binding diagram of LDH, GAPDH, pyruvate kinase and DHODH to the derivatized BIAcore chip at different protein concentrations (x-axis: protein concentration, y-axis: resonance units).

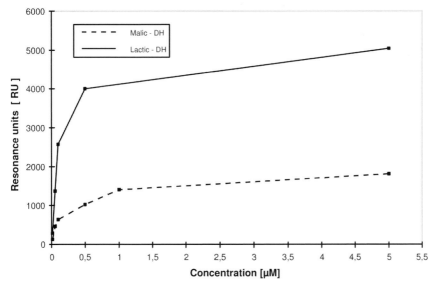

Figure 10.5 Binding diagram of MDH and LDH to the derivatized BIAcore chip at different protein concentrations (x-axis: protein concentration, y-axis: resonance units).

All proliferating cells, i.e., tumor cells as well as cells under normal cell growth conditions have to adapt their energy production by coordination of the different metabolic pathways. Tumor cells however, are also capable of continuous growth under unfavorable conditions such as poor supply of oxygen and/or glucose. Since they still are able to supply the daughter cells with building blocks for the DNA and structural proteins, alternative routes for energy regeneration are employed. Tumor cells are able to regenerate energy by glycolysis under hypoxic conditions or by the catabolic conversion of glutamine to lactate in the presence of oxygen. This process is called glutaminolysis. For those means, the composition of proteins forming the glycolytic enzyme complex can be changed. Concentration of pyruvate kinase, as the last regulatory enzyme in glycolysis, increases with tumor weight (Eigenbrodt et al., 1998). Four pyruvate kinase isoenzymes are found in mammals. These isoenzymes are named for their present tissues. L-type isoenzyme is found mostly from liver, R-type from erythrocytes and M1-type from muscles. M2-type isoenzyme of pyruvate kinase is present in many tissues such as kidney, intestine, lung, and stomach. All of these isoenzymes express their own unique electrophoretic, kinetic, and immunological properties (Ashizawa et al., 1991).

To coordinate the different metabolic pathways with oxygen and glucose supply, proliferating tumor cells express pyruvate kinase type M2 (M2-PK) (EC 2.7.1.40). The M2-PK itself can be transformed from a tetrameric into a dimeric form. In normal cells the tetrameric form is predominantly present; it exhibits high affinity for pyruvate whereas the dimeric from has low affinity and is inactive under physiological conditions. Therefore, in normal cells only a limited amount of the C atoms of glucose are fed into the generation of, e.g., nucleic acids. This also implies that the glucose is metabolized in normal cells to lactate to generate energy equivalents. In cancer cells the dimeric form (dim-M2-PK or tumor M2-PK) is the dominating isoform. This inactive form of the M2-PK in the cancer and all highly proliferating cells causes a blockade of the gycolysis above the PK and the usage of all glycolytic phospho-metabolites into the synthesis of DNA, phospholipids, and amino acids, building blocks needed in order to build new daughter cells after cellular fission.

Zwerschke et al. (1999) reported that tumor formation is linked to a constant dimeric–tetrameric interconversion of the M2-PK and Hacker et al. (1998) detected an isoenzyme shift from pyruvate kinase L-type to M2-type in late stage hepatoma carcinogenesis in rats. The isoenzyme shift seems to be characteristic for all cells that adapt their metabolism from normal to proliferation and can be detected particularly in tumor cells (Yilmaz et al., 2003). Typical cells containing high amounts of M2-PK are cancer cells, dividing cells in general, activated cells of the immune system, or stem cells. Therefore, the determination of pyruvate kinase isoenzyme type M2 was proposed as a marker for early onset in tumorgenenis (Eigenbrodt et al., 1998).

Besides pyruvate kinase, phosphoglyceromutase was already identified as a key regulatory element in the metabolic switch (Mazurek and Eigenbrodt, 2003). Therefore, it seemed to be worthwhile to evaluate the glycolytic complex and its susceptibility to dynamic change in general, and the process of glutaminolysis in

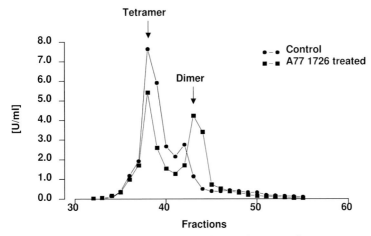

Figure 10.6 Generation of the M2-PK dimeric form by the action of 60 µm A77 1726.

particular as target for drug interaction. The presented affinity chromatography and BIAcore experiments strongly suggest that the active metabolite of Leflunomide interacts with high affinity to proteins of the glycolytic enzyme complex. The next step, then, was to show in cell assays that Leflunomide has a direct effect on glucose and glutamine metabolism.

In order to elucidate the influence of Leflunomide A77 1726 on the generation or stabilization of the different isoforms of the M2-PK (dim- or tetra-M2-PK) 60 µM of A77 1726 was incubated with cell extracts of Novikoff cells for 1 h on ice. The cell extracts were separated via isoelectric focusing. The experiment showed the dimerization of the M2-PK is favored under the influence of A77 1726 (Figure 10.6).

In a second experiment, the influence of A77 1726 on the flux of the major metabolites in the macrophage cell line RAW264.7 has been performed. The flux of glucose, glutamine, and lactate in the culture medium was measured in order to obtain data for the consumption or the generation of these metabolites, respectively. Figure 10.7 illustrates that A77 1726 inhibits strongly the generation of lactate from glucose. This indicates that glutaminolysis is preferred under such conditions by inhibition of the glycolysis.

Direct comparison of the chemical structures of glutamine and A77 1726 as well as other substrates in the glycolytic pathway with structural features of some of the common RA drugs finally lead to the synthesis of methyl 4-cyano-4-oxobuturate or carbomethoxypropionylcyanide (CMPC), a structural lead for a new class of low molecular weight compounds (Müllner and Eigenbrodt, 2001) with interesting *in vitro* activity (Figure 10.8). CMPC was synthesized (Tang and Sen, 1998) and the activity was compared with A77 1726 in Novikoff hepatoma cells. Novicoff is a cancer cell line from rat liver and it has been reported to have the highest pyruvate kinase activity compared to other cell lines (Mazurek et al., 1998), and therefore would make an ideal *in vitro* screening model for compounds interfering with pyruvate kinase or other members of the glycolytic complex.

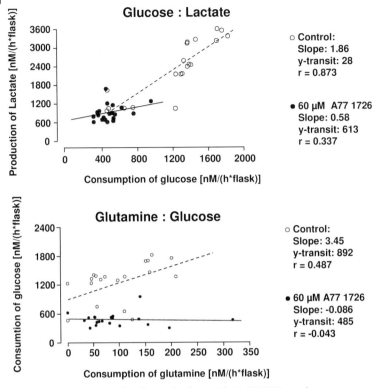

Figure 10.7 Direct metabolic effects of Leflunomide (A 77 1726) on glucose consumption and glycolysis. The *upper chart* shows that the drug suppresses both lactate production and glucose consumption. The *lower chart* shows that glutaminolysis is induced.

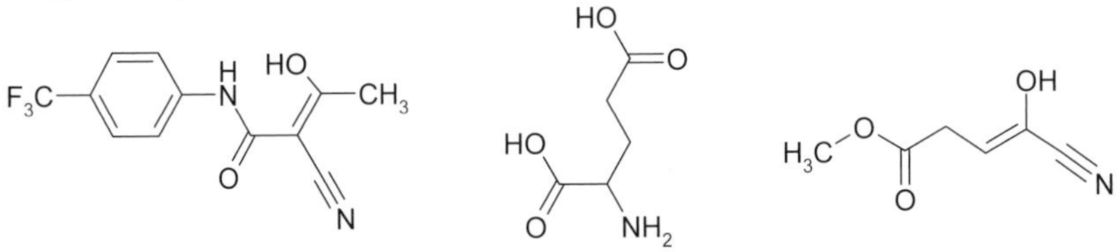

Figure 10.8 Structural comparison of **A** Leflunomide (A 77 1726), **B** glutamic acid, and **C** carbo-methoxy-proprionyl-cyanide.

Consequently, this cellular model was employed to evaluate the structure–activity relationship of some new acyl cyanides deduced from CMPC (Table 10.1). The derivatives with the highest biological activity compared to Leflunomide (A 77 1726) are shown with their respective IC_{50} values. Despite the fact that the activity increases slightly with chain length (CMPC < methyl-5-cyano-3-methyl-5-oxopentanoate ≤ methyl-6-cyano-6-oxohexanoate), branching of the C-chain reduces the activity of methyl 5-cyano-3-methyl-5-oxopentanoate to less than 10% of CMPC.

Table 10.1 Structure–activity relationship.

Compound	Structure	IC$_{50}$ Novikoff (mM)	Synonyms
A 77 1726		0.2–0.3	2-Cyano-3-oxo-*N*-(4-trifluoromethylphenyl)-butyramide
Methyl-4-cyano-4-oxobuturate		0.2–0.3	CMPC Carbo-methoxy-proprionyl-cyanide
Methyl-5-cyano-3-methyl-5-oxopentanoate		< 0.3	
Methyl-6-cyano-6-oxohexanoate		< 0.3	
Methyl-5-cyano-3-methyl-5-oxopentanoate		5–10	

10.4
Discussion

The results described in recent studies with differential proteomics approaches (Dax 1996; Dax et al., 1998; Ortlepp et al., 1998; Kirschbaum et al., 1999; Mangold et al., 1999) finally lead to additional insight into the mode of action of Leflunomide (Arava), a new antiproliferative and immonoregulatory drug for the treatment of RA. In the present paper we show that affinity chromatography led to the identification of ten intracellular potential binding partners of the active Leflunomide metabolite A 77 1726.

A subsequent BIAcore analysis revealed that binding of pyruvate kinase, GAPDH, MDH and LDH to A 77 1726 is due to a direct and strong interaction. Furthermore, the binding constants of GAPDH, LDH and pyruvate kinase are in a similar molar range as the binding constant for DHODH. It is known that enzymes of the glycolytic pathway form a protein complex which can adapt to different metabolic cellular states (Eigenbrodt et al., 1994). Phosphoglycerate mutase is part of this complex and might coelute through protein–protein interaction with a direct drug-binding protein. Actin, a ubiquitous protein in eukaryotes and major constituent of thin filaments is also known to be associated with the glycolytic enzyme GAPDH (Mejean et al., 1989). This interaction leads to a connection between the cytoskeleton and the glycolytic pathway and allows the organized transport of metabolites and ATP. Also, HSP90 is known to be a potential binding partner of actin (Koyasu et al., 1989). Binding of actin to the affinity column can therefore be explained by an interaction with GAPDH or through binding to a member of the heat shock family.

LDH and MDH are interesting targets in cancer therapy. It is a general principal that cell death occurs when ATP and NAD concentrations falls below a critical threshold. Because consumption of NAD for ADP ribosylation during DNA repair contributes significantly to energy depletion, anticancer drugs that have DNA as their primary target ultimately exert their cytotoxic effects by lowering ATP and NAD concentrations. Fast-proliferating cells either use glycolysis or glutaminolysis for energy production, so every enzyme of both pathways is a potential target for cancer therapy (Eigenbrodt et al., 1994) and inhibitors of LDH, for instance gossypol, have been proposed as therapeutic agents in cancer therapy already (Wu et al., 1989). Because inhibition of cell proliferation is a mechanism to treat autoimmune diseases, both enzymes could be interesting therapeutic targets. Additionally, in rheumatoid tissues, glucose consumption and lactic production is elevated and the enzymatic activity of LDH and its isoenzymes is used as an inflammatory marker of disease progression (Pejovic et al., 1992).

Two members of the HSP family HSP84 (HSP90) and HSC73 (HSP70) were also identified as potential binding partners of A 77 1726. In general, the expression of HSPs is increased under environmental stress or exposure to high temperature (Lindquist, 1986). Increased HSP synthesis has also been found following T-lymphocyte activation or during cellular differentiation (Craig, 1993). In general, infection represents stress for both the microorganism and the host, and increased synthesis of an extremely similar set of autologous and foreign stress proteins occurs at a time of active immune response. It is therefore not surprising that members of the HSP family are involved in several autoimmune diseases, like systemic lupus erythematosus (SLE) and RA (Winfield and Jarjour, 1991). Whether elution of HSP84 and HSC73 from the affinity column by A 77 1726 is due to a specific or unspecific interaction remains to be verified. The murine macrophage 23 kD stress protein (MSP23) was originally characterized from peritoneal macrophages (Ishii et al., 1993). It is a member of a novel family of highly conserved proteins with antioxidant activity implicated in the cellular response to oxidative stress and in control of cell proliferation and differentiation. MSP23 is closely related to the human PAG protein, to the murine MER5 family (Kawai et al., 1994) and to the natural killer-cell-enhancing factor A, NKEF-A (Shau et al., 1994) which is known to increase the cytotoxity of natural killer cells.

As discussed above, not all of the proteins identified by mode-of-action studies employing the SMART approach are suitable drug targets. However, the identification of enzymes of the glycolytic complex by affinity chromatography and their validation by BIAcore analysis finally led to the working hypothesis that Leflunomide (A77 1726) has a strong influence on energy production by interacting with enzymes responsible for glutaminolysis. One key element in glutaminolysis is the pyruvate kinase isoform M2 and the intracellular ratio between their tetrameric and dimeric form. It could be shown that Leflunomide directly affects this ratio. Depending on their metabolic state, normal nonneoplastic cells can also switch from glycolysis to glutaminolysis as a response to stimulation, diseased states, etc. This switch is functionally regulated by intracellular glutamine concentration. This leads us to compare the structure of glutamine with Leflunomide (A 77 1726) and the design of a small molecule CMPC. Subsequent synthesis of this compound and *in vitro* testing revealed that CMPC can serve as a very interesting scaffold for medicinal chemistry programs. In addition, the glycolytic pathway in general and pyruvate kinase M2 in particular proved to be a very important target for the development of antiproliferative and anti-inflammatory drugs.

References

ABE, H., OBINATA, T., MINAMIDE, L. S., BAMBURG, J. R. (**1996**). *Xenopus laevis* actin-depolymerizing factor cofilin: a phophorylation-regulated protein essential for development. *J. Cell Biol.* **132**, 871–885.

ASHIZAWA, K., WILLINGHAM, M. C., LIANG, C. M., CHENG, S. Y. (**1991**). *In vivo* regulation of monomer–tetramer conversion of pyruvate kinase subtype M2 by glucose is mediated via fructose 1,6–biphosphate. *J. Biol. Chem.* **266**, 16842–16846.

BAHL, A. K., DALE, M. M., FOREMAN, J. C. (**1994**). The effect of non-steroidal anti-inflammatory drugs on the accumulation and release of interleukin-1-like activity by peritoneal macrophages from the mouse. *Br. J. Pharmacol.* **113**, 809–814.

BARTLETT, R. R., DIMITRIJEVIC, M., MATTAR, T., ZIELINSKI, T., GERMANN, T., RUEDE, E., THOENES, G. H., KÜCHLE, C. C. A., SCHORLEMMER, H. U., BREMER, E., FINNEGAN, A., SCHLEYERBACH, R. (**1991**). Leflunomide (HWA 486), a novel immunomodulating compound for the treatment of autoimmune disorders and reactions leading to transplantation rejection. *Agents Actions* **31(1/2)**, 10–21.

BREEDVELD, F. C., DAYER, J. M. (**2000**). Leflunomide: mode of action in the treatment of rheumatoid arthritis. *Ann. Rheum. Dis.* **59**, 841–849.

BURGER, D., BEGUE-PASTOR, N., BENAVENT, S., GRUAZ, L., KAUFMANN, M. T., CHICHEPORTICHE, R., DAYER, J. M. (**2003**). The active metabolite of Leflunomide, A77 1726, inhibits the production of prostaglandin E(2), matrix metalloproteinase 1 and interleukin 6 in human fibroblast-like synoviocytes. *Rheumatology (Oxford)* **42**, 89–96.

CAO, W. W., KAO, P. N., CHAO, A. C., GARDNER, P., NG, J., MORRIS, R. E. (**1995**). Mechanism of the antiproliferative action of Leflunomide. A77 1726, the active metabolite of Leflunomide does not block T-cell receptor-mediated signal transduction but its antiproliferative effects are antagonized by pyrimidine nucleosides. *J. Heart Lung Transplant.* **14**, 1016–1030.

CHERWINSKI, H. M., BYARS, N., BALLARON, S. J., NAKANO, G. M., YOUNG, J. M., RANSOM, J. T. (**1995**). Leflunomide interferes with pyrimidine nucleotide biosynthesis. *Inflamm. Res.* **44**, 317–322.

CRAIG, E. A. (**1993**). Chaperones: helpers along the pathways to protein folding. *Science* **260**, 1902–1903.

DAX, C. I. (**1996**). *Proteinchemisch Methoden zur Identifizierung und Charakterisierung entzündungsrelevanter Polypeptide.* PhD Thesis, University of Frankfurt/M., Germany.

DAX, C. I., LOTTSPEICH, F., MÜLLNER, S. (1998). *In vitro* model system for the identification and characterization of proteins involved in inflammatory processes. *Electrophoresis* **19**, 1841–1847.

DEAGE, V., BURGER, D., DAYER, J. M. (1998). Exposure of T lymphocytes to Leflunomide but not to dexamethasone favors the production by monocytic cells of interleukin-1 receptor antagonist and the tissue-inhibitor of metalloproteinases-1 over that of interleukin-1beta and metalloproteinases. *Eur. Cytokine Netw.* **9**, 663–668.

EDWARDS, K. A., MONTAGUE, R. A., SHEPARD, S., EDGAR, B. A., ERIKSON, R. L., KIEHART, D. P. (1994). Identification of *Drosophila* cytoskeletal proteins by induction of abnormal cell shape in fission yeast. *Proc. Natl. Acad. Sci. USA* **91**, 4589–4593.

EIGENBRODT, E., GERBRACHT, U., MAZUREK, S., PRESEK, P., FRIIS, R. (1994). Carbohydrate metabolism and neoplasia: New perspectives for diagnosis and therapy. *Biochem. Mol. Aspects Sel. Cancers* **2**, 311–385.

EIGENBRODT, E., KALLINOWSKI, F., OTT, M., MAZUREK, S., VAUPEL, P. (1998). Pyruvate kinase and the interaction of amino acid and carbohydrate metabolism in solid tumors. *Anticancer Res.* **18**, 3267–3274.

ELDER, R. T., XU, X., WILLIAMS, J. W., GONG, H., FINNEGAN, A., CHONG, A. S. (1997). The immunosuppressive metabolite of Leflunomide, A77 1726 affects murine T cells through two biochemical mechanisms. *J. Immunol.* **159**, 22–27.

EMERY, P., BREEDVELD, F. C., LEMMEL, E. M., KALTWASSER, J. P., DAWES, P. T., GOMOR, B., VAN DEN BOSCH, F., NORDSTROM, D., BJORNEBOE, O., DAHL, R., HORSLEV-PETERSEN, K., RODRIGUEZ DE LA SERNA, A., MOLLOY, M., TIKLY, M., OED, C., ROSENBURG, R., LOEW-FRIEDRICH, I. (2000). A comparison of the efficacy and safety of Leflunomide and methotrexate for the treatment of rheumatoid arthritis. *Rheumatology (Oxford)* **39**, 655–665.

FAIRBANKS, L. D., BOFILL, M., RUCKEMANN, K., SIMMONDS, H. A. (1995). Importance of ribonucleotide availability to proliferating T-Lymphocytes from healthy humans. *J. Biol. Chem.* **270**, 29682–29689.

FELDMANN, M., BRENNAN, F. M., MAINI, R. N. (1996). Rheumatoid arthritis. *Cell* **85**, 307–310.

GESSANI, S., TESTA, U., VARANO, B., DiMARZIO, P., BORGHI, P., CONTI, L., BARBED, T., TRITARELLE, E., MARTUCCI, R., SERIPA, D., PESCHLE, C., BELARDELLI, F. (1993). Enhanced production of LPS-induced cytokines during differentiation of human monocytes to macrophages. Role of LPS receptors. *J. Immunol.* **151**, 3758–3766.

GUNSALUS, K. C., BONACCORSI, S., WILLIAMS, E., VERNI, F., GATTI, M., GOLDBERG, M. L. (1995). Mutations in *twinstar*, a DrosIophila gene encoding a cofilin/ADF homologue, result in defects in centrosome migration and cytokinesis. *J. Cell Biol.* **131**, 1243–1259.

HACKER, H. J., STEINBERG, P., BANNASCH, P. (1998). Pyruvate kinase isoenzyme shift from L-type to M2-type is a late event in hepatocarcinogenesis induced in rats by a choline-deficient/DL-ethionine-supplement diet. *Carcinogenesis* **19**, 99–107.

HAMILTON, L. C., VOJNOVIC, I., WARNER, T. D. (1999). A77 the active metabolite of Leflunomide, directly inhibits the activity of cyclo-oxygenase-2 *in vitro* and *in vivo* in a substrate-sensitive manner. *Br. J. Pharmacol.* **127**,1589–1596.

HERRMANN, M. L., SCHLEYERBACH, R., KIRSCHBAUM, B. J. (2000). Leflunomide: an immunomodulatory drug for the treatment of rheumatoid arthritis and other autoimmune diseases. *Immunopharmacology* **47**, 273–289.

HEUKESHOVEN, J., DERNICK, R. (1985). Simplified method for silver staining of proteins in polyacrylamide gels and the mechanism of silver staining. *Electrophoresis* **6**, 103–112.

HÜNIG, S., SCHALLER, R. (1982). Zur Chemie der Acylcyanide. *Angew. Chem. Int. Ed. Engl.* **21**, 36–49.

ISHII, T., YAMADA, M., SATO, H., MATSUE, M., TAKETANI, S., NAKAYAMA, K., SUGITA, Y., BANNAI, S. (1993). Cloning and characterization of a 23-kDa stress-induced mouse peritoneal macrophage protein. *J. Biol. Chem.* **268**, 18633–18636.

KANE, C. D., COE, N. R., VANLANDINGHAM, B., KRIEG, P., BERNLOHR, D. A. (**1996**). Expression purification, and ligand-binding analysis of recombinant keratinocyte lipid-binding protein (MAL-1), an intracellular lipid-binding found overexpressed in neoplastic skin cells. *Biochemistry* **35**, 2894–2900.

KAWAI, S., TAKESHITA, S., OKAZAKI, M., KIKUNO, R., KUDO, A., AMANN, E. (**1994**). Cloning and characterization of OSF-3, a new member of the MER5 family, expressed in mouse osteoplas-matic cells. *J. Biochem.* **115**, 641–643.

KIRCHNER, H., KRUSE, A., NEUSTOCK, P., RINK, L. (**1996**). *Cytokine and Interferone.* Spektrum Akademischer Verlag, Heidelberg.

KIRSCHBAUM, B., MÜLLNER, S., BARTLETT, R. (**1999**). *DEAH-Box Proteins.* US Patent 5,942,429.

KNECHT, W., BERGJOHANN, U., GONSKI, S., KIRSCHBAUM, B., LOEFFLER, M. (**1996**). Functional expression of a fragment of human dihydroorotate dehydrogenase by means of the baculovirus expression vector system, and kinetic investigation of the purified recombinant enzyme. *Eur. J. Biochem.* **240**, 292–301.

KOYASU, S., NISHIDA, E., KADOWAKI, T., MATSUZAKI, F., IIDA, K., HARADA, F., KASUGA, M., SAKAI, H., YAHARA, I., (**1989**). Two mammalian heat shock proteins, HSP90 and HSP100, are actin-binding proteins. *Proc. Natl. Acad. Sci. USA*, **83**, 8054–8058.

KRIEG, P., FEIL, S., FURSTENBERGER, G., BOWDEN, G. T. (**1993**). Tumor specific overexpression of a novel keratinocyte lipid-binding protein. Identification and characterization of a cloned sequence activated during multistage carcino-genesis in mouse skin. *J. Biol. Chem.* **268**, 17362–17369.

LAEMMLI, U. K. (**1970**). Cleavage of structural proteins during the assembly of the head of bacteriophage T4. *Nature* **227**, 680–685.

LAPPALAINEN, P., DRUBIN, D. G. (**1997**). Cofilin promotes rapid actin filament turnover *in vivo. Nature* **388**, 78–82.

LINDQUIST, S. (**1986**). The heat shock response. *Ann. Rev. Biochem.* **55**, 1151.

MANGOLD, U., DAX, C. I., SAAR, K., SCHWAB, W., KIRSCHBAUM, B.,

MÜLLNER, S. (**1999**). Identification and characterization of potential new therapeutic targets in inflammatory and autoimmune disease. *Eur. J. Biochem.* **266**, 1184–1191.

MANNA, S. K., AGGARWAL, B. B. (**1999**). Immunosuppressive Leflunomide metabolite (A77 1726) blocks TNF-dependent nuclear factor–kappa B activation and gene expression. *J. Immunol.* **162**, 2095–2102.

MANNA, S. K., MUKHOPADHYAY, A., AGGARWAL, B. B. (**2000**). Leflunomide suppresses TNF-induced cellular responses: effects on NF–kappa B, activator protein-1, c-Jun N-terminal protein kinase, and apoptosis. *J. Immunol.* **165**, 5962–5969.

MATTAR, T., KOCHHAR, K., BARTLETT, R., BREMER, E. G., FINNEGAN, A. (**1993**). Inhibition of the epidermal growth factor receptor tyrosine kinase activity by Leflunomide. *FEBS Lett.* **334**, 161–164.

MAZUREK, S., GRIMM, H., WILKER, S., LEIB, S., EIGENBRODT, E. (**1998**). Metabolic characteristics of different malignant cancer cell lines. *Anticancer Res.* **18**, 3275–3282.

MAZUREK, S., EIGENBRODT, E. (**2003**). The tumor metabolome. *Anticancer Res.* **23**, 1149–1154.

MAZUREK, S., EIGENBRODT, E., FAILING, K., STEINBERG, P. (**1999**). Alterations in the glycolytic and glutaminolytic pathways after malignant transformation of rat liver oval cells. *J. Cell Physiol.* **181**, 136–146.

MÉJEAN, C., PONS, F., BENYAMIN, Y., ROUSTAN, C. (**1989**). Antigenic probes locate binding sites for the glycolytic enzymes glyceralehyde-3-phosphate dehydrogenase, aldolase and phospho-fructokinase on the actin monomer in microfilaments. *Biochem. J.* **264**, 671–677.

MÜLLER, E., BADEL, K., MÜLLER, A., HERBERT, S., SEILER, A. (**1993**). Offenlegungsschrift DE 43 10 964 A1: *Aktivierte Trägermaterialien, ihre Herstellung und Verwendung.* Deutsches Patentamt.

MÜLLNER, S., EIGENBRODT, E. (**2001**). *1-Butyric acid derivatives such as carbo-methoxypropionyl cyanide or Leflunomide derivatives and the therapeutic use thereof.* European Patent Application (EP 1 377 291).

ORTLEPP, D., LAGGERBAUER, B., MÜLLNER, S., ACHSEL, T., KIRSCHBAUM, B., LÜHRMANN, R. (1998). The mammalian homologue of Prp16p is overexpressed in a cell line tolerant to Leflunomide, a new immunoregulatory drug effective against rheumatoid arthritis. *RNA* **4**, 1007–1018.

PEJOVIC, M., STANKOVIC, A., MITROVIC, D. R. (1992). Lactate dehydrogenase activity and its isoenzymes in serum and synovial fluid of patients with rheumatoid arthritis and osteoarthritis. *J. Rheumatol.* **19**, 529–533.

SAMSTAG, Y., DREIZLER, E. M., AMBACH, A., SCZAKIEL, G., MEUER, S. C. (1996). Inhibition of constitutive serine phosphate activity in T lymphoma cells result in phosphorylation of pp19/cofilin and induces apoptosis. *J. Immunol.* **156**, 4167–4173.

SAMSTAG, Y., ECKERSKORN, C., WESSELBORG, S., HENNING, S., WALLICH, R., MEUER, S. C. (1994). Costimulatory signals for human T-cell activation induce nuclear translocation of pp19/cofilin. *Proc. Natl. Acad. Sci. USA* **91**, 4494–4498.

SCHREIBER, S. L., CRABTREE, G. R. (1992). The mechanism of action of cylosporin A and FK506. *Immunol. Today* **13**, 136–142.

SHAU, H., BUTTERFIELD, L. H., CHIU, R., KIM, A. (1994). Cloning and sequence analysis of candidate human natural killer-enhancing factor genes. *Immunogenetics* **40**, 129–134.

SIEMASKO, K. F., CHONG, A. S., WILLIAMS, J. W., BREMER, E. G., FINNEGAN, A. (1996). Regulation of B cell function by the immunosuppressive agent Leflunomide. *Transplantation* **61**, 635–642.

SILVA, H. T., CAO, W., SHORTHOUSE, R. A., MORRIS, R. E. (1996). Mechanism of action of Leflunomide: *in vivo* uridine administration reverses its inhibition of lymphocyte proliferation. *Transplant. Proc.* **28**, 3082–3084.

SILVA, H. T., CAO, W., SHORTHOUSE, R. A., LOFFLER, M., MORRIS, R. E. (1997). *In vitro* and *in vivo* effects of Leflunomide, Brequinar and Cyclosporine on pyrimidine biosynthesis. *Transplant. Proc.* **29**, 1292–1293.

SMITH, A. B., FUKUI, M., VACCARO, H. A., EMPFIELD, J. R. (1991). Phyllanthoside-phyllanthostatin synthetic studies. 7. Total synthesis of (+)-Phyllanthocin and (+)-Phyllanthocindiol. *J. Am. Chem. Soc.* **113**, 2071–2092.

SMOLEN, J. S., KALDEN, J. R., SCOTT, D. L., ROZMAN, B., KVIEN, T. K., LARSEN, A., LOEW-FRIEDRICH, I., OED, C., ROSENBURG, R. (1999). Efficacy and safety of Leflunomide compared with placebo and sulphasalazine in active rheumatoid arthritis: a double-blind, randomised, multi-centre trial. *Lancet* **353**, 259–266.

STRAND, V., COHEN, S., SCHIFF, M., WEAVER, A., FLEISCHMANN, R., CANNON, G., FOX, R., MORELAND, L., OLSEN, N., FURST, D., CALDWELL, J., KAINE, J., SHARP, J., HURLEY, F., LOEW-FRIEDRICH, I. (1999). Treatment of active rheumatoid arthritis with Leflunomide compared with placebo and methotrexate. *Arch. Intern. Med.* **159**, 2542–2550.

STOSCHECK, C. M. (1990). Quantitation of Protein Methods in Enzymology. *Pierce*, Rockford, IL, USA, **182**, 50–69.

SUZUKI, K., YAMAGUCHI, T., TANAKA, T., KAWANISHI, T., NISHIMAKI-MOGAMI, T., YAMAMOTO, K., TSUJI, T., IRIMURA, T., HAYAKAWA, T., TAKAHASHI, A. (1995). Activation induces dephosphorylation of cofilin and its translocation to plasma membranes in neutrophil-like differentiated HL-60 cells. *J. Biol. Chem.* **270**, 19551–19556.

TANG, Q., SEN, S. E. (1998). Carbomethoxypropionyl cyanide: a regioselective C-acylation reagent for the preperation of β-dicarbonyl compounds. *Tetrahedron Lett.* **39**, 2249–2252.

VYSE, T. J., TODD, J. A. (1996). Genetic analysis of autoimmune disease. *Cell* **85**, 311–318.

WEN, S.-T., VAN ETTEN, R. A. (1997). The PAG gene product, a stress-induced protein with antioxidant properties, is an Abl SH3–binding protein and a physiological inhibitor of c-Abl tyrosine kinase activity. *Genes Dev.* **11**, 2456–2467.

WILLIAMSON, R. A., YEA, C. M., ROBSON, P. A., CURNOCK, A. P., GADHER, S., HAMBLETON, A. B., WOODWARD, K., BRUNEAU J-M,

HAMBLETON, P., MOSS, D., THOMSON, T. A., SPINELLA-JAEGLE, S., MORAND, T., COURTIN, O., SAUTÉS, C., WESTWOOD, R., HERCEND, T., KUO, E. A., RUUTH, E. (**1995**). Dihydroorotate dehydrogenase is a high-affinity binding protein of A77 1726 and mediator of a range of biological effects of the immunomodulatory compound. *J. Biol. Chem.* **270**, 22467–22472.

WILLIAMSON, R. A., YEA, C. M., ROBSON, P. A., CURNOCK, A. P., GADHER, S., HAMBLETON, A. B., WOODWARD, K., BRUNEAU, J. M., HAMBLETON, P., SPINELLA-JAEGLE, S., MORAND, P., COURTIN, O., SAUTES, C., WESTWOOD, R., HERCEND, T., KUO, E. A., RUUTH, E. (**1996**). Dihydroorotate dehydrogenase is a target for the biological effects of Leflunomide. *Transplant. Proc.* **28**, 3088–3091.

WINFIELD, J., JARJOUR, W. (**1991**). Do stress proteins play a role in arthritis and auto-immunity? *Immunol. Rev.* **121**, 121–220.

WRIGHT, S. D., RAMOS, R. A., TOBIAS, P. S., ULEVITCH, R. J., MATHISON, J. C. (**1990**). CD 14, a receptor for complexes of lipopolysaccharide (LPS) and LPS binding protein. *Science* **249**, 1431–1433.

WU, Y. W., CHIK, C. L., KNAZEK, R. A. (**1989**). An *in vitro* and *in vivo* study of antitumor effects of gossypol on human SW-13 adrenocortical carcinoma. *Cancer Res.* **49**, 3754–3758.

XU, X., WILLIAMS, J. W., GONG, H., FINNEGAN, A., CHONG, A. S. (**1996**). Two activities of the immunosuppressive metabolite of Leflunomide, A77 1726. Inhibition of pyrimidine nucleotide synthesis and protein tyrosine phosphorylation. *Biochem. Pharmacol.* **52**, 527–534.

YILMAZ, S., OZAN, S., OZERCAN, I. H. (**2003**). Comparison of pyruvate kinase variants from breast tumor and normal breast. *Arch. Med. Res.* **34**, 315–324.

ZIELINSKI, T., ZEITTER, D., MULLER, S., BARTLETT, R. R. (**1995**). Leflunomide, a reversible inhibitor of pyrimidine biosynthesis? *Inflamm. Res.* [Suppl. 2], S207–208.

ZWERSCHKE, W., MAZUREK, S., MASSIMI, P., BANKS, L., EIGENBRODT, E., JANSEN-DURR, P. (**1999**)Modulation of type M2 pyruvate kinase activity by the human papillomavirus type 16 E7 oncoprotein. *Proc. Natl. Acad. Sci. USA* **96**, 1291–1296.

11
Differential Phosphoproteome Analysis in Medical Research

Elke Butt and Katrin Marcus

Abstract

The field of proteomics has undergone tremendous advances over the past few years and enabled the detailed analysis of proteins in very complex systems, such as cells, tissues, or body fluids (at a given time point), including the capture of structural diversity of proteins, splicing, processing, and posttranslational modification (e.g., phosphorylation, glycosylation). Several developments in the area of proteome research have revolutionized science, as biological processes can be probed at the molecular level using sophisticated analytical tools including two-dimensional polyacrylamide gel electrophoresis (2D PAGE) and mass spectrometry (MS). The purpose of this review is to highlight how proteomics techniques have been used to answer specific questions related to signal trans-duction in platelets although the methods can be adopted in a wide variety of systems. Here, we restrict our discussion to the study of phosphoproteins. Our basic approach is to examine only those proteins that differ by some variable from the control sample. In this way, the number of proteins to be processed is greatly reduced. In eukaryotic cells, proteins are mainly phosphorylated on serine, threonine and tyrosine residues. Three examples of typical experiments to measure changes in protein phosphorylation in response to specific treatment will be discussed. In these studies the protein samples were separated by 2D PAGE and the phosphorylated proteins were directly assigned by detection of $[^{32}P]$-ortho-phosphate via autoradiography. Analysis of the phosphorylated proteins was done by MS. To concurrently identify the phosphoserine and phosphothreonine substrates of cAMP- and cGMP-dependent protein kinases associated with the inhibition of platelet aggregation, human platelets were analyzed by incubating the cells with $[^{32}P]$ before separation on 2-D gels. A similar approach was used to characterize the substrate specificities of different MAPK kinases by comparing the phosphoproteome of knock out mice with wild type animals after extracellular stress. The use of antibodies directed against phosphoamino acids represents an alternative procedure to detect phosphorylated proteins. This method was used for the additional detection of tyrosine-phosphorylated proteins after the activation

Proteomics in Drug Research
Edited by M. Hamacher, K. Marcus, K. Stühler, A. van Hall, B. Warscheid, H. E. Meyer
Copyright © 2006 Wiley-VCH Verlag GmbH & Co. KGaA, Weinheim
ISBN: 3-527-31226-9

of platelets by thrombin. The data will help to build the basis for future identification of new drug targets and therapeutic strategies against thrombotic diseases such as stroke and myocardial infarction.

11.1
Introduction

Proteomics has entirely shifted the paradigm for drug discovery and novel diagnostics, as it is an effective means to rapidly identify new biomarkers and therapeutic targets for the diagnosis and treatment of various diseases, while increasing our understanding of biological processes. Clinical applications of proteomics involve the use of proteomic technologies with the final goal being to characterize the information flow through the intra- and extracellular molecular protein networks that interconnect organ and circulatory systems.

Protein phosphorylation, a reversible posttranslational modification, occurs principally on serine, threonine, and tyrosine residues and plays a central role in cellular regulation either by altering a protein's activity directly or by inducing specific protein–protein interactions, which in turn affect localization, binding specificity or activity. Abnormal kinase or phosphatase function plays a role in several pathologies and tumorigenesis. The phosphoproteome especially will play a key role in clinical proteomics as the aberrant function of protein kinases and phosphatases are often the center of many diseases, e.g., cancer, stroke, inflammatory and viral pathologies. The new focus of narrowly focused molecular targeted therapeutics addresses this concept (Petricoin et al., 2004).

A comprehensive analysis of protein phosphorylation includes the identification of the phosphoprotein or phosphopeptide, the localization of the modified amino acid, and if possible, the quantification of phosphorylation.

Several methods have been used to detect and identify phosphorylated residues in a protein. The first step in a proteome study is the separation of a highly complex protein mixture resulting from a cell, tissue or organ lysate. In general, methods applied to the analysis of the proteome and protein identification are also well suited for the analysis of the phosphoproteome, including the localization of specific phosphorylation sites. The most prominent method for this purpose today is two-dimensional polyacrylamide gel electrophoresis (2D PAGE), which separates the proteins according to their isoelectric points in the first and their molecular weights in the second dimension (Görg et al., 1985; Klose, 1999). The phosphoproteins can be visualized using different methods such as radiolabeling with [^{32}P]-orthophosphate combined with autoradiography (Garrison et al., 1993), immunodetection with phosphospecific antibodies, and/or staining using phosphospecific dyes (Schulenberg et al., 2004). The specific isolation of phosphorylated proteins/peptides from a complex protein mixture is facilitated by different methods such as immunoprecipitation (Marcus et al., 2000, 2003; Gronborg et al., 2002), immobilized metal affinity chromatography (Salomon et al., 2003; Ficarro et al., 2002), or chemical modification followed by specific purification (Oda et al., 2001; Zhou et al., 2001). Mass spectrometric identification and localization of the phosphorylated amino acid residue can be done by several different methods such as mass spectrometry (MS)/MS-based peptide sequencing, precursor ion or neutral loss scanning (Steen and Mann, 2002; Batemann et al., 2002; Mann et al., 2002)

Indeed, the analysis of phosphoproteins and the detection and localization of the phosphorylated residues are not straightforward for various reasons:

- The site of phosphorylation in a protein may vary resulting in a very hetero-geneous mixture of different phosphorylation forms.
- Phosphorylation in most instances emerges as a transient modification within signaling cascades and is accompanied by dephosphorylation. Hence, the stoichiometry is generally relatively low and the phosphoprotein detection method must be very sensitive.
- Most analytical techniques for studying protein phosphorylation have a limited dynamic range and only major phosphorylation sites may be detected, whereas minor sites might be difficult to localize.
- Phosphorylated peptides are generally suppressed in mass spectrometric analysis because of their low ionization rates.

In consequence, the localization of the phosphorylated residue often requires the specific isolation, detection and analysis of the modified region, which is complicated by the physicochemical conditions and complexity of the analyzed protein sample.

The following three sections show concrete examples of the analysis of phosphoproteins as potential drug targets and demonstrate the applicability and enormous power of proteomics in this field.

11.2
Phosphoproteomics of Human Platelets

Blood platelets arise as pinch-offs of megakaryocytes in the bone marrow, playing a critical role in wound healing. In their inactive, nonadhesive state they circulate freely in the blood. Interactions with structures in the subendothelial matrix caused by injuries of the blood vessels initiate a rapid platelet activation resulting in the formation of vascular plugs and release of intracellular substances (e.g., thrombin) that initiate repair processes. Platelet adhesion at the vessel wall is the first step in platelet activation in vivo. Starting from this adhesion, intracellular signaling cas-cades are initiated resulting in secretion of activating substances and aggregation of the platelets. One important event within these intracellular signaling cascades is the phosphorylation/dephosphorylation of multiple proteins on various tyro-sine, serine and threonine residues (Watson, 1997; Presek and Martinson, 1997).

Platelets are important components of hemostasis underlying cardiovascular diseases such as myocardial infarction or stroke (Brown et al., 1997; Glass and

Witzum, 2001). Genetic defects may result in bleeding disorders like Glanzmann thrombasthenia and Bernard-Soulier syndrome (Nurden, 1999). The high clinical relevance of arterial thrombosis, the limited knowledge about the underlying mechanisms on the molecular level, and the small number of available antiplatelet drugs (Fitzgerald, 2001) show the necessity for gaining more profound insights into the system. Although in recent years inhibitors of platelet aggregation have been established as the outstanding principle for secondary, and partially for primary prevention of cardiovascular diseases (Lefkowitz and Willerson, 2001; Bennett, 2001), only a few effective drugs exist so far (Fitzgerald, 2001). Hence, the search for better and perhaps safer antiplatelet drugs is of particular clinical interest, and it is therefore imperative to understand the exact molecular mechanisms in platelet activation, to find new potential components of the signal transduction pathways and to determine the role of protein phosphorylation.

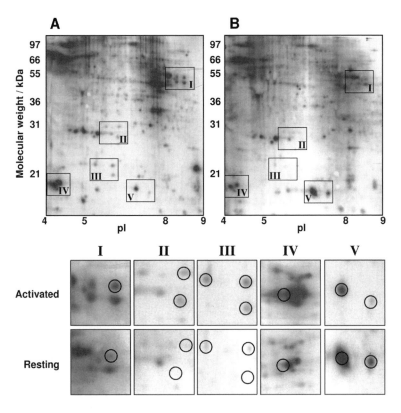

Figure 11.1 Autoradiography of a two-dimensional separation of phosphorylated human platelet proteins after labeling with [^{32}P] in the pH range 3–10.
Top: Autoradiography of the 2D gel with pH 3–10 after application of 400 μg protein lysate of activated (**A**) and resting (**B**) human platelets.
Below: Scaled-up sections I–V of the 2D gels. This figure clearly demonstrates the power of autoradiography in the differential phosphoproteome analysis.

In the study described, a differential phosphoproteomic approach using the combination of several methods – 2D PAGE, 1D PAGE, autoradiography, immunoblotting, and nano high-performance liquid chromatography (LC)–electrospray ionization (ESI) MS/MS – was employed, resulting in a comprehensive analysis of the platelet phosphoproteome.

After ex vivo stimulation with thrombin and radiolabeling of the platelet (resting platelets served as control), proteins were separated using 2D PAGE and phosphorylated proteins were directly assigned by detection of $[^{32}P]$ via autoradiography (Figure 11.1). Discrimination of serine/threonine-phosphorylated proteins on the one hand and tyrosine-phosphorylated proteins on the other was done by detection of phosphotyrosine-containing proteins via immunoblotting with an antiphosphotyrosine-specific antibody. Phosphorylated protein spots/bands were analyzed by nano-LC MS/MS.

Several of the identified proteins were known to be involved in signal transduction and platelet function, e.g., cortactin, myosin regulatory light chain (myosin RLC), and protein disulfide isomerase (PDI).

11.2.1
Cortactin

In several spots of the immunoblots and the autoradiogram (Figure 11.2, white arrow) the actin-binding protein cortactin was identified. The human form of cortactin is known as an oncogene, often amplified in human colon carcinomas and tumor cells (Schuuring et al., 1993) and seems to be involved in the signal transduction caused by extracellular stimuli such as thrombin, integrins, or the EGF (Huang et al., 1997). On the basis of the MS/MS spectra of the nonphosphorylated and the phosphorylated peptide, a phosphorylation of Ser-418 was detected (Figure 11.3A, B). Additionally, two further phosphorylation sites were localized at Thr-401 and Ser-405 (results not shown). These sites were already described to be phosphorylated by the extracellular signal-regulated kinase (ERK) in vivo and in vitro (Campbell et al., 1999). Although the exact role of this phosphorylation is not yet clear, there is evidence for an important function in the regulation of the subcellular localization of cortactin (van Damme et al., 1997). Additionally, a tyrosine-phosphorylation of cortactin by the tyrosine kinase pp60c-src at different tyrosine residues (Tyr-421, Tyr-466, and Tyr-482) is known to be responsible for its reduced actin-crosslink activity (Raffel et al., 1996).

11.2.2
Myosin Regulatory Light Chain

The regulatory light chains of myosin are known to be phosphorylated at different serine and threonine residues. Thr-18 and Ser-19 are substrates of the myosin light chain-kinase (MLCK) whereas Ser-1, Ser-2 und Ser-9 are phosphorylated by the protein kinase C (PKC) (Naka et al., 1988; Kawamoto et al., 1989; Moussavi et al., 1993). As the first slow phosphorylation by the MLCK is already present in the

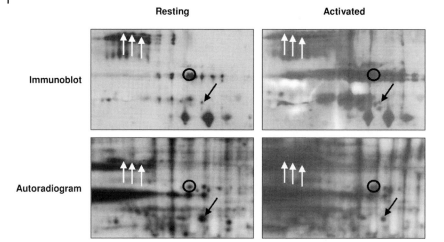

Figure 11.2 Extracts of an anti-phosphotyrosine immunoblot (*top*) and autoradiography (*bottom*) after two-dimensional separation in the pH range 4–7 of resting and activated human platelets. (For better orientation a *black circle* is drawn in.) The extracts of the regions between pI 5.4–5.9 and 40–55 kDa clearly demonstrate the differences in the phosphorylation patterns of resting and thrombin-activated platelets. An intensity increase of different protein spots identified as, e.g., cortactin (three different phosphorylated forms; *white arrows*) and protein disulfide-isomerase (*black arrow*) could be detected after stimulation with thrombin in both the immunoblot and the autoradiogram. Several of the identified proteins were known to be involved in signal transduction and platelet function.

resting platelets the second fast phosphorylation by the PKC occurred after thrombin stimulation. This second Ca^{2+}-dependent phosphorylation plays an important role in the secretion of substances from platelet intracellular granules (Nishikawa et al., 1980). In 1988, Naka and coworkers (Naka et al., 1988) described the existence of three different forms of myosin RLC in thrombin-activated platelets (non-, mono-, and bis-phosphorylated RLC) by nonproteomic studies. These results were confirmed unequivocally and were even augmented by this proteomic study (Figure 11.3C, D) (Marcus et al., 2003).

11.2.3
Protein Disulfide Isomerase

PDI is a noncovalent homodimer with a molecular weight of 57 kDa for each subunit. The protein is generally involved in the formation, reduction and rearrangement of disulfide bridges and plays an important role in platelet function. PDI is localized at the extracellular side of the plasma membrane and is secreted after platelet activation (Chen et al., 1992). In platelets, PDI catalyzes the formation of disulfide-bridged complexes of thrombospondin 1 and thrombin–antithrombin III and is thus involved in hemostasis and wound healing (Milev and Essex, 1999).

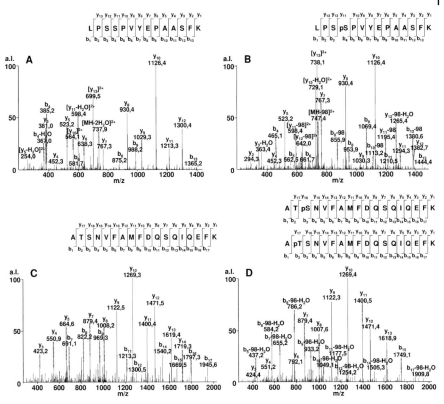

Figure 11.3 Localization of the phosphorylation sites of cortactin (**A** and **B**) and the myosin regulatory light chain (RLC; **C**, **D**).

A: Fragment ion spectrum of the nonphosphorylated peptide LPSpSPVYEDAASFK of cortactin.

B: Fragment ion spectrum of the nonphosphorylated peptide LPSSPVYEDAASFK of cortactin. Due to the secessions of –98 Da from y₁₁ Ser-418 (Ser-4 in the spectrum) was unambiguously detected to be phosphorylated.

C: Fragment ion spectrum of the nonphosphorylated peptide ATSNVFAMFDQSQIQEFK of the RLC.

D: Fragment spectrum of the phosphorylated peptide ApTSNVFAMFDQSQIQEFK or ATpSNVFAMFDQSQIQEFK, respectively. This peptide contained three possible phosphorylation sites (Thr-2, Ser-3, and Ser-12). The comparison of the two spectra revealed Thr-2 or Ser-3 as phosphorylated residues. As no secession of –98 Da was observed till y_{13} a phosphorylation of Ser-13 could be excluded. On the basis of the MS/MS spectrum the exact site of phosphorylation was not determinable.

In addition, it is part of the integrin-mediated platelet adhesion (Lahav et al., 2000). A threonine phosphorylation has been described for the PDI localized in the endoplasmatic reticulum (Quemeneur et al., 1994).

All proteins identified could be classified into different categories based on their function: 23 cytoskeletal proteins, 25 signaling molecules, 19 metabolic proteins, 5 vesicular proteins, 9 mitochondrial proteins, 6 proteins with extracellular function, 12 proteins involved in protein processing, 23 unknown proteins, and 12 miscellaneous. Overall, this proteome study resulted in a successful identification of 134 proteins and the localization of eight partly unknown phosphorylation sites, clearly demonstrating the power of proteomics. Methods, procedures and results are described in detail in Marcus et al., 2003.

The identification of phosphorylated proteins from such global approaches is only the first step in characterizing these signaling molecules. It is important to verify the involvement of the proteins identified and to prove the localized phosphorylation sites in regard to their functionality. Such studies may be performed by using antibodies against the protein of interest and/or over-expression of wild-type and mutant forms of epitope-tagged proteins to examine their role in signal transduction pathways.

11.3
Identification of cAMP- and cGMP-Dependent Protein Kinase Substrates in Human Platelets

The activation process of human platelets and vessel wall–platelet interactions are tightly regulated under physiological conditions and are often impaired in thrombosis, arteriosclerosis, hypertension, and diabetes. Platelet activation can be inhibited by a variety of agents, including aspirin and Ca^{2+} antagonists, as well as cGMP- and cAMP-elevating agents such as sodium nitroprusside (NO) and prostaglandin I_2 (Koesling and Nuernberg, 1997). The inhibitory effects of cGMP and cAMP are mediated by cAMP- and cGMP-dependent protein kinase (PKA and PKG), respectively (Münzel et al., 2003) For example, platelets from certain patients with chronic myelocytic leukemia showed decreased expression of PKG but a normal level of PKA. Phosphorylation of vasodilator-stimulated phospho-protein (VASP) a specific substrate of both PKA and PKG, was severely impaired in these cGMP-dependent protein-kinase-deficient platelets, despite an exaggerated cGMP response to sodium nitroprusside, whereas the cAMP-mediated VASP phosphorylation was functionally intact (Eigenthaler et al., 1993). In PKG-deficient mice cGMP-mediated inhibition of platelet aggregation is impaired (Massberg et al., 1999). However, the molecular mechanisms of platelet inhibition by cGMP signaling downstream of cGK activation are only partially understood and to date, only a few substrates for PKA and PKG have been identified and characterized in intact human platelets. In this respect, we used differential phosphoproteomic display of radiolabeled human platelets to identify specific substrates of PKG and PKA.

Therefore, human platelets were labeled with ^{32}P, stimulated with 500 µM of the specific PKG activator 8-pCPT-cGMP, and proteins of the resulting platelet lysate were separated by 2D PAGE. The two-dimensional phosphoproteomes resulting from this experiment (Figure 11.4) demonstrate the phosphorylation/ dephosphorylation of several proteins after stimulation with 8-pCPT-cGMP. For identification of these proteins, the spots were excised from the gel, digested with trypsin, and the resulting peptides were analyzed by nano LC-ESI MS/MS (Butt et al., 2003). Alongside the increase of phosphorylation in several spots, dephosphorylation is also observed (Spots 4, 5 and 8) indicating the activation of phosphatases by activated PKG (Table 11.1).

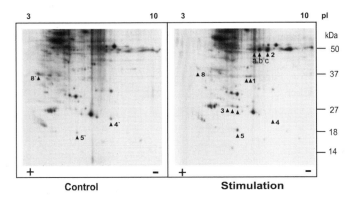

Figure 11.4 PKG-dependent phosphorylation in intact human platelets. Human platelets were incubated in the presence of [^{32}P]-orthophosphate and treated with buffer alone (control) or with 500 µM PKG activator 8pCPT-cGMP for 30 min (stimulation).

Platelet homogenate was separated by two-dimensional gel electrophoresis and an autoradiogram was obtained. Spot 1: LASP; spots 2a, b and c: VASP; spot 3: rap 1b; spots 4, 5 and 8 have not yet been identified. The pH gradient is indicated at the top of the gel.

Table 11.1 Identification and function of cAMP- and cGMP-dependent protein kinase substrates in human platelets.

Spot No.	Identification	Function
1	LIM and SH3 domain protein LASP	Phosphorylation of LASP at Ser-146 leads to a redistribution of the actin-bound protein from the tips of the cell membrane to the cytosol accompanied with a reduced cell migration (Butt et al., 2003; Keicher et al., 2004)
2a, b and c	VASP	Three distinct phosphorylation sites (Ser-157, Ser-239 and Thr-278) with different specificities for PKA and PKG (Butt et al., 1994). VASP phosphorylation is thought to be involved in the negative regulation of integrin $\alpha_{IIb}b_{III}$ (Horstrup et al., 1994). Experiments *in vitro* revealed reduced F-actin binding and actin polymerization of phosphorylated VASP (Harbeck et al., 2000).
3	Rap 1b	Phosphorylation of Rap 1b at Ser-179 is associated with translocation of the protein from the membrane to the cytosol (Torti et al., 1994)
4, 5, 8	Not identified yet	
Gel not shown	Heat shock protein 27 Hsp27	Phosphorylation of Hsp27 at Thr-143 by PKG reduced the stimulatory effect of MAPKAP kinase 2-phosphorylated Hsp27 on actin polymerization (Butt et al., 2001)

All PKG substrates identified so far in human platelets interact directly or indirectly with the actin filament and the cytoskeleton of the cell and are involved in the negative regulation of platelet activation after phosphorylation by PKG. The proteomic approach in combination with functional studies is a very successful tool for studying phosphorylation-dependent effects in platelets. Ongoing proteome studies will now focus on membrane proteins for new therapeutic targets to develop antithrombotic drugs.

11.4
Identification of a New Therapeutic Target for Anti-Inflammatory Therapy by Analyzing Differences in the Phosphoproteome of Wild Type and Knock Out Mice

Genetically modified mice either overexpressing or knocking out specific components have provided exciting advances in our knowledge of diseases. Recently, we focused on the p38 mitogen-activated protein kinase (MAPK) pathway and its physiological effect on inflammation.

An intact cytokine response is essential for efficient host defense against invading pathogens. Inhibition of the p38 MAPK pathway that is involved in a number of cellular stress responses has been used successfully to decrease cytokine production in vitro (Lee et al., 1994), suggesting a key role for this MAPK pathway in inflammation. However, the central role of p38 MAPK, which is responsible for the activation of a number of kinases and transcription factors, limits the use of p38 inhibitors as a selective anti-inflammatory strategy. Thus, downstream substrates of the MAPK pathway could represent promising targets for a specific suppression of cytokine production. In mouse and human, three MAPKAP kinases (MK2, MK3, and MK5) have been identified and are activated by p38. The substrate specificities of these kinases also seem very similar and the small heat shock protein HSP25/27 is an efficient substrate for all three kinases in vitro (Stokoe et al., 1992). Therefore, it was rather surprising that lack of MK2 resulted in a defect of cytokine biosynthesis and decreased levels of p38 MAPK. This was unexpected since MK3 and MK5 should be able to compensate for the loss of MK2, at least to a certain degree. To analyze these significant differences, we deleted the MK2 and MK5 gene from mice and compared the phenotypes of the resulting MK5 knockout (KO) mice and the MK2-deficient mice with the wild type (WT) (Shi et al., 2003).

For this approach WT, MEK2 –/– and MEK5 –/– mouse embryonic cardiac fibroblasts were labeled with ^{32}P, stimulated by 100 µM arsenite and proteins of the cell lysates were separated by 2D PAGE. The differential phosphoproteomes (Figure 11.5) demonstrate the phosphorylation of a protein identified by MS as HSP25 (arrow) in wild type and MK5-deficient cells indicating that in vivo, MK2 was the only kinase phosphorylating HSP25 in these cells stimulated by arsenite (Shi et al., 2003).

Obviously, MK5 is not able to compensate for this effect, making MK2 a unique pharmacological target for anti-inflammatory therapy.

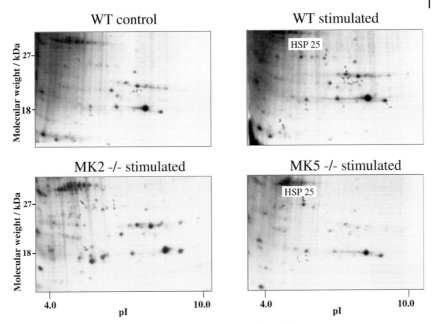

Figure 11.5 Autoradiograms of 2D phosphoproteomics from wildtype (WT), MAPKAP kinase 2 deficient (MK2 –/–) and MAPKAP kinase 5 deficient (MK5 –/–) mouse cardiac fibroblasts before and after stimulation with 100 µM arsenite. The spot for phospho-HSP 25 was identified by mass spectrometry and is indicated by an *arrow*.

11.5
Concluding Remarks and Outlook

In summary, modern proteomics approaches offer unprecedented opportunities to identify novel phosphorylated proteins for diagnosis and therapy of major illnesses. The development of new clinical proteomic tools away from ^{32}P labeling methods and the monitoring of protein expression by two-dimensional gels towards MS-based methods using phosphospecific isotope-tagged labeled peptides followed by one-dimensional electrophoresis will gain in importance. In the future, better protein enrichment strategies and quantification of phosphorylation will allow the clinical transfer to functional application and early detection of pathological organ conditions, as a supplement to existing and coevolving advances in diagnostic imaging.

Acknowledgements

Parts of the authors' work were funded by the German Ministry of Education and Research (BMBF).

References

BATEMAN, R. H., CARRUTHERS, R., HOYES, J. B., JONES, C., LANGRIDGE, J. I., MILLAR, A., VISSERS, J. P. (2002). A novel precursor ion discovery method on a hybrid quadrupole orthogonal acceleration time-of-flight (Q-TOF) mass spectrometer for studying protein phosphorylation. *J. Am. Soc. Mass Spectrom.* **13**, 792–803.

BENNETT, J. S. (2001). Platelet-fibrinogen interactions. *Annu. Rev. Med.* **52**, 161–184.

BROWN, A. S., ERUSALIMSKY, J. D., MARTIN, J. F. (1997). Megakaryocytopoiesis: the megakaryocyte/platelet haemistatic axis. In VON BRUCHHAUSEN, F., WALTER, U. (Eds.), *Platelets and their Factors.* Springer-Verlag, Berlin, pp. 3–19.

BUTT, E., ABEL, K., KRIEGER, M., PALM, D., HOPPE, V., HOPPE, J., WALTER, U. (1994). cAMP- and cGMP-dependent protein kinase phosphorylation sites of the focal adhesion vasodilator-stimulated phosphoprotein (VASP) *in vitro* and in intact human platelets. *J. Biol. Chem.* **269**, 14509–14517.

BUTT, E., GAMBARYAN, S., GÖTTFERT, N., GALLER, A., MARCUS, K., MEYER, H. E. (2003). Actin binding of human LIM and SH3 protein is regulated by cGMP- and cAMP-dependent protein kinase phosphorylation on serine 146. *J. Biol. Chem.* **278**, 15601–15607.

BUTT, E., IMMLER, D., MEYER, H. E., KOTLYAROV, A., LAASS, K., GAESTEL, M. (2001). Heat shock protein 27 is a substrate of cGMP-dependent protein kinase in intact human platelets: phosphorylation-induced actin polymerization caused by HSP27 mutants. *J. Biol. Chem.* **276**, 7108–7113.

CAMPBELL, D. H., SUTHERLAND, R. L., DALY, R. J. (1999). Signaling pathways and structural domains required for phosphorylation of EMS1/cortactin. *Cancer Res.* **59**, 5376–5385.

CHEN, K., LIN, Y., DETWILER, T. C. (1992). Protein disulfide isomerase activity is released by activated platelets. *Blood* **79**, 2226–2228.

EIGENTHALER, M., ULLRICH, H., GEIGER, J., HORSTRUP, K., HONIG-LIEDL, P., WIEBECKE, D., WALTER, U. J. (1993). Defective nitrovasodilator-stimulated protein phosphorylation and calcium regulation in cGMP-dependent protein kinase-deficient human platelets of chronic myelocytic leukemia. *Biol. Chem.* **268**, 13526–13531.

FICARRO, S. B., MCCLELAND, M. L., STUKENBERG, P. T., BURKE, D. J., ROSS, M. M., SHABANOWITZ, J., HUNT, D. F., WHITE, F. M. (2002). Phosphoproteome analysis by mass spectrometry and its application to *Saccharomyces cerevisiae. Nat. Biotechnol.* **20**, 301–305.

FITZGERALD, D. J. (2001). Vascular biology of thrombosis: the role of platelet–vessel wall adhesion. *Neurology* **57**, S1–S4.

GARRISON, J. C. (1993). Protein phosphorylation. In HARDIE, D. G. (Ed.), *A Practical Approach.* Oxford University Press, Oxford, pp. 1.

GLASS, C. K., WITZUM, J. L. (2001). Atherosclerosis: The Road Ahead. *Cell* **104**, 503–516.

GÖRG, A., POSTEL, W., GÜNTHER, S., WESER, J. (1985). Improved horizontal two-dimensional electrophoresis with hybrid isoelectric focusing in immobilized pH gradients in the first dimension and laying-on transfer to the second dimension. *Electrophoresis* **6**, 599–604.

GRONBORG, M., KRISTIANSEN, T. Z., STENSBALLE, A., ANDERSEN, J. S., OHARA, O., MANN, M., JENSEN, O. N., PANDEY, A. (2002). A mass spectrometry-based proteomic approach for identification of serine/threonine-phosphorylated proteins by enrichment with phospho-specific antibodies: identification of a novel protein, Frigg, as a protein kinase A substrate. *Mol. Cell Proteomics* **1**, 517–527.

HARBECK, K., HÜTTELMAIER, S., SCHLÜTER, K., JOCKUSCH, B. M., ILLENBERGER, S. (2000). Phosphorylation of the vasodilator-stimulated phosphoprotein regulates its interaction with actin. *J. Biol. Chem.* **275**, 30817–30825.

HORSTRUP, K., JABLONKA, B., HÖNIG-LIEDL, P., JUST, M., KOCHSIEK, K., WALTER, U. (1994). Phosphorylation of focal adhesion vasodilator-stimulated phosphoprotein at Ser157 in intact human platelets correlates with fibrinogen receptor inhibition. *Eur. J. Biochem.* **225**, 21–27.

HUANG, C., NI, Y., WANG, T., GAO, Y., HAUDENSCHILD, C. C., ZHAN, X. (**1997**). Down-regulation of the filamentous actin cross-linking activity of cortactin by Src-mediated tyrosine phosphorylation. *J. Biol. Chem.* **272**, 13911–13915.

KAWAMOTO, S., BENGUR, A. R., SELLERS, J. R., ADELSTEIN, R. S. (**1989**). *In situ* phosphorylation of human platelet myosin heavy and light chains by protein kinase C. *J. Biol. Chem.* **264**, 2258–2265.

KEICHER, C., GAMBARYAN, G., SCHULZE, E., MARCUS, K., MEYER, H. E., BUTT, E. (**2004**). Phosphorylation of mouse LASP-1 on threonine 156 by cAMP- and cGMP-dependent protein kinase. *Biochem. Biophys. Res. Commun.* **324**, 308–316.

KLOSE, J. (**1999**). Large-gel 2-D electrophoresis. *Methods Mol. Biol.* **112**, 147–172.

KOESLING, D., NUERNBERG, B. (**1997**). Platelet G proteins and adenylyl and guanylyl cyclases. In VON BRUCHHAUSEN, F., WALTER, U. (Eds.), *Platelets and their Factors.* Springer-Verlag, Berlin, pp. 181–218.

LAHAV, J., GOFER-DADOSH, N., LUBOSHITZ, J., HESS, O., SHAKLAI, M. (**2000**). Protein disulfide isomerase mediates integrin-dependent adhesion. *FEBS Lett.* **475**, 89–92.

LEE, J. C., LAYDON, J. T., MCDONNEL, P. C., GALLAGHER, T. F., KUMAR, S., GREEN, D., MCNULTY, D., BLUMENTHAL, M. J., HEYS, J. R., LANDVATTER, S. W. (**1994**). A protein kinase involved in the regulation of inflammatory cytokine biosynthesis. *Nature* **372**, 739–746.

LEFKOWITZ, R. J., WILLERSON, J. T. (**2001**). Prospects for cardiovascular research. *JAMA* **285**, 581–587.

MANN, M., ONG, S. E., GRONBORG, M., STEHEN, H., JENSEN, O. N., PANDEY, A. (**2002**). Analysis of protein phosphorylation using mass spectrometry: deciphering the phosphoproteome. *Trends Biotechnol.* **20**, 261–268.

MARCUS, K., IMMLER, D., STERNBERGER, J., MEYER, H. E. (**2000**). Identification of platelet proteins separated by two-dimensional gel electrophoresis and analyzed by matrix assisted laser desorption/ionization-time of flight-mass spectrometry and detection of tyrosine-phosphorylated proteins. *Electrophoresis* **21**, 2622–2636.

MARCUS, K., MOEBIUS, J., MEYER, H. E. (**2003**). Differential analysis of phosphorylated proteins in resting and thrombin-stimulated human platelets. *Anal. Bioanal. Chem.* **276**, 973–993.

MASSBERG, S., SAUSBIER, M., KLATT, P., BAUER, M., PFEIFER, A., SIESS, W., FASSLER, R., RUTH, P., KROMBACH, F., HOFMANN, F. (**1999**). Increased adhesion and aggregation of platelets lacking cyclic guanosine 3′,5′-monophosphate kinase I. *J. Exp. Med.* **189**, 1255–1264.

MILEV, Y., ESSEX, D. W. (**1999**). Protein disulfide isomerase catalyzes the formation of disulfide-linked complexes of thrombospondin-1 with thrombin–antithrombin III. *Arch. Biochem. Biophys.* **361**, 120–126.

MOUSSAVI, R. S., KELLEY, C. A., ADELSTEIN, R. S. (**1993**). Phosphorylation of vertebrate nonmuscle and smooth muscle myosin heavy chains and light chains. *Mol. Cell Biochem.* **127–128**, 219–227.

MÜNZEL, T., FEIL, R., MÜLSCH, A., LOHMANN, S. M., HOFMANN, F., WALTER, U. (**2003**). Physiology and pathophysiology of vascular signaling controlled by guanosine 3′,5′-cyclic monophosphate-dependent protein kinase. *Circulation* **108**, 2172–2183.

NAKA, M., SAITOH, M., HIDAKA, H. (**1988**). Two phosphorylated forms of myosin in thrombin-stimulated platelets. *Arch. Biochem. Biophys.* **261**, 235–240.

NISHIKAWA, M., TANAKA, T., HIDAKA, H. (**1980**). Ca^{2+}-calmodulin-dependent phosphorylation and platelet secretion. *Nature* **287**, 863–865.

NURDEN, A. T. (**1999**). Inherited abnormalities of platelets. *Thromb Haemost* **82**, 468–480.

ODA, Y., NAGASU, T., CHAIT, B. T. (**2001**). Enrichment analysis of phosphorylated proteins as a tool for probing the phosphoproteome. *Nat. Biotechnol.* **19**, 379–382.

PRESEK, P., MARTINSON, E. A. (1997). Platelet protein tyrosine kinases cyclases. In VON BRUCHHAUSEN, F., WALTER, U. (Eds.), *Platelets and their Factors.* Springer-Verlag, Berlin, pp. 263–296.

PETRICOIN, E., WULFKUHLE, J., ESPINA, V., LIOTTA, L. A. (2004). Clinical proteomics: revolutionizing disease detection and patient tailoring therapy. *J. Proteome Res.* 3, 209–217.

QUEMENEUR, E., GUTHAPFEL, R., GUEGUEN, P. (1994). *J. Biol. Chem.* 269, 5485–5488.

RAFFEL, G. D., PARMAR, K., ROSENBERG, N. (1996). *In vivo* association of v-Abl with Shc mediated by a non-phosphotyrosine-dependent SH2 interaction. *J. Biol. Chem.* 271, 4640–4645.

SALOMON, A. R., FICARRO, S. B., BRILL, L. M., BRINKER, A., PHUNG, Q. T., ERICSON, C., SAUER, K., BROCK, A., HORN, D. M., SCHULTZ, P. G., et al. (2003). Profiling of tyrosine phosphorylation pathways in human cells using mass spectrometry. *Proc. Natl. Acad. Sci. USA* 100, 443–448.

SCHULENBERG, B., GOODMAN, T. N., AGGELER, R., CAPALDI, R. A., PATTON, W. F. (2004). Characterization of dynamic and steady-state protein phosphorylation using a fluorescent phosphoprotein gel stain and mass spectrometry. *Electrophoresis* 15, 2526–2532.

SCHUURING, E., VERHOEVEN, E., LITVINOV, S., MICHALIDES, R. J. (1993). The product of the EMS1 gene, amplified and over-expressed in human carcinomas, is homologous to a v-src substrate and is located in cell-substratum contact sites. *Mol. Cell Biol.* 13, 2891–2898.

SHI, Y., KOTLYAROV, A., LAASS, K., GRUBER, A. D., BUTT, E., MARCUS, K., MEYER, H. E., FRIEDRICH, A., VOLK, H. D., GAESTEL, M. (2003). Elimination of protein kinase MK5/PRAK activity by targeted homologous recombination. *Mol. Cell Biol.* 23, 7732–7741.

STEEN, H., MANN, M. (2002). A new derivatization strategy for the analysis of phosphopeptides by precursor ion scanning in positive ion mode. *J. Am. Soc. Mass Spectrom.* 13, 996–1003.

STOKOE, D., ENGEL, K., CAMPBELL, D. G., COHEN, P., GAESTEL, M. (1992). Identification of MAPKAP kinase 2 as a major enzyme responsible for the phosphorylation of the small mammalian heat shock proteins. *FEBS Lett.* 313, 307–313.

TORTI, M., LAPETINA, E. G. (1994). Structure and function of rap proteins in human platelets. *Thromb Haemost* 71, 533–543.

VAN DAMME, H., BROK, H., SCHUURING-SCHOLTES, E., SCHUURING, E. (1997). The redistribution of cortactin into cell-matrix contact sites in human carcinoma cells with 11q13 amplification is associated with both overexpression and posttranslational modification. *J. Biol. Chem.* 272, 7374–7380.

WATSON, S. P. (1997). Protein phosphatases in platelet function. In VON BRUCHHAUSEN, F., WALTER, U. (Eds.), *Platelets and their Factors.* Springer-Verlag, Berlin, pp. 297–325.

ZHOU, H., WATTS, J. D., AEBERSOLD, R. (2001). A systematic approach to the analysis of protein phosphorylation. *Nat. Biotechnol.* 19, 375–378.

12
Biomarker Discovery in Renal Cell Carcinoma Applying Proteome-Based Studies in Combination with Serology

Barbara Seliger and Roland Kellner

Abstract

Multiparametric biomarker analysis in combination with artificial neural networks and pattern recognition represent a most promising technology for the identification and monitoring of cancer. This might lead to early detection of disease which generally also provides a better outcome for tumor patients. In addition, the implementation of this multivariable approach will result in a better understanding of the molecular basis of cancer, as well as the definition of novel cancer drugs and their development. The most promising benefits are the design of molecular targeted cancer therapies leading to individually tailored therapies for cancer patients. The translation of molecular insights into useful therapeutic approaches represents a challenge due to its high complexity, which needs an integrative collaboration of pharmaceutical, biotechnological, academic, governmental and patient advocacy groups in order to translate experimental data into rational therapies. Thus, one major goal is the integration of complementary efforts including molecular diagnostics and therapeutic technologies that require the development and implementation of a tissue bank consisting of tumor lesions, autologous normal tissues, cell lines, serum, blood samples and urine as well as the ethics commission to work with this material. These collaborative efforts will lead to both the discovery and validation of mechanism-based biomarkers for improved application in prognosis, diagnosis and therapeutic intervention. So far, only a few markers of limited value are available for monitoring of renal cell carcinoma (RCC) patients. In addition, an effective therapy for this metastatic disease is currently not available. Therefore biomarkers are urgently needed for the monitoring and treatment of this disease. The identification of biomarkers in kidney cancer might result not only in the early detection of localized high-risk patients or early disease recurrence after nephrectomy, but also in their use as therapeutic targets. These RCC-specific biomarkers will be discovered employing a combination of molecular, cellular and protein technologies, and/or functional analysis focused on disease mechanisms using RCC cell lines, primary and/or metastatic tumor lesions and biological fluids such as serum, peripheral blood

Proteomics in Drug Research
Edited by M. Hamacher, K. Marcus, K. Stühler, A. van Hall, B. Warscheid, H. E. Meyer
Copyright © 2006 Wiley-VCH Verlag GmbH & Co. KGaA, Weinheim
ISBN: 3-527-31226-9

mononuclear cells and urine obtained from RCC patients. The resulting candidate markers will be carefully validated in independent patient populations and on a broad number of RCC lesions of distinct subtypes, grading and staging, as well as in a healthy control group to ensure their specificity. This review focuses on the identification of RCC-specific biomarkers using proteome-based methods, in particular classical two-dimensional gel-electrophoresis-based proteome analysis and PROTEOMEX, representing a combination of proteomics and serology.

12.1
Introduction

12.1.1
Renal Cell Carcinoma

Renal cell carcinoma (RCC) is the most common kidney malignancy and accounts for approximately 2% of all cancers worldwide. Patients are diagnosed typically between the ages of 50 and 75 years. Approximately 20–30% of patients have already developed metastases at the time of diagnosis. Patients with metastatic disease have a 5-year survival rate of 5%, whereas patients with localized disease have a survival rate ranging between 50 and 90% (Motzer et al., 1996). In this context it is noteworthy that the mortality rate of RCC patients is higher than in any other urological malignancy.

According to the Union Internationale Contre le Cancer (UICC) and the American Joint Committee on Cancer (AJCC), RCC can be histopathologically classified into the most prominent clear cell type, which is associated with the occurrence of highly specific deletion of chromosome 3p and mutation of the von Hippel Lindau (*VHL*) gene as well as the chromophobe, papillary/chromophil or collecting duct types (Kovacs, 1999).

RCC is resistant to standard radiation and chemotherapy; only nephrectomy is successfully implemented at the early stages of the disease. Immunotherapy with cytokines such as interleukin-2 and interferon (IFN)-α increases survival of patients and in some cases can lead to tumor regression (Nathan and Eisen, 2002). These findings have therefore stimulated interest in the effects of the immune system on RCC (Mancuso and Sternberg, 2005). Currently, there exists no satisfactory tumor marker to screen for RCC. Early diagnosis relies mainly on unrelated radiographic screening (Kessler et al., 1994). Owing to the low 5-year survival rate of RCC patients, the limited success of conventional therapies and radiographic evaluation being the only screening approach, novel diagnostic and prognostic markers as well as therapeutic targets are urgently needed for the management of this malignancy.

12.2
Rational Approaches Used for Biomarker Discovery

Recent technological advances have permitted the development and identification of biomarkers important for the initiation and progression of malignancies. A number of molecular profiling approaches have been implemented that are based on genomics, transcriptomics, epigenetics and/or proteomics in order to define the molecular signature of an individual cancer, such as gene and protein expression patterns, DNA alterations like methylation, single nucleotide polymorphism, microsatellite instability and chromosomal aberrations (Petricoin et al., 2002; Petricoin and Liotta, 2004).

A multimodal approach was first successfully implemented for screening of ovarian cancer patients (Jacobs et al., 1999). Based on these results it is suggested that the identification and application of early candidate biomarkers in populations of high risk for RCC could present an efficient approach for the determination of screening markers.

A major goal for the use of biomarkers is their histological specificity and their efficient measurement, which enables determination of the onset as well as following of the progression of tumors (Hanash, 2004). At the genomic level, comparative genomic hybridization and spectral karyotyping for detection of chromosomal aberrations have been employed (Veldman et al., 1997). cDNA microarrays have been widely used to determine the transcriptional profiles associated with various tumor types, stages and grading. A comprehensive catalogue of changes in gene expression accompanying the development progression of RCC has been performed and at least partially correlated with histopathological, cytogenetic and clinical findings (Huber et al., 2002; Wang et al., 2004; Shi et al., 2005; Yao et al., 2005). In addition, cell-based technologies have been implemented to detect and quantify circulating tumor and endothelial cells (Shaked et al., 2004). However, the development of various malignancies is frequently associated with posttranscriptional or posttranslational modifications that can only be detected at the protein level. Therefore, proteome-based technologies have evolved to be a significant tool and are becoming increasingly important for the determination of protein expression profiles in cancer studies.

Single-protein biomarkers, the general protein signature or protein patterns containing multiple biomarkers are rapidly progressing toward becoming clinically relevant tools. In this context, three major proteome-based strategies were used for the identification of biomarkers in tumor biopsies or in body fluids such as serum, blood samples, lavage or plasma (Petricoin and Liotta, 2004). These include classical two-dimensional gel electrophoresis (2DE)-based proteomics, PROTEOMEX and the ProteinChip system manufactured by Ciphergen Biosystems (Figure 12.1) (Lichtenfels et al., 2003; Seliger et al., 2003; Merchant and Weinberger, 2000). The latter system employs surface-enhanced laser desorption and ionization time-of-flight (SELDI–TOF).

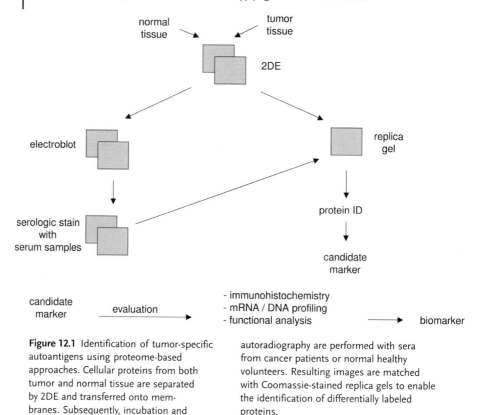

Figure 12.1 Identification of tumor-specific autoantigens using proteome-based approaches. Cellular proteins from both tumor and normal tissue are separated by 2DE and transferred onto membranes. Subsequently, incubation and autoradiography are performed with sera from cancer patients or normal healthy volunteers. Resulting images are matched with Coomassie-stained replica gels to enable the identification of differentially labeled proteins.

12.3
Advantages of Different Proteome-Based Technologies for the Identification of Biomarkers

Classical 2DE in combination with chromatographic separation is still the most powerful tool for fractionating and subsequently identifying complex protein mixtures. It allows the detection of the differential expression pattern of proteins in solid tumor samples, the direct identification of pathological changes in specific organs or tissues and the comprehensive display of distinctive proteomes in fluids, including urine (Pieper et al., 2003a, 2004; Yoshida et al., 2005).

Since tumor tissue is the primary manifestation of cancer, it contains the most immediate information and is the preferred material to investigate molecular targets. Experimental approaches will benefit from the enhanced protein concentration compared to body fluids. However, one major drawback of analyzing tumor biopsies is their limited availability and their heterogenicity which often requires microdissection. Alternatively, body fluids are the preferred clinical sources because they are most readily available in a noninvasive and continuous manner.

Serum proteomics is an attractive technology that allows the characterization of a huge number of individual serum proteins/peptides, which may enable the discovery of an increasing number of reliable disease markers (Park et al., 2004; Petricoin et al., 2002; Pieper et al., 2003a; Rapkiewicz et al., 2004). Furthermore, the urinary proteome detects changes in the protein pattern as a result of diseases or drug toxicity, particular of those affecting the kidney and the urogenital tract (Pieper et al., 2004; Wu et al., 2004). Disease-related information from human plasma and urine on the proteomic level is essential for the examination of diagnostic markers or therapeutic targets because relevant posttranscriptional alterations cannot be detected at the transcript level (Anderson and Anderson, 2002; Anderson et al., 2004). Unfortunately, body fluids are extremely complex protein mixtures and investigations are hampered by several facts: (1) regarding serum proteomics, the high protein content of serum is dominated by a small number of serum components such as albumin, transferrin, immunoglobulins, lipoproteins, etc., comprising more than 90% of total protein; (2) an extraordinarily dynamic range of protein concentrations exceeds the analytical capabilities of traditional proteomics and the detection of low-abundance serum proteins becomes extremely challenging; and (3) sensitivity is crucial to identifying these low-abundance proteins that represent the most promising marker candidates. For example, serum albumin ($\sim$50 mg mL^{-1}) exceeds the serum concentration of a well known marker, the prostate specific antigen (1 ng mL^{-1}) by a factor of 50 000 000. Depletion of high-abundance proteins and reducing the complexity of serum and/or plasma protein mixtures is a mandatory part of the strategy to analyze the serum proteome (Tam et al., 2004; Pieper et al., 2003b). Similarly, a protein fractionation strategy enriching proteins in the urine is required for the identification of low-abundance proteins in the urine. High levels of posttranslational modifications were found in most urinary proteins. Only one-third of the proteins identified in the urinary proteome are described as classical plasma proteins in circulation (Pieper et al., 2004).

In addition to classical proteomics, an attractive alternative for biomarker discovery is PROTEOMEX, a combination of classical proteome analysis and serology. This method also termed serological proteome analysis (SERPA) has been successfully employed for the characterization of a number of markers in various cancers including RCC (Kellner et al., 2002; Seliger et al., 2003; Lichtenfels et al., 2003; Klade et al., 2001). Furthermore, SELDI allows the detection of proteins and the analysis of complex protein patterns directly from crude extracts due to specific capture of proteins on a functionalized surface (protein chip assay) with subsequent detection of retained proteins by matrix-assisted laser desorption ionization–time-of-flight mass spectrometry (MALDI–TOF). This technology has also been successfully implemented for the identification of diagnostic markers in RCC (Won et al., 2003; Fetsch et al., 2002; Tolson et al., 2004; Junker et al., 2005; Walther et al., 2004).

12.4
Type of Biomarker

Biomarker determination can help to explain empirical results of preclinical research or clinical trials by relating the effects of intervention or progression on molecular and cellular pathways. An expert working group was convened by the National Institutes of Health and proposed terms, definitions and a conceptual model for the use of biomarkers (Biomarkers Definitions Working Group, 2001). Stratification of biomarkers allow their division to different stages of disease, development and/or progression. Monitoring of the early period of disease development at the time at which diagnosis is performed and during the initiated treatment has to be accompanied by the evaluation of the follow-up period of patients to diagnose and treat any local recurrence, distant relapse or metastases (Park et al., 2004). This might allow the identification of biomarkers which could be employed for the detection of individuals at risk for developing the disease (markers of risk), of individuals with early signs of disease or recurrence (early detection markers), and for the prediction of tumor biology or individual therapy response in patients (prognostic or selection markers). Furthermore, some of these markers might be developed as therapeutic targets and used for the design of novel therapies (therapeutic markers) (Park et al., 2004).

Molecular risk biomarkers could detect systemic or local changes indicating that the carrier of this specific biomarker is at higher risk for disease development in the future. Examples include the presence of human papillomavirus in the cervix, which is associated with a higher risk for the development of cervical cancer, the association between *Helicobacter pylori* and gastric cancer as well as EBV and nasopharyngeal carcinoma. The drawback of these markers is that they are not helpful for the detection of early disease.

Early disease markers allow the identification of patients who have already developed the clinically detectable disease at its early stages, implying an easier treatment (Adam et al., 2002; Iwaki et al., 2004). These include, for example, the prostate-specific antigen CA-125 as a predictor of prognosis in both prostate and ovarian cancer (Santala et al., 2004) and HER-2, a prognostic marker identified in many tumor specimens of distinct histology. In all cases, the prognostic and predictive markers must be associated with the histological examination of tumor tissue. Early detection biomarkers often determined in accessible fluids like serum or urine represent specific changes in urine or plasma proteins. In contrast, RNA and/or protein profiles of tumor lesions in comparison to normal kidney epithelium that correlate with local and/or distant disease relapse or progression would be very useful for the detection of therapeutic or preventive markers in RCC (Sarto et al., 1997; Ferguson et al., 2004). These markers may serve either as surrogate end-points for preliminary studies of treatment efficacy or may guide early chemotherapeutic and chemopreventive intervention.

Some serological markers have been used as prognostic determinants, such as ferritin, erythropoetin, calcium and renin. These markers include the proliferating cell nuclear antigen Ki-67, specific cytogenetic alterations, P-glycoprotein, p53 and myc mutations and elevated levels of β_2-microglobulin, interleukin-6, γ-enolase and E-cadherin. However, although these markers have some potential utility regarding the aggressive behavior of a tumor of distinct origin, their use is limited in diagnosing RCC (Shimazui et al., 1998), although Ki-67 has been demonstrated as a potential prognostic biomarker for RCC to improve survival prediction and classification of kidney cancer (Bui et al., 2004).

Recent comparison of gene expression profiles of von Hippel-Lindau+ versus von Hippel-Lindau-cell lines obtained from the clear cell RCC subtype identified proteins that might serve as candidate molecular markers. Inactivation of VHL is a hallmark in most sporadic clear cell RCC and it occurs early in renal carcinogenesis (Skates and Iliopoulus., 2004; Latif et al., 1993). Furthermore, the profile of secreted proteins could be directly ?AQ4?determined by analyzing comparatively the genomic and proteomic patterns of these cell lines as well as their conditioned tissue culture supernatants (Ferguson et al., 2004). A combination of cDNA microarray and proteome analyses appears reasonable since not every difference at the transcriptome level will translate into differences at the protein level, whereas posttranscriptional/posttranslational modifications were not detectable by transcriptomics.

12.5
Proteome Analysis of Renal Cell Carcinoma Cell Lines and Biopsies

Comprehensive proteome analysis of both cell lines and tumor specimens of RCC is just beginning (Unwin et al., 2003b; Ferguson et al., 2004). Some routine laboratory parameters are predictors of survival in RCC, such as serum calcium, alkaline phosphatase, hemoglobin and the erythrocyte sedimentation rate. In addition, some poor markers of RCC have been demonstrated including the proliferation antigen Ki-67, PCNA, topoisomerase II-a and p100. Caveolin-1 overexpression appears to predict poor disease survival in patients with RCC and may serve as a useful prognostic marker, but this marker has only been identified by immunohistochemistry and not by proteome-based analysis (Campbell et al., 2003). Furthermore, Konety and coworkers (1998) identified nuclear matrix proteins which are specifically up-regulated in cell lines and human renal cancer. The presence of these proteins can be determined quantitatively in the serum or urine of RCC patients, demonstrating that these proteins might serve as biomarkers for early detection of this disease. The appearance/overexpression of Ki-67 and the carboanhydrase 9 (CAIX) might improve the survival prediction and classification of kidney cancers as demonstrated by immunohistochemical analysis of a large series of RCC lesions obtained from RCC patients treated with nephrectomy (Bui et al., 2004). Interestingly, both markers have also been identified by classical proteome analysis (reviewed by Seliger et al., 2003). In addition, the molecular analysis of serum by SELDI mass spectrometry led to the characterization of other disease-related proteins such as the serum amyloid α-1 (SAA-1) and haptoglobin α, as well as yet unidentified markers (Tolson et al., 2004).

Regarding the classical prognostic factors of RCC, the combined evaluation of the histological grade, histological type, performance status, number and locations of metastatic sites and the time-to-appearance of metastasis still represents the best parameter. Therefore, the only striking advance during the last 5 years has been the proven contribution of radical nephrectomy for metastatic disease patients with good performance status (Mejean et al., 2003).

A novel strategy called random forest clustering was employed for tumor profiling based on tissue microarray data of RCC. By analyzing various tumor markers a final subclassification of clear cell carcinoma patients was demonstrated. A subgroup has been shown with long surviving clear cell patients carrying the distinct molecular profile and novel tumor subclasses in low grade clear cell patients that could not be explained by any clinical pathological variables (Shi et al., 2005). Recently, the gene expression profiles of 34 renal cell carcinomas and 9 corresponding kidney samples using oligonucleotide microarray demonstrated a high up-regulation of the adipose differentiation-related protein (ADFP) and nicotinamide *N*-methyltransferase (NNMT) expression in clear cell carcinoma (Yao et al., 2005). Validation by quantitative reverse transcription-polymerase chain reaction (RT-PCR) revealed a more frequent up-regulation of both markers in the clear cell RCC subtype than in other non-clear-cell RCC subtypes. It is noteworthy that ADFP, a lipid-storage-associated protein, might be regulated by the *VHL* hypoxy-inducible factor (HIF-1). In addition, clear cell RCC lesions contain abundant lipids and cholesterols, suggesting that the ADFP upregulation following *VHL* inactivation is involved in the morphological appearance of clear cell RCC. Thus, ADFP might be a useful prognostic biomarker for patients with RCC. This also holds for NNMT which is differentially expressed in kidney tumors as demonstrated by classical proteomics as well as PROTEOMEX. In addition, glutathione S-transferase (GST)-α is of diagnostic value in RCC. This biomarker was found by proteome analysis and is differentially expressed in RCC compared to kidney epithelium (Sarto et al., 1997, 1999; Lichtenfels et al., 2003). Immuno-histochemical analysis using a GST-α-specific antibody demonstrated a significant increase of GST expression in the majority of clear cell carcinoma, whereas other renal tumor subtypes did not exhibit a significant immunoreactivity. These data have been confirmed by Chuang and coworkers (2005), suggesting that GST-α represents a specific marker for clear cell carcinoma (Chuang et al., 2005).

Various epithelial adhesion molecules have been demonstrated to be differentially expressed in RCC by proteome analysis (Seliger et al., 2003). EpCam represents a marker that is overexpressed in a broad range of carcinomas, and in particular with a high frequency in clear cell RCC. Its expression might allow discrimination between clear cell carcinoma, chromophobe and oncocytoma (Went et al., 2005). In addition, as EpCam appears to be associated with poor prognosis it might be used as potential therapeutic target in RCC patients. Similarly, the adhesion molecule CD44 and its isoforms are of prognostic value in RCC (Lucin et al., 2004).

A combination of conventional proteome analysis with serology has been developed as a promising experimental approach for the discovery of serological markers in RCC (Seliger et al., 2003; Klade et al., 2001). This analysis is achieved by 2DE followed by gel blotting and incubation of the membrane with serum (Figure 12.1). Reactive proteins are identified by mass spectrometry. To identify proteins reacting with IgG molecules in sera, total protein extracts obtained from cell lines or primary/metastatic lesions of RCC and from cell lines/tissues obtained from autologous renal epithelium were subjected to 2DE combined with immuno-blotting using serum samples of RCC patients and healthy volunteers. Techno-logical advantages and limitations of PROTEOMEX and proteomics have recently been discussed in detail (Seliger et al., 2003). So far, more than 400 differentially expressed proteins have been identified by conventional proteomics (Table 12.1) and more than 50 by PROTEOMEX in RCC of different subtypes (Table 12.2). Differentially expressed proteins belong to various protein families including heat shock proteins, annexins, antigen processing genes, regulators of cell growth and cell cycle, chemotherapy resistance genes, oncogenes, tumor suppressor genes,

Table 12.1 RCC biomarkers identified by classical proteomics.
IHC Immunohistochemistry, *NB* northern blot, *WB* western blot.

RCC subtype	Protein name	Protein family	Validation	Literature
Not specified	Enoyl-CoA hydrolase	Metabolic enzymes	RT-PCR	Balabanov et al., 2001
	Aldelyde dehydrogenase I		–	Balabanov et al., 2001
	Aminoacylase-I		–	Balabanov et al., 2001
	α-Glycerol-3-phosphate dehydrogenase		–	Balabanov et al., 2001
Clear cell	Agmatinase		RT–PCR, NB, IHC	Dallmann et al., 2004
	Superoxide dismutase	Enzymes of the redox system	IHC, WB	Sarto et al., 1999, 2001
	Thioredoxin		WB	Lichtenfels et al., 2003
	Glutathione peroxidase		WB	Lichtenfels et al., 2003
	NADH-ubiquinone oxidoreductase		IHC	Sarto et al., 1997, 2001
	hsp27, 60, 75, 90, gp96	Heat shock proteins	WB, IHC	Lichtenfels et al., 2002; Sarto et al., 2004; Tremolada et al., 2005
	Annexin II	Annexins	–	Lichtenfels et al., personal communication
	Annexin IV		WB	
	er60	Chaperone	WB	Lichtenfels et al., 2001
	α-Enolase I	Enzyme		
	MECL 1	Antigen processing	WB	Lichtenfels et al., 2003

adhesion molecules, components of the cytoskeleton and signal transduction pathways, a number of different metabolic enzymes and some unknown proteins (Tables 12.1, 12.2). A functional relevance of distinct protein subgroups in tumor situations has been discussed (Seliger et al., 2003; Lichtenfels et al., 2002; Kellner et al., 2002; Ferguson et al., 2005; Balabanov et al., 2001; Dallmann et al., 2004; Sarto et al., 1999). Using these methodologies it became obvious that the tumor status of RCC is associated with changes in cellular metabolism (Unwin et al., 2003a; Lichtenfels et al., 2003). An example for a typical readout utilizing the PROTEOMEX strategy is given by a group of nine distinct metabolic enzymes

Table 12.2 Biomarkers of clear cell RCC identified by PROTEOMEX. IHC Immunohistochemistry, NB northern blot, WB western blot.

Protein name	Protein family	Validation	Literature
SM 22αtubulin		IHC, NB	Klade et al., 2001
Tropomyosin		–	Kellner et al., 2002
Tubulin	Cytoskeletal	IHC, WB	Kellner et al., 2002
Cytokeratin		IHC, WB	Kellner et al., 2002
Vimentin		IHC, WB	Kellner et al., 2002
Annexin I	Annexins	RT–PCR,	Unwin et al., 2003a;
Annexin IV	Annexins	IHC	Zimmermann et al., 2004
Major vault protein (MVP)	Transporter	IHC, WB	Unwin et al., 2003
GBB1 (G-protein)	Signal transducer	–	Unwin et al., 2003
Carbonic anhydrase I		IHC, WB	Klade et al., 2001; Unwin et al., 2003
Triosephosphate isomerase			Unwin et al., 2003
Mn superoxide dismutase		–	Unwin et al., 2003
α-Enolase		WB	Unwin et al., 2003
Enoyl-CoA-hydratase	Metabolic enzymes	–	Lichtenfels et al., 2003
Thioredoxin			Lichtenfels et al., 2003
α-Casein S1		–	Unwin et al., 2003
Thymidin phosphorylase		–	Unwin et al., 2003
Lactate dehydrogenase		–	Unwin et al., 2003
Triosephosphate isomerase			Lichtenfels et al., 2003
HSP27, 60, 75, 90, gp96	Heat shock protein	WB, IHC	Lichtenfels et al., 2002
Ubiquitin-carboxy terminal hydrolase	Antigen processing	IHC, WB	Lichtenfels et al., 2003
Bip/grp78	Antigen processing	WB	Lichtenfels et al., 2002
Unidentified	–		Unwin et al., 2003

that could be identified. At least one candidate marker, namely thioredoxin, was shown to be twofold up-regulated in RCC cell lines when compared to autologous normal kidney epithelium (Lichtenfels et al., 2003). Furthermore, annexins, which belong to a family of proteins binding phospholipids in a calcium-dependent way, were identified by both classical proteomics and PROTEOMEX (Table 12.2). As shown representatively in Figure 12.2, annexin is differentially expressed in RCC lesions compared to normal kidney epithelium. This suggests a role for annexin IV in tumor dissemination (Zimmermann et al., 2004).

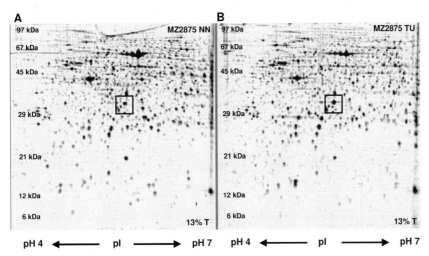

Figure 12.2 Representative 2DE map of RCC biopsy and its corresponding normal kidney epithelium. Proteins obtained from a primary RCC lesion of clear cell subtype of patient MZ2875 (B) as well as from the corresponding normal kidney epithelium (A) were extracted and subjected to 2DE separation. Gels were silver-stained and exposed to image analysis. The marked region is zoomed in Figure 12.3.

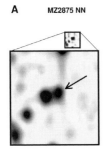
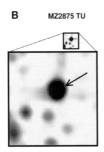

Figure 12.3 Zoomed out window of RCC biopsy MZ2875RC and corresponding normal kidney epithelium. The window shown in Figure 12.2 was zoomed out representing the matching 2DE spot pattern of MZ2875NN (A) and MZ2875RC (B). The protein spot representing annexin IV is marked by *arrows*. Other differentially expressed protein spots are currently investigated.

12.6
Validation of Differentially Expressed Proteins

Alterations in the protein pattern are immanent for mammalian body fluids or tissues. Differential expression profiles for the comparison of distinct samples are to be expected due to the biological variability related to patient age, gender or lifestyle. In order to distinguish disease-related features from false-positive differences the critical validation and assessment of putative markers is of fundamental importance (Zolg and Langen, 2004).

One key issue is the validation of the differential protein expression identified in kidney tumors by proteome-based technologies using RT-PCR analyses and if monoclonal antibodies were available, immunohistochemistry (IHC) and/or western blot analysis. The development of a tissue array comprising RCC lesions and normal kidney epithelium is mandatory for monitoring protein expression. A prerequisite is the availability of samples from both RCC lesions and RCC cell lines in conjunction with corresponding normal kidney epithelium. So far, more than 40 differentially expressed proteins, so-called candidate markers, have been validated by IHC, western blot and/or RT-PCR. One example is representatively shown for annexin IV. A differential expression pattern of this putative marker was also detected in RCC cell lines compared to normal kidney epithelium cell cultures using western blot analysis (Figure 12.4). Often, the validation is hampered by the lack of available antibodies directed against the respective biomarkers. Therefore, RCC biomarker-specific polyclonal or monoclonal antibodies have to be raised for potent immunohistological screens.

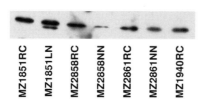

MZ1851RC MZ1851LN MZ2858RC MZ2858NN MZ2861RC MZ2861NN MZ1940RC

Figure 12.4 Heterogeneous annexin IV expression in RCC lesions compared to normal kidney epithelium. Proteins from various RCC cell lines as well as corresponding normal kidney epithelium were subjected to western blot analysis and probed with an anti-annexin-IV-specific antibody. The results demonstrate a higher annexin expression in RCC than in normal kidney epithelium. Furthermore, annexin expression is more pronounced in a metastatic lesion (MZ1851LN) than in the primary tumor (MZ1851RC).

12.7
Conclusions

Many disease processes are associated with quantitative and functional changes in proteins in body fluids (Anderson et al., 2002). The challenge of biomarker research is exemplified by the fact that in the last 10 years only ten new plasma protein diagnostic tests have been approved (Tam et al., 2004). Low-abundance components have to be detected and identified out of a complex mixture of circa 3700 plasma proteins (Pieper et al., 2003b). Detection and quantification of putative markers requires optimized protocols (e.g., for sample storage and handling) as well as the latest technologies (e.g., to gain maximum sensitivity and statistical significant throughput) (Zolg and Langen, 2004).

Proteome-based technologies facilitate the analysis of the entire proteome. So far, there exist only a limited number of studies in kidney cancer, mainly focused on the clear cell subtype of RCC. 2DE-PAGE and mass spectrometry have been successfully employed to identify some novel markers. Their relevance in terms of prognostic and diagnostic value and their implementation as therapeutic targets is currently being analyzed. If such marker candidates can be verified regarding their clinical sensitivity and specificity, a product might be developed that allows the rapid monitoring of RCC patients.

References

ADAM B-L, QU, Y., DAVIS, J. W., WARD, M. D., CLEMENTS, M. A., CAZARES, L. H., SEMMES, O. J., SCHELLHAMMER, P. F., YASUI, Y., FENG, Z., et al. (**2002**). Serum protein fingerprinting coupled with a pattern-matching algorithm distinguishes prostate cancer from benign prostate hyperplasia and healthy men. *Cancer Res.* **62**, 3609–3614.

ANDERSON, N. L., ANDERSON, N. G. (**2002**). The human plasma proteome: history, character, and diagnostic prospects. *Mol. Cell Proteomics* **1**, 845–867.

ANDERSON, N. L., POLANSKI, M., PIEPER, R., GATLIN, T., TIRUMALAI, R. S., CONRADS, T. P., VEENSTRA, T. D., ADKINS, J. N., POUNDS, J. G., FAGAN, R., et al. (**2004**). The human plasma proteome: a nonredundant list developed by combination of four separate sources. *Mol. Cell Proteomics* **3**, 311–326.

BALABANOV, S., ZIMMERMANN, U., PROTZEL, C., SCHARF, C., KLEBINGAT, K. J., WALTHER, R. (**2001**). Tumour-related enzyme alterations in the clear cell type

of human renal cell carcinoma identified by two-dimensional gel electrophoresis. *Eur. J. Biochem.* **268**, 5977–5980.

BIOMARKERS DEFINITIONS WORKING GROUP (**2001**). Biomarkers and surrogate endpoints: preferred definitions and conceptual framework. *Clin. Pharmacol. Ther.* **69**, 89–95.

BUI, M. H., VISAPAA, H., SELIGSON, D., KIM, H., HAN, K. R., HUANG, Y., HORVATH, S., STANBRIDGE, E. J., PALOTIE, A., FIGLIN, R. A., et al. (**2004**). Prognostic value of carbonic anhydrase IX and Ki67 as predictors of survival for renal clear cell carcinoma. *J. Urol.* **171**, 2461–2466.

CAMPBELL, L., GUMBLETON, M., GRIFFITHS, D. F. (**2003**). Caveolin-1 overexpression predicts poor disease-free survival of patients with clinically confined renal cell carcinoma. *Br. J. Cancer* **89**, 1909–1913.

CHUANG, S. T., CHU, P., SUGIMURA, J., TRETIAKOVA, M. S., PAPAVERO, V., WANG, K., TAN, M., LIN, F., TEH, B. T.,

YANG, X. (2005). Overexpression of glutathione S-transferase alpha in clear cell renal cell carcinoma. *Am. J. Clin. Pathol.* **123**, 421–429.

DALLMANN, K., JUNKER, H., BALABANOV, S., ZIMMERMANN, U., GIEBEL, J., WALTHER, R. (2004). Human agmatinase is diminished in the clear cell type of renal cell carcinoma. *Int. J. Cancer* **108**, 342–347.

FERGUSON, R. E., JACKSON, S. M., STANLEY, A. J., JOYCE, A. D., HARNDEN, P., MORRISON, E. E., PATEL, P. M., PHILLIPS, R. M., SELBY, P. J., BANKS, R. E. (2005). Intrinsic chemotherapy resistance to the tubulin-binding antimitotic agents in renal cell carcinoma. *Int. J. Cancer* **115**, 155–163.

FERGUSON, R. E., SELBY, P. J., BANKS, R. E. (2004). Proteomic studies in urological malignancies. *Contrib. Nephrol.* **141**, 257–279.

FETSCH, P. A., SIMONE, N. L., BRYANT-GREENWOOD, P. K., MARINCOLA, F. M., FILIE, A. C., PETRICOIN, E. F., LIOTTA, L. A., ABATI, A. (2002). Proteomic evaluation of archival cytologic material using SELDI affinity mass spectrometry: potential for diagnostic applications. *Am. J. Clin. Pathol.* **118**, 870–876.

HANASH, S. (2004). Integrated global profiling of cancer. *Nat. Rev. Cancer* **4**, 638–644.

HUBER, W., BOER, J. M., VON HEYDEBRECK, A., GUNAWAN, B., VINGRON, M., FUZESI, L., POUSTKA, A., SULTMANN, H. (2002). Transcription profiling of renal cell carcinoma. *Verh. Dtsch. Ges. Pathol.* **86**, 153–164.

IWAKI, H., KAGEYAMA, S., ISONO, T., WAKABAYASHI, Y., OKADA, Y., YOSHIMURA, K., TERAI, A., ARAI, Y., IWAMURA, H., KAWAKITA, M., et al. (2004). Diagnostic potential in bladder cancer of a panel of tumor markers (calreticulin, gamma-synuclein, and catechol-*o*-methyltransferase) identified by proteomic analysis. *Cancer Sci.* **95**, 955–961.

JACOBS, I. J., SKATES, S. J., MACDONALD, N., MENON, U., ROSENTHAL, A. N., DAVIES, A. R., WOOLAS, R., JEYARAJAH, A. R., SIBLEY, K., LOWE, D. G., et al. (1999). Screening for ovarian cancer: a pilot randomised controlled trial. *Lancet* **353**, 1207–1210.

JUNKER, K., GNEIST, J., MELLE, C., DRIESCH, D., SCHUBERT, J., CLAUSSEN, U., EGGELING, F. (2005). Identification of protein pattern in kidney cancer using ProteinChip arrays and bioinformatics. *Int. J. Mol. Med.* **15**, 285–290.

KELLNER, R., LICHTENFELS, R., ATKINS, D., BUKUR, J., ACKERMANN, A., BECK, J., BRENNER, W., MELCHIOR, S., SELIGER, B. (2002). Targeting of tumor-associated antigens in renal cell carcinoma using proteome-based analysis and their clinical significance. *Proteomics* **2**, 1743–1751.

KESSLER, O., MUKAMEL, E., HADAR, H., GILLON, G., KONECHEZKY, M., SERVADIO, C. (1994). Effect of improved diagnosis of renal cell carcinoma on the course of the disease. *J. Surg. Oncol.* **57**, 201–204.

KLADEM, C. S., VOSS, T., KRYSTEK, E., AHORN, H., ZATLOUKAL, K., PUMMER, K., ADOLF, G. R. (2001). Identification of tumor antigens in renal call carcinoma by serological proteome analysis. *Proteomics* **1**, 890–898.

KONETY, B. R., NANGIA, A. K., NGUYEN, T. S., VEITMEIER, B. N., DHIR, R., ACIERNO, J. S., BECICH, M. J., HREBINKO, R. L., GETZENBERG, R. H. (1998). Identification of nuclear matrix protein alterations associated with renal cell carcinoma. *J. Urol.* **159**, 1359–1363.

KOVACS, G. (1999). Molecular genetics and diagnosis of renal cell tumors. *Urol. A* **38**, 433–441.

LATIF, F., TORY, K., GNARRA, J., YAO, M., DUH, F. M., ORCUTT, M. L., STACKHOUSE, T., KUZMIN, I., MODI, W., GEIL, L., et al. (1993). Identification of the von Hippel-Lindau disease tumor suppressor gene. *Science* **260**, 1317–1320.

LICHTENFELS, R., ACKERMANN, A., KELLNER, R., SELIGER, B. (2001). Mapping and expression pattern analysis of key components of the major histo-compatibility complex class I antigen processing and presentation pathway in a representative human renal cell carcinoma cell line. *Electrophoresis* **22**, 1801–1809.

LICHTENFELS, R., KELLNER, R., ATKINS, D., BUKUR, J., ACKERMANN, A., BECK, J., BRENNER, W., MELCHIOR, S., SELIGER, B. (2003). Identification of metabolic enzymes in renal cell carcinoma utilizing PROTEOMEX analyses. *Biochim. Biophys. Acta* **1646**, 21–31.

LICHTENFELS, R., KELLNER, R., BUKUR, J., BECK, J., BRENNER, W., ACKERMANN, A., SELIGER, B. (2002). Heat shock protein expression and anti-heat shock protein reactivity in renal cell carcinoma. *Proteomics* **2**, 561–570.

LUCIN, K., MATUSAN, K., DORDEVIC, G., STIPIC, D. (2004). Prognostic significance of CD44 molecule in renal cell carcinoma. *Croat. Med. J.* **45**, 703–708.

MANCUSO, A., STERNBERG, C. N. (2005). New treatments for metastatic kidney cancer. *Can. J. Urol.* **12**, 351–355.

MEJEAN, A., OUDARD, S., THIOUNN, N. (2003). Prognostic factors of renal cell carcinoma. *J. Urol.* **169**, 821–827.

MERCHANT, M., WEINBERGER, S. R. (2000). Recent advancements in surface-enhanced laser desorption/ionization-time of flight-mass spectrometry. *Electrophoresis* **21**, 1164–1177.

MOTZER, R. J., BANDER, N. H., NANUS, D. M. (1996). Renal-cell carcinoma. *N. Engl. J. Med.* **335**, 865–875.

NATHAN, P. D., EISEN, T. G. (2002). The biological treatment of renal-cell carcinoma and melanoma. *Lancet Oncol.* **3**, 89–96.

PARK, J. W., KERBEL, R. S., KELLOFF, G. J., BARRETT, J. C., CHABNER, B. A., PARKINSON, D. R., PECK, J., RUDDON, R. W., SIGMAN, C. C., SLAMON, D. J. (2004). Rationale for biomarkers and surrogate end points in mechanism-driven oncology drug development. *Clin. Cancer Res.* **10**, 3885–3896.

PETRICOIN, E. F., ARDEKANI, A. M., HITT, B. A., LEVINE, P. J., FUSARO, V. A., STEINBERG, S. M., MILLS, G. B., SIMONE, C., FISHMAN, D. A., KOHN, E. C., et al. (2002). Use of proteomic patterns in serum to identify ovarian cancer. *Lancet* **359**, 572–577.

PETRICOIN, E. F., LIOTTA, L. A. (2004). SELDI–TOF-based serum proteomic pattern diagnostics for early detection of cancer. *Curr. Opin. Biotechnol.* **15**, 24–30.

PIEPER, R., GATLIN, C. L., MAKUSKY, A. J., RUSSO, P. S., SCHATZ, C. R., MILLER, S. S., SU, Q., McGRATH, A. M., ESTOCK, M. A., PARMAR, P. P., et al. (2003a). The human serum proteome: display of nearly 3700 chromatographically separated protein spots on two-dimensional electrophoresis gels and identification of 325 distinct proteins. *Proteomics* **3**, 1345–1364.

PIEPER, R., GATLIN, C. L., McGRATH, A. M., MAKUSKY, A. J., MONDAL, M., SEONARAIN, M., FIELD, E., SCHATZ, C. R., ESTOCK, M. A., AHMED, N., et al. (2004). Characterization of the human urinary proteome: a method for high-resolution display of urinary proteins on two-dimensional electrophoresis gels with a yield of nearly 1400 distinct protein spots. *Proteomics* **4**, 1159–1174.

PIEPER, R., SU, Q., GATLIN, C. L., HUANG, S. T., ANDERSON, N. L., STEINER, S. (2003b). Multi-component immunoaffinity subtraction chromatography: an innovative step towards a comprehensive survey of the human plasma proteome. *Proteomics* **3**, 422–432.

RAPKIEWICZ, A. V., ESPINA, V., PETRICOIN III, E. F., LIOTTA, L. A. (2004). Biomarkers of ovarian tumours. *Eur. J. Cancer* **40**, 2604–2612.

SANTALA, M., RISTELI, J., KAUPPILA, A. (2004). Comparison of carboxyterminal telopeptide of type I collagen (ICTP) and CA 125 as predictors of prognosis in ovarian cancer. *Anticancer Res.* **24**, 1057–1062.

SARTO, C., DEON, C., DORO, G., HOCHSTRASSER, D. F., MOCARELLI, P., SANCHEZ, J. C. (2001). Contribution of proteomics to the molecular analysis of renal cell carcinoma with an emphasis on manganese superoxide dismutase. *Proteomics* **1**, 1288–1294.

SARTO, C., FRUTIGER, S., CAPPELLANO, F., SANCHEZ, J. C., DORO, G., CATANZARO, F., HUGHES, G. J., HOCHSTRASSER, D. F., MOCARELLI, P. (1999). Modified expression of plasma glutathione peroxidase and manganese superoxide dismutase in human renal cell carcinoma. *Electrophoresis* **20**, 3458–3466.

SARTO, C., MAROCCHI, A., SANCHEZ, J. C., GIANNONE, D., FRUTIGER, S., GOLAZ, O., WILKINS, M. R., DORO, G., CAPPELLANO, F., HUGHES, G., et al. (1997). Renal cell carcinoma and normal kidney protein expression. *Electrophoresis* 18, 599–604.

SARTO, C., VALSECCHI, C., MAGNI, F., TREMOLADA, L., ARIZZI, C., CORDANI, N., CASELLATO, S., DORO, G., FAVINI, P., PEREGO, R. A., et al. (2004). Expression of heat shock protein 27 in human renal cell carcinoma. *Proteomics* 4, 2252–2260.

SELIGER, B., LICHTENFELS, R., KELLNER, R. (2003). Detection of renal cell carcinoma-associated markers via proteome- and other 'ome'-based analyses. *Brief Funct. Genomic Proteomic* 3, 194–212.

SHAKED, Y., BERTOLINI, F., MAN, S., ROGERS, M. S., CERVI, D., FOUTZ, T., RAWN, K., VOSKAS, D., DUMONT, D. J., BEN-DAVID, Y., et al. (2005). Genetic heterogeneity of the vasculogenic phenotype parallels angiogenesis; implications for cellular surrogate marker analysis of antiangiogenesis. *Cancer Cell* 7, 101–111.

SHI, T., SELIGSON, D., BELLDEGRUN, A. S., PALOTIE, A., HORVATH, S. (2005). Tumor classification by tissue microarray profiling: random forest clustering applied to renal cell carcinoma. *Mod. Pathol.* 18, 547–557.

SHIMAZUI, T., OOSTERWIJK, E., AKAZA, H., BRINGUIER, P., RUIJTER, E., van BERKEL, H., WAKKA, J. O., AN BOKHOVEN, A., DEBRYUNE, F. M., SCHALKEN, J. A. (1998). Expression of cadherin-6 as a novel diagnostic tool to predict prognosis of patients with E-cadherin-absent renal cell carcinoma. *Clin. Cancer Res.* 4, 2419–2424.

SKATES, S., ILIOPOULOS, O. (2004). Molecular markers for early detection of renal carcinoma. *Clin. Can. Res.* 10, 6296–6301.

TAM, S. W., PIRRO, J., HINERFELD, D. (2004). Depletion and fractionation technologies in plasma proteomic analysis. *Expert Rev. Proteomics* 1, 411–420.

TOLSON, J., BOGUMIL, R., BRUNST, E., BECK, H., ELSNER, R., HUMENY, A., KRATZIN, H., DEEG, M., KUCZYK, M., MUELLER, G. A., et al. (2004). Serum

protein profiling by SELDI mass spectrometry: detection of multiple variants of serum amyloid alpha in renal cancer patients. *Lab. Invest.* 84, 845–856.

TREMOLADA, L., MAGNI, F., VALSECCHI, C., SARTO, C., MOCARELLI, P., PEREGO, R., CORDANI, N., FAVINI, P., GALLI KIENLE, M., SANCHEZ, J. C., et al. (2005). Characterization of heat shock protein 27 phosphorylation sites in renal cell carcinoma. *Proteomics* 5, 788–795.

UNWIN, R. D., CRAVEN, R. A., HARNDEN, P., HANRAHAN, S., TOTTY, N., KNOWLES, M., EARDLEY, I., SELBY, P. J., BANKS, R. E. (2003a). Proteomic changes in renal cancer and co-ordinates demonstration of both the glycolytic and mitochondrial aspects of the Warburg effect. *Proteomics* 3, 1620–1632.

UNWIN, R. D., HARNDEN, P., PAPPIN, D., RAHMAN, D., WHELAN, P., CRAVEN, R. A., SELBY, P. J., BANKS, R. E. (2003b). Serological and proteomic evaluation of antibody responses in the identification of tumor antigens in renal cell carcinoma. *Proteomics* 3, 45–55.

VELDMAN, T., VIGNON, C., SCHROCK, E., ROWLEY, J. D., RIED, T. (1997). Hidden chromosome abnormalities in haemotological malignancies detected by multicolour spectral karyotyping. *Nat. Genet.* 15, 406–410.

WALTHER, R., BALABANOV, S., JUNKER, H., SCHARF, C., ZIMMERMANN, U. (2004). Proteome analysis of renal call carcinoma to develop new strategies in diagnosis and therapy. *Int. J. Clin. Pharmacol. Ther.* 42, 635–636.

WANG, E., LICHTENFELS, R., BUKUR, J., NGALAME, Y., PANELLI, M. C., SELIGER, B., MARINCOLA, F. M. (2004). Ontogeny and oncogenesis balance the transcriptional profile of renal cell cancer. *Cancer Res.* 64, 7279–7287.

WENT, P., DIRNHOFER, S., SALVISBERG, T., AMIN, M. B., LIM, S. D., DIENER, P. A., MOCH, H. (2005). Expression of epithelial cell adhesion molecule (EpCam) in renal epithelial tumors. *Am. J. Surg. Pathol.* 29, 83–88.

WON, Y., SONG, H. J., KANG, T. W., KIM, J. J., HAN, B. D., LEE, S. W. (2003). Pattern analysis of serum proteome distinguishes renal cell carcinoma from other urologic diseases and healthy persons. *Proteomics* **3**, 2310–2316.

Wu, D. L., Wang, W. J., Guan, M., Jin, S. B., Jin, C. R., Zhang, Y. F. (**2004**). Screening urine markers of renal cell carcinoma using SELDI-TOF-MS. *Zhonghua Yi Xue Za Zhi* **84**, 1092–1095.

Yao, M., Tabuchi, H., Nagashima, Y., Baba, M., Nakaigawa, N., Ishiguro, H., Hamada, K., Inayama, Y., Kishida, T., Hattori, K., et al. (**2005**). Gene expression analysis of renal carcinoma: adipose differentiation-related protein as a potential diagnostic and prognostic biomarker for clear-cell renal carcinoma. *J. Pathol.* **205**, 377–387.

Yoshida, Y., Miyazaki, K., Kamiie, J., Sato, M., Okuizumi, S., Kenmochi, A., Kamijo, K., Nabetani, T., Tsugita, A.,

Xu, B., et al. (**2005**). Two-dimensional electrophoretic profiling of normal human kidney glomerulus proteome and construction of an extensible markup language (XML)-based database. *Proteomics* **5**, 1083–1096.

Zimmermann, U., Balabanov, S., Giebel, J., Teller, S., Junker, H., Schmoll, D., Protzel, C., Scharf, C., Kleist, B., Walther, R. (**2004**). Increased expression and altered location of annexin IV in renal clear cell carcinoma: a possible role in tumour dissemination. *Cancer Lett.* **209**, 111–118.

Zolg, J. W., Langen, H. (**2004**). How industry is approaching the search for new diagnostic markers and biomarkers. *Mol. Cell Proteomics* **3**, 345.

13
Studies of Drug Resistance Using Organelle Proteomics

Catherine Fenselau and Zongming Fu

Abstract

This chapter discusses examples of methodology and considerations for examining the changes in protein abundances that cause and accompany acquired drug resistance in human breast cancer cells. The strategy includes isolation and examination of subcellular organelles, which provide smaller mixtures of proteins, and thus more total identifications; consideration of function within the context of organelle biology; and in many cases, new observations about which proteins are present or translocated in which cellular addresses. In the present work quantitative comparisons are made of the relative abundances of proteins in susceptible MCF-7 cells and a subline selected for resistance to mitoxantrone. Metabolic labeling with isotope-labeled arginine and lysine, and isotope ratio measurements made by matrix-assisted laser desorption ionization (MALDI) or electrospray mass spectrometry, were especially useful for difficult studies of plasma membrane proteins. A novel pellicle method was used to isolate enriched membrane proteins. Relative abundances of nuclear proteins were measured using metabolic labeling, and also by comparative densitometric analysis of two-dimensional gel electrophoretic arrays. Around 5% of the proteins analyzed in each organelle were found to have abundances altered by more than twofold. Many of these were associated with the normal function of the organelle. Some correlations could be made between protein changes in the nucleus and membrane, notably in inhibition of apoptosis.

Proteomics in Drug Research
Edited by M. Hamacher, K. Marcus, K. Stühler, A. van Hall, B. Warscheid, H. E. Meyer
Copyright © 2006 Wiley-VCH Verlag GmbH & Co. KGaA, Weinheim
ISBN: 3-527-31226-9

13.1
Introduction

13.1.1
The Clinical Problem and the Proteomics Response

Currently the major obstacle to successful treatment of cancer is the development by the patient of resistance to the chemotherapeutic agents being used. Often this acquired resistance also extends to other drugs with similar chemical structures or mechanisms. Acquired drug resistance is also a serious consideration in other illnesses that require long term therapy, such as diabetes, autoimmune deficiency syndrome, and epilepsy. The acquisition of drug resistance involves selection of target cells – cancer cells in the present chapter – that can survive the xenobiotic challenge because genetic mutations and other cellular changes lead to changes in protein expression, degradation and structures. A few of the major protein contributors to resistance have been identified, for example, ATP-dependent membrane pumps such as P-glycoprotein (Pgp) (Gottesman et al., 2002) and breast cancer resistance protein (BCRP) (Ross et al., 1999), and the general mechanisms are predicted to be metabolic inactivation, increased DNA repair, enhanced efflux, decreased drug uptake, inhibition of apoptosis, and selective loss of cell cycle checkpoints (Gottesman et al., 2002; Ross et al., 1999; Pervaiz and Clement, 2004; Litman et al., 2001; Stein et al., 2004). Some of these mechanisms are illustrated in Figure 13.1, which also illustrates proteins critical to drug resistance function throughout the cell. In most of the cases which have been examined carefully, it has been reported that no single protein can account for the level of resistance

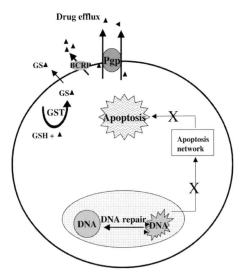

Figure 13.1 Acquired drug resistance in cancer involves a complex mechanism mediated by many proteins. Some of these are illustrated in this schematic cell. Figure provided by Y. Hathout.

observed. Thus a global differential analysis of proteins in drug-resistant and nonresistant cancer cells is needed.

It seems both appropriate and necessary to apply proteomic strategies to learn more about the proteins involved and the mechanisms implicated in acquired resistance. Proteins of interest are identified by quantitative comparison of their abundances and structures in drug-resistant cell lines or tissue samples and drug-susceptible or control samples (Hathout et al., 2002, 2004; Brown and Fenselau, 2004; Fu and Fenselau, in press; Rahbar and Fenselau, in press; Castagna et al., 2004; Lage, 2004). Ultimate objectives may include increased understanding of cell biology, identification of new drug targets, or identification of markers to guide individualized therapy.

13.2
Objectives and Experimental Design

Toward the above objectives, and also that of developing new methods for proteomic investigation, this laboratory has undertaken the identification of proteins whose abundances or primary structures are modified in drug-resistant strains of the human breast cancer cell line MCF-7. The parental cell line and several strains selected by exposure to different anti-cancer drugs were provided by K. H. Cowan (Eppley Cancer Center, University of Nebraska), P. Gutierrez and D. D. Ross, (Greenebaum Cancer Center, University of Maryland), and J. A. Moscow, (Department of Pediatrics, University of Kentucky).

13.2.1
The Cell Lines

In this chapter we will discuss comparative proteomic studies of an MCF-7 human breast cancer cell line (Soule et al., 1973) resistant to mitoxantrone (Nakagawa et al., 1992) as an example of the application of organelle proteomics to study acquired drug resistance. Mitoxantrone is used to treat breast (and other) cancer and was designed to act in the nucleus by inhibiting topoisomerase II. MCF-7 cells selected for resistance to mitoxantrone have also been found to exhibit resistance to some other therapeutic agents, including etoposide, doxorubicin (Nakagawa et al., 1992), and topotecan (Erlichman et al., 2001). The structures of these drugs are shown in Figure 13.2, and all contain a fused planar ring moiety. Like mitoxantrone, etoposide acts by inhibiting topoisomerase II (Volk et al., 2000), while doxorubicin is thought to induce apoptosis by introducing oxidative stress (Schneider et al., 1994), and topotecan is an inhibitor of topoisomerase I. Despite these different modes of action, these four drugs are all substrates for the overexpressed BCRP pump, as one of the mechanisms contributing to their resistance. MCF-7 cells selected for resistance to mitoxantrone have also been shown to be resistant to methotrexate, which is reported not to be a substrate for BCRP (Gewirtz, 1999), but to be detoxified by other cellular proteins.

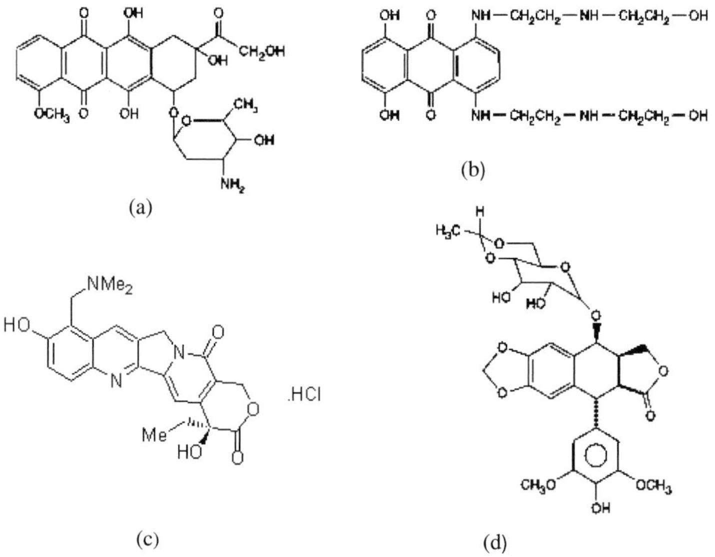

Figure 13.2 Structures of a doxorubicin, b mitoxantrone, c topotecan and d etoposide.

13.2.2
Organelle Isolation

The strategy employed is to fractionate cells into organelles and examine the proteins in each organelle. This approach provides smaller mixtures of proteins to be examined, with the prospect of analyzing more proteins overall, allows observations to be correlated within the frame of reference of organelle biology, and occasionally allows the contribution of new observations about which proteins occur in which part of the cell, that is, annotation of the human genome.

13.2.2.1 Criteria for Isolation
The literature contains many methods for isolation of cytosol, mitochondria, and other organelles. In our ongoing work, several procedures are evaluated for isolation of each organelle, using both physiological (e.g., electron microscopy), and biochemical criteria such as western blots. In addition, protein identifications by mass spectrometry (MS) help to evaluate the purity of the isolated organelles and the suitability of the isolation method. The presence of many proteins that have been classically assigned to other organelles, e.g., cytosolic proteins in a membrane preparation, would suggest that the isolation has not provided a pure organelle sample. Alternatively, the identification of a mitochondrial protein in a nuclear fraction, for example, could indicate that the protein is present in the nucleus as well as in mitochondria. When this kind of observation is made by several different laboratories, consideration should probably be given to reassigning the locations of the protein. Similarly, observations of isoforms and modified proteins using proteomics strategies can augment the protein atlas.

13.2.2.2 Plasma Membrane Isolation

The plasma membrane is one of the most important subcellular organelles to study when considering drug–protein interactions, with a multiplicity of proteins that are potential drug targets, and also a cohort of proteins that are already known to contribute to drug resistance. It is also one of the most difficult subcellular organelles to isolate in pure form. In part this reflects its centrifugal similarity to other membranous parts of the cell, in part the diffuse boundary on the cytosolic side, to which structural, signaling and other proteins are attached. As a consequence, strategies are under development, which take advantage of the unique accessibility of the exterior surface of cells. Alkylating agents that react with functional groups on the exterior side of the plasma membrane have been linked to biotin affinity tags to permit subsequent purification of alkylated proteins and chunks of membrane on, e.g., beads carrying immobilized streptavidin. One strategy incorporating biotin tagging is illustrated in Figure 13.3. This approach has been realized with reagents for lysine side chains (Shin et al., 2003; Peirce et al., 2004; Zhao et al., 2004) and sulfhydryl groups (Laragione et al., 2003). Although it has not provided pure membranes, its application has provided enriched plasma membranes and has expanded the inventory of plasma membrane proteins to include, for example, several chaperone proteins (Shin et al., 2003; Peirce et al., 2004).

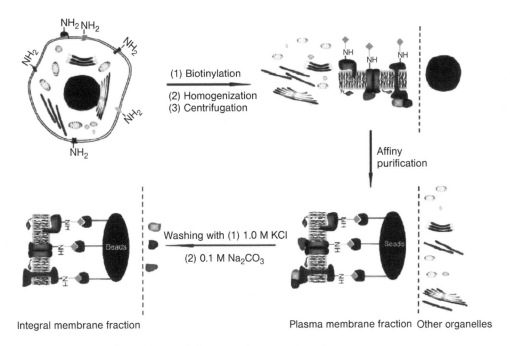

Figure 13.3 A strategy for enrichment of plasma membrane proteins using biotinylated alkylating reagents. Reprinted with permission from the American Chemical Society (Zhao et al., 2004).

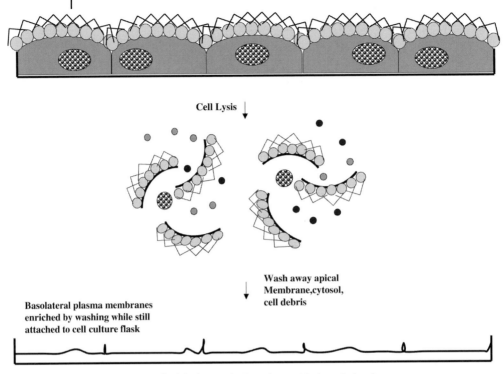

Figure 13.4 The colloidal silica method used to enrich the apical and basolateral plasma membranes recovered from cells growing on a surface. Adapted by A. Rahbar from Spector and Leinwand (1998).

This laboratory has incorporated into proteomics strategies an older procedure, in which isolation of the plasma membrane is facilitated by coating it with colloidal silica (Rahbar and Fenselau, 2004). The silica coating is subsequently stabilized by cross-linking with polyacrylic acid. When the cell is lysed, heavy, coated pieces of the plasma membrane may be separated from other cellular membranes by centrifugation. Like the biotinylation method discussed above, coating with colloidal silica allows the recovery of an enriched sample of plasma membrane. The colloidal silica method is summarized in Figure 13.4, as it was applied to MCF-7 cells growing on a surface. The coated apical membrane may be recovered, based on its increased mass. The basolateral membrane can be washed well and, finally, released from the surface in the usual way. Based on western blotting with the antibody to ATPase, the basolateral membrane was recovered, enriched 20-fold (Rahbar and Fenselau, 2004).

13.2.3
Protein Fractionation and Identification

For fractionation and characterization of proteins from isolated organelles, the optimum set of tools depends on the properties of the proteins in the organelle being examined. Thus, membrane proteins, often very large with hydrophobic domains, require a different approach than cytosolic proteins, which are mostly water-soluble. We have evaluated several different strategies to fractionate proteins and peptides from different organelles for identification by MS: the two-dimensional (2D) gel-based method (Henzel et al., 1993, Jensen et al., 1999), a shotgun approach (Eng et al., 1994) employing tandem high-pressure liquid chromatography for peptides from cytosolic proteins (Brown and Fenselau, 2004), a combination of one-dimensional (1D) SDS gel electrophoresis, *in situ* proteolysis and liquid chromatography (LC) MS/MS for membrane proteins (Rahbar and Fenselau, 2004), and solution electrochromatography (Zuo and Speicher, 2000; Wall et al., 2000; Herbert and Righetti, 2000) in combination with other separation methods (An et al., 2005). Peptide and protein identifications are obtained using bioinformatics, based on experimental sequence information from tandem MS experiments in which either matrix-assisted laser desorption ionization (MALDI) or electrospray ionization (ESI) is used, as discussed by experts elsewhere in this book.

The membrane proteins isolated using the colloidal silica method described above are especially difficult to solubilize, and difficult to separate by 2D gel electrophoresis. Consequently the membranes were solubilized in Laemmli buffer with sonication in a 60 °C water bath, and separated by 1D SDS gel electrophoresis. This 1D gel was cut into 1 mm bands, each of which was subjected to trypsin digestion. The peptides were recovered, separated and analyzed by LC/ESI MS/MS. This strategy is illustrated in Figure 13.5. More than 3200 peptides were

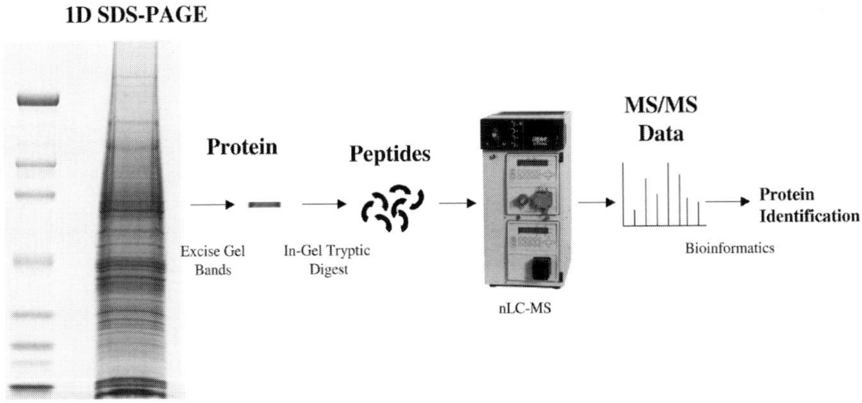

1D SDS-PAGE

Protein   Peptides   MS/MS Data

Excise Gel Bands   In-Gel Tryptic Digest   nLC-MS   Bioinformatics   Protein Identification

Figure 13.5 Schematic representation of the fractionation and analysis of plasma membrane proteins isolated using the colloidal silica method. Figure provided by A. Rahbar.

identified, corresponding to 540 proteins, of which about 50% were already classified as plasma membrane proteins (Rahbar and Fenselau, 2004, 2005). Only identifications based on microsequences of two or more peptides were accepted. As many as 30–50 tryptic peptides were identified from individual proteins in the most favorable cases. Among the proteins identified, 73 were based on predicted amino acid sequences not previously observed.

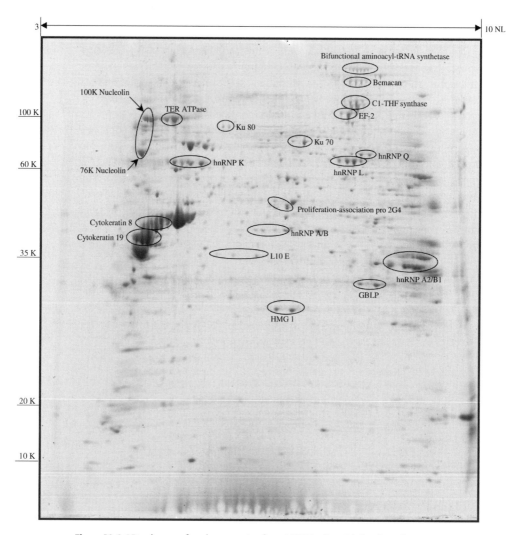

Figure 13.6 2D gel array of nuclear proteins from MCF-7 cells, with families of isoforms indicated (Fu, 2004). *EF-2* Elongation factor 2, *HMG-1* high mobility group protein 1, *hnRNP* heterogeneous nuclear ribonucleoprotein, *GBLP* guanine nucleotide-binding protein beta, *Cl-THF synthase* C-1-tetrahydrofolate synthase.

In a related study, nuclei were isolated from MCF-7 cells and three drug-resistant strains by standard methods (Fu and Fenselau, 2005). The nuclear pellet was lysed and centrifuged and the supernatant was collected for proteomic analysis. These soluble nuclear proteins were separated by 2D gel electrophoresis. The 160 spots from the drug-susceptible cell line that were excised were incubated with trypsin and identified either by peptide mapping from MALDI mass spectra, or microsequencing by ESI MS/MS. Two or more sequence tags were required for identification, and typically, peptide maps covered more than 30% of the protein sequence. More than 90% of the proteins identified were proteins classically catalogued as nuclear proteins. The 160 identifications yielded 123 different proteins, and more than 40 isoforms. The isoforms characterized are illustrated in Figure 13.6. Structural differences were not defined for all isoforms in this study; however it is clear that the nucleus is a good source of modified and processed proteins.

There are some important differences between the protein inventories developed for the MCF-7 plasma membrane using the shot-gun strategy, and for the MCF-7 nucleus using 2D gel arrays. The shot-gun strategy identified many more proteins. The gel array allowed many isoforms to be recognized.

13.2.4
Quantitative Comparisons of Protein Abundances

A major objective of our ongoing proteomics investigation is to identify proteins with modified abundances or structures in drug-resistant strains of human breast cancer MCF-7 cells, which can be further evaluated as potential contributors to cell survival. For comparisons of abundances on a proteomic scale, we have relied on both isotope ratios measured by MS, and comparative densitometry of digitized 2D gel maps. Isotopic labels have been introduced in our work by metabolic labeling (Gerhmann et al., 2004) or enzyme-catalyzed incorporation of ^{18}O atoms (Yao et al., 2001), depending on the organelle under study. The use of synthetic internal standards (Szeto et al., 2001; Gerber et al., 2003) and isotope labeling by various chemical derivatization reactions (Aebersold and Mann, 2003) could also be used. Recently, quantification has been proposed based on the relative peak areas of reproducible liquid chromatograms (Chelius et al., 2003; Nakamura et al., 2004).

Metabolic labeling, the introduction of isotope-labeled amino acids into cells in culture (Oda et al., 1999; Jiang and English, 2002; Ong et al., 2002; Gu et al., 2002), is illustrated for comparative proteomics in Figure 13.7. This strategy is especially appropriate for studies of plasma membrane proteins, because they are difficult to solubilize and maintain in solution, e.g., for isotope labeling by chemical reactions. In a variation on this method in our laboratory (Gerhmann et al., 2004), both arginine and lysine were replaced with $^{13}C_6$-arginine and $^{13}C_6$-lysine in proteins synthesized by drug-susceptible MCF-7 cells, which could then be mixed with various drug-resistant cell lines whose proteins contained unlabeled arginine and lysine. This is a relatively expensive method, which provides, however,

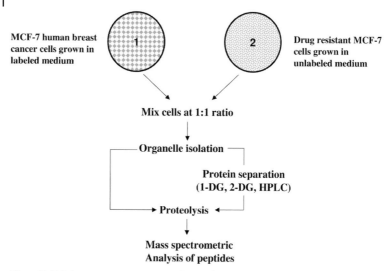

Figure 13.7 Schematic representation of quantification using metabolic labeling of relative abundances of proteins from two cell populations.

an isotope pair at the carboxyl-terminus of every tryptic peptide. The isotope ratio reveals the relative abundances of all peptides (except that which contains the original C-terminus), and the isotope pair tags all "y" ions in MS/MS spectra, assisting manual interpretation of the sequences. Corrections have to be made for incomplete labeling as well as for the ratio of cells in the mixture. In this estrogen-sensitive mammalian cell line, accommodation had to be made to provide low molecular weight hormones in order for the cells to grow (Gerhmann et al., 2004).

Relative quantitation of MCF-7 nuclear protein abundances was carried out using two methods. In one, the densities of Coomassie-stained protein spots in digitized 2D gel arrays from drug-susceptible and -resistant cell lines were compared with computer support (Fu and Fenselau, 2005). In the second, relative abundances of the nuclear proteins were determined by MS following metabolic labeling, with the isotopic protein mixture separated by 2D gel electrophoresis. In either approach relative quantification was combined with 2D gel separations, which allowed protein isoforms to be quantified. For example, two forms of nucleolin were identified (see Figure 13.6), with molecular masses of approximately 100 kDa and 76 kDa. The lighter protein was found to have two to three times higher abundance in MCF-7 cells resistant to mitoxantrone and etoposide, while the heavier isoform was only modestly more abundant. A comparison between the two methods of quantification indicates that they provide comparable ratios for most proteins, as illustrated in Figure 13.8 and Table 13.1. Exceptions are attributed to the presence of more than one protein in a spot, or to the tendency of densitometry to underestimate very abundant proteins.

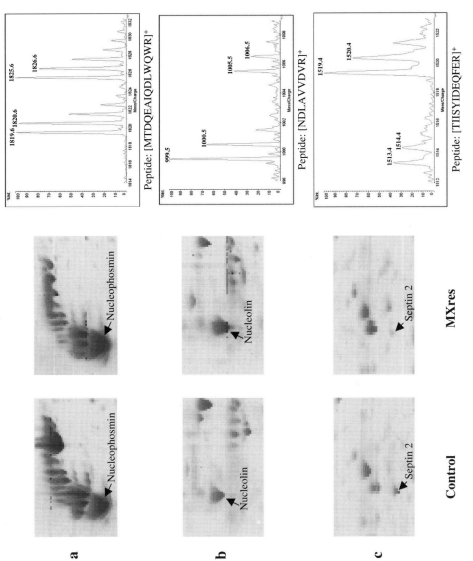

Figure 13.8 Examples of differential abundances of nuclear proteins in mitoxantrone-resistant and susceptible cells measured by densitometry and MS (Fu, 2004).
a Nucleophosmin, determined to be unchanged and used to correct for the ratio of cells in the mixture.

b Nucleolin 76 kDda, determined to be more abundant in the resistant cell line labeled with light isotopes.
c Septin 2, determined to be less abundant in the resistant cell line. Mass spectra were obtained using MALDI on a Kratos Axima time-of-flight (TOF) instrument.

Table 13.1 Abundance changes in nuclear proteins from mitoxantrone-resistant MCF-7 cells (Fu and Fenselau, 2005).

Protein name	Accession number	Isotope ratio MXres/control	Densitometry MXres/control
78K GRP	P11021	4.910.46	5.53±0.69
Prohibitin	P35232	3.69±0.33	2.81±0.31
HMG-1	P09429	2.56±0.18	1.52±0.16
40S ribosomal protein SA	P08861	2.35±0.18	1.74±0.19
Cyclophilin B	P23284	2.22±0.16	1.66±0.45
Nucleolin(76K)	P19338	2.18±0.16	1.98±0.24
Nucleolin (100K)	P19388	1.58±0.18	1.47±0.26
Cytokeratin 19	P08727	0.53±0.09	0.58±0.14
Cytokeratin 8 (truncated)	P05787	0.49±0.05	0.43±0.06
PARP-1	P09874	0.46±0.06	0.30±0.13
EF-1-beta	P24534	0.36±0.06	0.28±0.09
Septin 2	Q15019	0.11±0.03	0.22±0.03
Alpha tropomyosin	P09493	0.10±0.02	Not detected in MXres cells

It should be noted that these measurements report the relative abundance of each protein in the cell extracts. Isotope ratios and comparative densitometry do not distinguish changes in expression, regulation, or degradation. Follow-up experiments are required to make these distinctions.

13.3
Changes in Plasma Membrane and Nuclear Proteins in MCF-7 Cells Resistant to Mitoxantrone

Using the criteria that abundances are altered by at least twofold, 12 proteins of interest were identified in the nucleus and 15 in the plasma membrane of mitoxantrone-resistant MCF-7 cells. These are listed in Tables 13.1 and 13.2. Experimental variation in Table 13.1 is based on isotope ratio determinations of peptides from at least three gels, each from three separate cultures, thus incorporating both biological variability and experimental uncertainty. Experimental variation in Table 13.2 is based on isotope ratios measured for multiple peptides for each protein, from three separate cultures.

The proteins whose abundances are altered in the plasma membrane in drug-resistant cells (Table 13.2) include those involved in increased amino acid uptake (4F2 cell-surface antigen heavy chain, large neutral amino acids transporter small subunit-1), reduced glucose uptake (stomatin, facilitated glucose transport

Table 13.2 Abundance changes in plasma membrane proteins from mitoxantrone resistant MCF-7 cells (Rahbar and Fenselau, 2005).

Protein name	Accession number	Isotope ratio MXres/control
ATP-binding cassette protein ABCG2	Q8lX16	Not detected in control cells
Transferrin receptor protein 1	P02786	Not detected in control cells
Large neutral amino acids transporter small subunit 1	Q01650	7.9±0.6
Dihydropyridine receptor alpha 2 subunit	Q9UlU0	6.8±0.4
4F2 cell-surface antigen heavy chain	P08195	4.4±0.3
Clathrin heavy chain 1	Q00610	3.7±0.4
Guanine nucleotide-binding protein G(S), alpha subunit	P04895	3.5±0.3
Erythrocyte band 7 integral membrane protein (stomatin)	P27105	2.6±0.3
Ephrin type-B receptor 4 precursor	P54760	2.3±0.3
Integrin alpha-2 precursor	P17301	0.33±0.03
Hypothetical protein DKFZp686D0452	Q7Z3V1	0.3±0.02
Solute carrier family 2, facilitated glucose transporter member 1	P11166	0.25±0.02
Tumor-associated calcium signal transducer 2	Q7Z7Q4	0.24±0.08
Guanine nucleotide-binding protein, alpha 13 subunit	Q14344	0.085±0.01
Integrin alpha-3 precursor	P26006	0.073±0.006

member-1), an activator of apoptosis (guanine nucleotide-binding protein alpha-13 subunit), and three integrins, which function in adhesion and metastasis (Rahbar and Fenselau, 2005). Most interesting, perhaps, are the appearances of the ATP-dependent pump breast cancer resistant protein (BCRP) and the transferrin receptor in the drug-resistant cell line, while they are undetectable in the drug-susceptible parent line. This observation of BCRP is consistent with previous reports by Doyle, Ross and colleagues based on genetic analysis and immunochemistry (Ross et al., 1999). BCRP has been shown to pump mitoxantrone and several other drugs out of the cell. This was the first time that the transferrin receptor had been associated with acquired drug resistance, and the observation suggests further study.

Abundances are either increased or decreased in the nucleus (Table 13.1) for proteins thought to participate in DNA repair (the nucleolin isoforms, high mobility group protein 1), in protein folding (cyclophilin B, also known as proline isomerase), in apoptosis (78K glucose-regulated protein, PARP-1), and cell cycle regulation (septin-2) (Fu and Fenselau, 2005). PARP-1 activates a caspase-independent cascade that leads to cell death, and its decreased abundance would reduce the incidence of that endpoint (Yu et al., 2002). Increased abundances of nucleolin stabilizer bcl-2, which inhibits caspase-dependent apoptosis (Sengupta et al., 2004). Several of these proteins may be translocated into the nucleus in response to genotoxic stress (e.g., glucose-regulated protein and prohibitin). Abundances of several cytoskeletal proteins were observed to be decreased, perhaps as part of broader cellular cytoskeletal reorganization. Changes in abundances of proteins that manage protein folding and degradation have been observed widely in cells under stress (Hampton, 2002; Kaneko and Nomura, 2003; Bianchi et al., 2005), as have abundances of proteins affecting apoptosis. However, the involvement of nuclear proteins in the suppression of apoptosis is a rather novel observation.

Many of the altered proteins in these two organelles are associated with the functions of the organelle. In the nucleus, modifications in protein abundances enhance DNA repair, and in the plasma membrane up-regulated BCRP pumps the drug out of the cell. In both organelles proteins are modified to accommodate reduced glucose uptake (stomatin, facilitated glucose transport member-1, and 78K glucose-regulated protein). And in both organelles protein mechanisms are activated to inhibit or down-regulate apoptosis, perhaps reflecting a synchronized transcellular response. Ongoing studies of the cytoplasm and mitochondria will test that possibility further.

Taken together, these and other observations strongly support the hypothesis that drug resistance in cancer cells is a complex mechanism, a multifactorial process mediated by many proteins. Many proteins with modified abundances and/or structures are involved directly or indirectly, even though one or a few proteins may provide the dominant mechanisms in a given case. Organelle-based proteomic studies indicate that different sets of proteins play important roles in cells selected to survive genotoxic or cytotoxic assaults by different drugs) (Fu and Fenselau, 2005). Applied to other cancers, other drugs, and with ever-improving methodology, proteomics studies will continue to illuminate the changes in cell biology associated with, and responsible for, acquired drug resistance. Future research will include hypothesis-driven studies, studies that focus on posttranslational modifications, and studies that address protein–protein interactions and complexes.

References

AEBERSOLD, R., MANN, M. (2003). Mass spectrometry-based proteomics. *Nature* **422**, 198–207.

AN, Y., FU, Z., GUTIERREZ, P., FENSELAU, C. (2005). Solution isoelectric focusing for peptide analysis: comparative investigation of an insoluble nuclear protein fraction. *J. Proteome Res.* **4**.

BIANCHI, L., CANTON, C., BINI, L., ORLANDI, R., MENARD, S., ARMINI, A., CATTANEO, M., PALLINI, V., BERNARDI, L. R., BIUNNO, I. (2005). Protein profile changes in the human breast cancer cell line MCF-7 in response to SELIL gene induction. *Proteomics* **5**, 2433–2442.

Brown, K. J., Fenselau, C. (**2004**). Investigation of doxorubicin resistance in MCF-7 breast cancer cells using shotgun comparative proteomics with proteolytic 18O labeling. *J. Proteome Res.* **3**, 455–462.

Castagna, A., Antonioli, P., Astner, H., Hamdan, M., Righerri, S. C., Perego, P., Zunino, F., Righerri, P. G. (**2004**). A proteomic approach to cisplatin resistance in the cervix squamous cell carcinoma cell line A431. *Proteomics* **4**, 3246–3267.

Chelius, D., Zhang, T., Wang, G., Shen, R. F. (**2003**). Global protein identification and quantification technology using two-dimensional liquid chromatography nanospray mass spectrometry. *Anal. Chem.* **75**, 6658–6665.

Eng, J. K., McCormack, A. L., Yates, J. R. (**1994**). An approach to correlate tandem mass spectral data of peptides with amino acid sequences in a protein database. *J. Am. Soc. Mass Spectrom.* **5**, 976–989.

Erlichman, C., Boerner, S. A., Hallgren, C. G., Spieker, R., Wang, X. Y., James, C. D., Scheffer, G. L., Maliepaard, M., Ross, D. D., Bible, K. C., Kaufmann, S. H. (**2001**). The HER tyrosine kinase inhibitor CI1033 enhances cytotoxicity of 7-ethyl-10-hydroxycamptothecin and topotecan by inhibiting breast cancer resistance protein-mediated drug efflux. *Cancer Res.* **61**, 739–748.

Fu, Z. (**2004**). PhD thesis, University of Maryland.

Fu, Z., Fenselau, C. (**2005**). Proteomic evidence for roles for nucleation and poly(ADP-ribosyl)transferase in drug resistance. *J. Proteome Res.* **4**.

Gehrmann, M. L., Hathout, Y., Fenselau, C. (**2004**). Evaluation of metabolic labeling for comparative proteomics in breast cancer cells. *J. Proteome Res.* **3**, 1063–1068.

Gerber, S. A., Rush, J., Stemman, O., Kirschner, M. W., Gygi, S. P. (**2003**). Absolute quantification of proteins and phosphoproteins from cell lysates by tandem MS. *Proc. Natl. Acad. Sci. USA* **100**, 6940–6945.

Gewirtz, D. A. (**1999**). A critical evaluation of the mechanisms of action proposed for the antitumor effects of the anthracycline antibiotics adriamycin and daunorubicin. *Biochem. Pharmacol.* **57**, 727–741.

Gottesman, M. M. (**2002**). Mechanisms of cancer drug resistance. *Annu. Rev. Med.* **53**, 615–627.

Gu, S., Pan, S., Bradbury, E. M., Chen, X. (**2002**). Use of deuterium-labeled lysine for efficient protein identification and peptide de novo sequencing. *Anal. Chem.* **74**, 5774–5785.

Hampton, R. Y. (**2002**). ER-associated degradation in protein quality control and cellular regulation. *Curr. Opin. Cell Biol.* **14**, 476–482.

Hathout, Y., Gehrmann, M. L., Chertov, A., Fenselau, C. (**2004**). Proteomic phenotyping: metastatic and invasive breast cancer. *Cancer Lett.* **210**, 245–253.

Hathout, Y., Riordan, K., Gehrmann, M., Fenselau, C. (**2002**). Differential protein expression in the cytosol fraction of an MCF-7 breast cancer cell line selected for resistance toward melphalan. *J. Proteome Res.* **1**, 435–442.

Henzel, W. J., Billeci, T. M., Stults, J. T., Wong, S. C., Grimley, C., Watanabe, C. (**1993**). Identifying proteins from two-dimensional gels by molecular mass searching of peptide fragments in protein sequence databases. *Proc. Natl. Acad. Sci. USA* **90**, 5011–5015.

Herbert, B., Righetti, P. G. (**2000**). A turning point in proteome analysis: sample prefractionation via multi-compartment electrolyzers with iso-electric membranes. *Electrophoresis* **21**, 3639–3648.

Jensen, O. N., Wilm, M., Shevchenko, A., Mann, M. (**1999**). Sample preparation methods for mass spectrometric peptide mapping directly from 2–DE gels. *Methods Mol. Biol.* **112**, 513–530.

Jiang, H., English, A. M. (**2002**). Quantitative analysis of the yeast proteome by incorporation of isotopically labeled leucine. *J. Proteome Res.* **1**, 345–350.

Kaneko, M., Nomura, Y. (**2003**). ER signaling in unfolded protein response. *Life Sci.* **74**, 199–205.

Lage, H. (**2004**). *Proteomics* in cancer cell research: an analysis of therapy resistance. *Pathol. Res. Pract.* **200**, 105–117.

Laragione, T., Bonetto, V., Casoni, F., Massignan, T., Bianchi, G., Gianazza, E., Ghezzi, P. (**2003**). Redox regulation of surface protein thiols: identification of integrin alpha-4 as a molecular target by using redox proteomics. *Proc. Natl. Acad. Sci. USA* **100**, 14737–14741.

Litman, T., Druley, T. E., Stein, W. D., Bates, S. E. (**2001**). From MDR to MXR: new understanding of multidrug resistance systems, their properties and clinical significance. *Cell Mol. Life Sci.* **58**, 931–959.

Nakagawa, M., Schneider, E., Dixon, K. H., Horton, J., Kelley, K., Morrow, C., Cowan, K. H. (**1992**). Reduced intracellular drug accumulation in the absence of P-glycoprotein (mdr1) overexpression in mitoxantrone-resistant human MCF-7 breast cancer cells. *Cancer Res.* **52**, 6175–6181.

Nakamura, T., Dohmae, N., Takio, K. (**2004**). Characterization of a digested protein complex with quantitative aspects: an approach based on accurate mass chromatographic analysis with Fourier transform-ion cyclotron resonance mass spectrometry. *Proteomics* **4**, 2558–2566.

Oda, Y., Huang, K., Cross, F. R., Cowburn, D., Chait, B. T. (**1999**). Accurate quantitation of protein expression and site-specific phosphorylation. *Proc. Natl. Acad. Sci. USA* **96**, 6591–6596.

Ong, S. E., Blagoev, B., Kratchmarova, I., Kristensen, D. B., Steen, H., Pandey, A., Mann, M. (**2002**). Stable isotope labeling by amino acids in cell culture, SILAC, as a simple and accurate approach to expression proteomics. *Mol. Cell Proteomics* **1**, 376–386.

Peirce, M. J., Wait, R., Begum, S., Saklatvala, J., Cope, A. P. (**2004**). Expression profiling of lymphocyte plasma membrane proteins. *Mol. Cell Proteomics* **3**, 56–65.

Pervaiz, S., Clement, M. V. (**2004**). Tumor intracellular redox status and drug resistance – serendipity or a causal relationship? *Curr. Pharm. Des.* **10**, 1969–1977.

Rahbar, A. M., Fenselau, C. (**2004**). Integration of Jacobson's pellicle method into proteomic strategies for plasma membrane proteins. *J. Proteome Res.* **3**, 1267–1277.

Rahbar, A. M., Fenselau, C. (**2005**). Unbiased examination of changes in plasma membrane proteins in drug resistant cancer cells. *J. Proteome Res.* **4**.

Ross, D. D., Yang, W., Abruzzo, L. V., Dalton, W. S., Schneider, E., Lage, H., Dietel, M., Greenberger, L., Cole, S. P., Doyle, L. A. (**1999**). Atypical multidrug resistance: breast cancer resistance protein messenger RNA expression in mitoxantrone-selected cell lines. *J. Natl. Cancer Inst.* **91**, 429–433.

Schneider, E., Horton, J. K., Yang, C. H., Nakagawa, M., Cowan, K. H. (**1994**). Multidrug resistance-associated protein gene overexpression and reduced drug sensitivity of topoisomerase II in a human breast carcinoma MCF7 cell line selected for etoposide resistance. *Cancer Res.* **54**, 152–158.

Sengupta, T. K., Bandyopadhyay, S., Fernandes, D. J., Spicer, E. K. (**2004**). Identification of nucleolin as an AU-rich element binding protein involved in bcl-2 mRNA stabilization. *J. Biol. Chem.* **279**, 10855–10863.

Shin, B. K., Wang, H., Yim, A. M., LeNaour, F., Brichory, F., Jang, J. H., Zhao, R., Puravs, E., Tra, J., Michael, C. W., Misek, D. E., Hanash, S. M. (**2003**). Global profiling of the cell surface proteome of cancer cells uncovers an abundance of proteins with chaperone function. *J. Biol. Chem.* **278**, 7607–7616.

Soule, H. D., Vazguez, J., Long, A., Albert, S., Brennan, M. (**1973**). A human cell line from a pleural effusion derived from a breast carcinoma. *J. Natl. Cancer Inst.* **51**, 1409–1416.

Spector, D. L., Leinwand, R. D. (**1998**). *Cells: A Laboratory Manual*. Cold Spring Harbor Laboratory Press, USA.

Stein, W. D., Bates, S. E., Fojo, T. (**2004**). Intractable cancers: the many faces of multidrug resistance and the many targets it presents for therapeutic attack. *Curr. Drug Targets* **5**, 333–346.

Szeto, H. H., Lovelace, J. L., Fridland, G., Soong, Y., Fasolo, J., Wu, D., Desiderio, D. M., Schiller, P. W. (2001). *In vivo* pharmacokinetics of selective mu-opioid peptide agonists. *J. Pharmacol. Exp. Ther.* **298**, 57–61.

Volk, E. L., Rohde, K., Rhee, M., McGuire, J. J., Doyle, L. A., Ross, D. D., Schneider, E. (2000). Methotrexate cross-resistance in a mitoxantrone-selected multidrug-resistant MCF7 breast cancer cell line is attributable to enhanced energy-dependent drug efflux. *Cancer Res.* **60**, 3514–3521.

Wall, D. B., Kachman, M. T., Gong, S., Hinderer, R., Parus, S., Misek, D. E., Hanash, S. M., Lubman, D. M. (2000). Isoelectric focusing nonporous RP HPLC: a two-dimensional liquid-phase separation method for mapping of cellular proteins with identification using MALDI-TOF mass spectrometry. *Anal. Chem.* **72**, 1099–1111.

Yao, X., Freas, A., Ramirez, J., Demirev, P. A., Fenselau, C. (2001). Proteolytic ^{18}O labeling for comparative proteomics: model studies with two serotypes of adenovirus. *Anal. Chem.* **73**, 2836–2842.

Yu, S. W., WangH, Poitras, M. F., Coombs, C., Bowers, W. J., Federoff, H. J., Poirier, G. G., Dawson, T. M., Dawson, V. L. (2002). Mediation of poly(ADP-ribose) polymerase-1-dependent cell death by apoptosis-inducing factor. *Science* **297**, 259–263.

Zhao, Y., Zhang, W., Kho, Y., Zhao, Y. (2004). Proteomic analysis of integral plasma membrane proteins. *Anal. Chem.* **76**, 1817–1823.

Zuo, X., Speicher, D. W. (2000). A method for global analysis of complex proteomes using sample prefractionation by solution isoelectrofocusing prior to two-dimensional electrophoresis. *Anal. Biochem.* **248**, 266–278.

14

Clinical Neuroproteomics of Human Body Fluids: CSF and Blood Assays for Early and Differential Diagnosis of Dementia

Jens Wiltfang and Piotr Lewczuk

Abstract

An aging population and increasing life expectancy result in a growing number of patients with dementia. This symptom can be a part of a completely curable disease of the central nervous system (e.g., neuroinflammation), or a disease currently considered irreversible [e.g., Alzheimer's disease (AD)]. In the latter case, several potentially successful treatment approaches are being tested now, demanding reasonable standards of early diagnosis. Cerebrospinal fluid and serum analysis (CSF/serum analysis), whereas routinely performed in neuroinflammatory diseases, still requires standardization to be used as an aid to the clinically based diagnosis of AD. Several AD-related CSF parameters (total tau, phosphorylated forms of tau, amyloid β peptides, ApoE genotype, p97, etc.) tested separately or in a combination provide sensitivity and specificity of about 85%, the figure commonly expected from a good diagnostic tool. In this review, recently published reports regarding progress in neurochemical durante vitam diagnosis of dementias are discussed with a focus on an early and differential diagnosis of AD. Novel perspectives offered by recently introduced technologies, e.g., surface-enhanced laser desorption/ionization time-of-flight mass spectrometry (SELDI-TOF MS), differential gel electrophoresis (DIGE), and multiplexing technologies are briefly discussed.

14.1
Introduction

Along with the aging of population and increasing life expectancy, dementia will become a serious problem for the health care system. Regarding Alzheimer's disease (AD), however, the increasing number of patients has not resulted in achieving accurate standards of *durante vitam* diagnosis so far. Although sensitivity of clinical diagnosis is relatively high (93%), specificity may be lower; for instance it was reported as 55% in a multicenter clinical autopsy study (Mayeux, 1998).

Proteomics in Drug Research
Edited by M. Hamacher, K. Marcus, K. Stühler, A. van Hall, B. Warscheid, H. E. Meyer
Copyright © 2006 Wiley-VCH Verlag GmbH & Co. KGaA, Weinheim
ISBN: 3-527-31226-9

Although in expert hands the clinical diagnosis of AD is predictive of AD pathology in 80–90% of cases, very early diagnosis of AD, and differential diagnosis of unusual presentations of patients with dementia remains difficult on clinical grounds.

With the introduction of potentially successful treatment against dementia hitherto considered irreversible, like acetylocholinesterase inhibitors in AD (recently reviewed in (Bullock, 2002; Knopman, 2001), the need for an early and differential diagnosis of dementia becomes even more urgent (The Ronald and Nancy Reagan Research Institute, 1998). Biomarkers such as neuroimaging studies and cerebrospinal fluid (CSF) analysis have been developed as aids to clinical diagnosis. The aim of this review, therefore, is to summarize the current state-of-the-art in the field of neurochemical diagnosis of dementias with a special focus on biomarkers for AD. Finally, new perspectives on the search for dementia biomarkers offered by novel sophisticated techniques, like multiplexing, differential gel electrophoresis, and mass spectrometry are briefly discussed.

Since CSF is in direct contact with the environment of the central nervous system (CNS), it is obvious that any changes in biochemical composition of brain parenchyma should be predominantly reflected in the CSF. A recent review by Reiber (2001) presents a complete concept of distinguishing diffusion of brain-derived proteins into CSF from diffusion of proteins from blood into CSF, allowing proteins that originate in the brain to be prioritized. Lumbar puncture is an easy procedure, with a low incidence of complications. In a large study (Andreasen et al., 2001), only 4.1% of all patients experienced postlumbar headache, and an even smaller proportion of 2% was reported by Blennow et al. (1993). Thus, it is reasonable to postulate that lumbar puncture (LP) is a feasible and only moderately invasive procedure, and that CSF analysis could possibly improve current clinical and neuroimaging-based approaches to diagnosis.

14.2
Neurochemical Markers of Alzheimer's Disease

The major cause of dementia in the elderly is AD. Moreover, in older subjects, AD is a major overall cause of morbidity and death. Neuropathological markers of AD are extracellular β-amyloid deposits (plaques), and deposits of neurofibrillary tangles (NFT) [reviewed in (Wiltfang et al., 2001b)]. Neuronal loss with brain atrophy, disturbances of neurotransmission as well as local inflammatory reactions of glia are also commonly found. Recently, requirements to be met for a test to become an acceptable diagnostic parameter in AD have been proposed (The Ronald and Nancy Reagan Research Institute, 1998). Ideally, such a test should be: able to detect a fundamental feature of AD pathology, validated in neuropathologically confirmed cases, and be precise, reliable, noninvasive, simple to perform and inexpensive. A common consensus is that the sensitivity and specificity of such a test should not be lower than approximately 85%, and 75–85%, respectively (The Ronald and Nancy Reagan Research Institute, 1998).

14.2.1
β-Amyloid Precursor Protein (β-APP): Metabolism and Impact on AD Diagnosis

β-Amyloid plaques are composed mainly of peptides coming from the enzymatic cut of β-amyloid precursor protein (β-APP) (Kang et al., 1987). This trans-membrane protein is encoded in man by a gene on chromosome 21, and its alternative splicing results in at least eight forms. The form known as β-APP 695 (i.e., one consisting of 695 amino acid residues) is expressed predominantly in the brain (Panegyres, 1997). The physiological role of β-APP is not yet clear; however, an involvement in cell-to-cell and matrix interactions is postulated. β-APP is enzymatically processed by α-, β-, and γ-secretases to release several forms of amyloid β (Aβ) peptides. This process is schematically presented in Figure 14.1. Interestingly, the discovery of Aβ peptides ending at different C termini leads to a conclusion that different γ-secretase activities may exist. If so, perhaps a specific inhibitor might be synthesized to prevent the generation of Aβ1-42 by blocking the cut at the position 42, thus preventing the formation of β-amyloid plaques.

Several recently published studies including these from our group have reported decreased concentration of Aβ1-42 in the CSF (Lewczuk et al., 2004d; Wiltfang et al., 2002; and reviews in: Blennow et al., 2001; Wiltfang et al., 2001b), whereas the total level of Aβ peptides remained unchanged (Motter et al., 1995). Even more importantly, CSF Aβ1-42 seems to decrease in the course of the disease before a

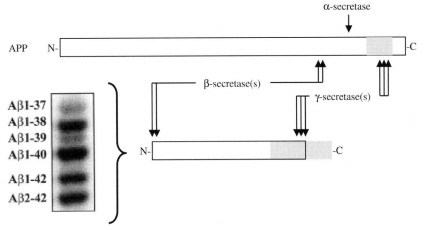

Figure 14.1 Common cut of amyloid precursor protein (APP) by β-, and γ-secretases leads to a formation of at least six different Aβ peptides. The "classical" position of the cut by β-secretase is defined as position 1 of Aβ peptides, however, an alternative cut by β-secretase leads to a release of peptides beginning at position 2 and also, probably, position 3. γ-Secretase(s) cut(s) ahead of amino acid isoleucine (position 40) leads to the generation of Aβ1-40, and the cut ahead of the amino acid threonine (position 42) leads to the generation of Aβ1-42. Recently, peptides ending at C-terminus positions 37, 38, and 39 have also been found (Wiltfang et al., 2002). This may indicate this enzyme having more than one activity, resulting in a searching for a possible specific inhibitor. The *gray box* indicates the position of the transmembrane domain of APP.

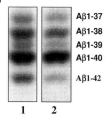

Aβ1-37
Aβ1-38
Aβ1-39
Aβ1-40

Aβ1-42

1   2

Figure 14.2 Quantitative Aβ-SDS-PAGE/Immunoblot shows a decrease in Aβ1-42 in CSF of a patient with AD (*band 2*) compared to a control case (*band 1*). According to Wiltfang et al. (2001a) with modifications.

severe collapse of patients' cognitive functions appears (Riemenschneider et al., 2000). As an illustration, decreased CSF Aβ1-42 in AD is clearly shown in the Aβ-SDS-PAGE/immunoblot method in Figure 14.2. It must be stressed that CSF concentration of Aβ1-42 strongly depends on preanalytical sample treatment (Wiltfang et al., 2002). Results obtained with methods involving incubation with detergent or heat (e.g., immunoprecipitation in the presence of detergent or SDS heat denaturation) are up to threefold higher than those obtained without such pretreatment [reviewed in (Wiltfang et al., 2002]. This finding indicates the presence of a fraction of Aβ peptides not accessible to antibodies, most probably due to binding to carrier proteins.

Mechanisms leading to decreased concentrations of Aβ1-42 in CSF of patients with AD are not yet clear. Accumulation of the peptide in the plaques is suggested by some investigators. This hypothesis, however, cannot explain our results (Otto et al., 2000) showing decreased concentration of the peptide in the CSF of a subgroup of patients with Creutzfeldt–Jakob disease (CJD) who did not develop any amyloid plaques. Similarly, decreased levels of CSF Aβ1-42 have been reported to occur in bacterial meningitis (Sjögren et al., 2001), a disease that can cause chronic memory deficits but does not present with beta-amyloid plaques. In fact, the lowest values found in a large multicenter study of Hulstaert et al. were observed in two cases of subacute sclerosing panencephalitis and in one case of bacterial meningitis (Hulstaert et al., 1999). As an alternative hypothesis, it might be postulated that the formation of Aβ1-42 binding complexes is responsible for decreased CSF concentration of the peptide (Wiltfang et al., 2002). Indeed, evidence for such complexes in CSF of patients with AD have been recently obtained by fluorescent correlation spectroscopy (Pitschke et al., 1998).

The sensitivity and specificity of Aβ1-42 alone to distinguish AD from elderly controls were 78 and 81%, respectively, in a study by Hulstaert et al. (1999), and Galasko et al. (1998) reported similar figures of 78 and 83% for sensitivity and specificity respectively. In our study (Lewczuk et al., 2004d), application of CSF Aβ42 alone resulted in a correct classification of 87% of subjects when non-Alzheimer's dementia and nondementia patients were treated as a control group for AD. Blennow et al. analyzed data from eight studies with a total of $n = 562$ AD patients and $n = 273$ controls, and reported mean sensitivity and specificity of 85 and 84%, respectively (Blennow et al., 2001).

We have recently found an additional amino-terminal-truncated Aβ peptide, i.e., Aβ2-42 in CSF of patients with AD (Wiltfang et al., 2001a), and this peptide is the second most abundant Aβ peptide in the frontal lobe of patients with AD, after Aβ1-42. In the CSF, it is present in a subset of 35% of AD cases. Experimental studies with knock-out mice showed that Aβ2-42 was most probably produced by an alternative β-secretase cut, and of particular pathophysiological importance is that this peptide may serve as a first nidus for β-amyloid plaque formation (Wiltfang et al., 2001a).

Application of a urea-based electrophoresis system (Wiltfang et al., 1991) resulted in finding other carboxyl-terminal-truncated Aβ peptides, i.e., Aβ1-37/38/39 in human CSF (Wiltfang et al., 2002) as well as in blood (Lewczuk et al., 2004b). This highly conserved pattern has been observed in all CSF samples investigated so far, with AD characterized by an elevated fraction (%) of Aβ1-40 (Aβ1-40%), reduced Aβ1-42%, and elevated Aβ1-38% in some cases (Wiltfang et al., 2002). Elevation of the latter peptide concentration, observed also in some cases of chronic neuro-inflammatory diseases, further supports the commonly accepted role of inflammation in developing AD (Eikelenboom et al., 2000; McGeer and McGeer, 2001).

14.2.2
Tau Protein and its Phosphorylated Forms

Tau proteins belong to the family of microtubule-associated proteins (MAPs) found in neuronal and nonneuronal cells (for a review, see Buee et al., 2000). The human tau gene is located on the long arm of chromosome 17. Its alternative splicing leads to formation of six forms of the protein ranging from 352 to 441 amino acids. Studies on the role of tau proteins have revealed that their main function is to promote neuronal microtubule stability and assembly. They are also involved in promoting microtubule nucleation, growth and bundling, and it is hypothesized that phosphorylation of the tau molecule is an important factor in regulating the tau–microtubule interaction (for a review, see Shahani and Brandt, 2002). The phosphorylation status of tau is considered to change during development, with a relatively high degree of phosphorylation during the fetal phase followed by a steady decrease with age, most probably as a result of phosphatase activation (Mawal-Dewan et al., 1994; Rosner et al., 1995). An increasing amount of evidence indicates that tau also interacts directly or indirectly with the actin cytoskeleton, playing an important role in regulating the cell's shape, motility and the interactions between microtubule and the plasma membrane (for a review, see Shahani and Brandt, 2002). Interestingly, as recently reported, tau regulates intracellular traffic of vesicles and inhibits transport of amyloid precursor protein into neuronal extensions, which leads to accumulation of APP in the cell body (Stamer et al., 2002). Studies with cell culture models indicate that tau is also involved in neurite outgrowth and stabilization (Baas et al., 1991; Knops et al., 1991). Moreover, primary cultures of neurones from genetically modified mice with knocked out tau gene show a significant delay in their axonal and dendritic extensions (Dawson et al., 2001); however, another group reported a completely normal phenotype in mice lacking the tau gene (Harada et al., 1994). These seemingly conflicting results have been recently explained by the finding that other MAPs possibly compensate for the missing tau gene (Takei et al., 2000). The role of tau and its phosphorylation by glycogen synthase kinase 3 (GSK-3) in anterograde transport and in neurite outgrowth has recently been shown by Tatebayashi et al. (2004).

Total tau protein concentration has been extensively studied as a nonspecific marker of neuronal destruction in AD. Recently published meta-analysis of Sunderland et al. (2003) was based on the data from 17 reports on Aβ42 and 34 reports on CSF tau in AD, and all studies included in this meta-analysis reported increased CSF total tau in AD. Since an age-related increase of tau concentration has been reported in nondementia controls by some investigators (Buerger et al., 2003), age-dependent reference values of total tau should be considered as recently suggested (Sjogren et al., 2001). Increased CSF total tau concentrations are observed in neuropsychiatric disorders other than AD, e.g., CJD and stroke (Hesse et al., 2001; Otto et al., 1997). Nevertheless, given the fact that tau most likely can be used to monitor the efficacy of neuron-protective drugs in AD patients, and that CJD and acute stroke are easily distinguishable from AD clinically, this should not dampen the value of this marker.

14.2.2.1 Hyperphosphorylation of Tau as a Pathological Event

It is suggested that some phosphorylation events change the conformational status of the tau molecules, leading to decreased microtubule binding, a reduced ability to promote microtubule assembly, and increased dynamic instability of microtubules. It has been reported that phosphorylation of serine 262 partially abolishes the ability of tau to bind to microtubules (Singh et al., 1996), whereas phosphorylation at threonine 231 and serine 235 markedly influence tau's binding to microtubules (Sengupta et al., 1998). In the brains of patients afflicted with Alzheimer's disease, hyperphosphorylated molecules of tau protein assemble to form intraneuronal filamentous inclusions, neurofibrillary tangles, one of the hallmarks of the disease, and it has been observed that the number of neurofibrillary tangles correlates closely with the degree of dementia (Alafuzoff et al., 1987; Arriagada et al., 1992; Braak and Braak, 1991). Intracellular deposits of hyperphosphorylated tau are observed in several neurodegenerative disorders, known as tauopathies; however, in Alzheimer's disease these inclusions are described only in neurons. Recently it has been suggested that the phosphorylation of tau proteins in AD happens in a form of evolution with the amino acid positions 153, 262, and 231 modified in an early stage of the disease and the positions 199, 202, 205, 396, and 404 relatively late (Augustinack et al., 2002). This evolution of phosphorylation sites of the tau molecule corresponds to the morphological evolution of the tangles in AD. Simultaneously, an increased concentration of phosphorylated forms of tau protein is measured in CSF of AD patients (Buerger et al., 2002; Itoh et al., 2001; Vanmechelen et al., 2000). Several kinases have been shown to phosphorylate tau molecules as recently reviewed (Buee et al., 2000), including stress-activated protein kinases, mitogen-activated protein kinases, GSK 3, and others.

14.2.2.2 **Phosphorylated Tau in CSF as a Biomarker of Alzheimer's Disease**

While the increase in the total CSF tau concentration is considered to reflect nonspecific disruption of nerve cells, abnormal hyperphosphorylation of tau is a hallmark of AD (Iqbal et al., 1986), and hyperphosphorylated molecules of tau form neurofibrillary tangles (Grundke-Iqbal et al., 1986). Tau can be phosphorylated at 79 putative positions, serine and threonine being predominant. Recently, independent groups have reported increased CSF concentrations of phosphorylated tau to distinguish AD from other dementia diseases. In studies available so far, mean sensitivity and specificity of tau phosphorylated at different positions varied in the ranges 44–94%, and 80–100%, respectively (Blennow et al., 2001). Interestingly, Hu et al. has shown recently that the pTau396/404 to total tau ratio in CSF could discriminate AD from other dementias and neurological disorders at a sensitivity of 96% and a specificity of 94% (Hu et al., 2002). These findings suggest that CSF analysis of tau phosphorylated at serine 396/404 might be more promising than some of the other sites reported to date. It should also be noted that while processes of hyperphosphorylation of tau dominate in AD, dephosphorylation of these molecules is supposed to happen as well, and the ratio of phosphorylated and dephosphorylated tau molecules must be considered dynamically. Moreover, dephosphorylation of hyperphosphorylated tau seems to be one of the most promising therapeutic targets in AD, as recently reviewed (Iqbal et al., 2002).

In our recent study, we found significantly increased CSF concentrations of pTau181 in the group of AD patients with clinical diagnosis supported neurochemically by decreased Aβ42 in CSF (Lewczuk et al., 2004a). This form seems to be particularly interesting, since pTau181 remains unchanged while total tau is increased after acute stroke (Hesse et al., 2001). This may suggest that pTau181 is not only a marker of simple neuronal loss. Similarly, Vanmechelen et al. (2001) reported significantly increased levels of CSF pTau181 in AD compared to all other groups studied [frontotemporal dementia (FTD), Lewy body dementia (LBD), Parkinson's disease, multiple system atrophy, and progressive supranuclear palsy] except for corticobasal degeneration. Parnetti et al. (2001) confirmed recently that pTau181 was a useful biomarker to distinguish AD from dementia with Lewy bodies. Moreover, in agreement with the results of our study, Vanmechelen et al. (2001) and Nagga et al. (2002) found a strong correlation between total tau and pTau181 independently of the diagnostic group. Similarly, increased CSF concentrations of pTau181 have been reported very recently in subjects with probable AD compared to controls (Nägga et al., 2002), and Papassotiropoulos et al. found increased CSF concentration of pTau181 associated with a polymorphism of CYP46. The group of C. Hock reported that this gene was associated with an increased risk of late-onset sporadic AD (Papassotiropoulos et al., 2003).

Regarding other phosphorylation sites, in an international multicenter study, Itoh et al. reported a significant overall increase of pTau199 in patients with AD compared to all other non-AD groups. In this study, both sensitivity and specificity of CSF pTau199 for discriminating AD from other studied groups yielded 85% at the cutoff level of 1.05 fmol mL^{-1} (Itoh et al., 2001). Tau phosphorylated at threonine 231 (pTau231) seems to help in the differentiation of AD from relevant diseases, i.e., FTD, vascular dementia (VD), and LBD (reviewed in Blennow et al., 2001). A follow up study revealed increased CSF concentration of pTau231 at the onset of the disease followed by decreasing concentrations of pTau231, but not total tau, in a group of untreated AD patients. This may suggest a possible role for this isoform in tracking the natural course of the disease (Hampel et al., 2001). Interestingly, tau protein phosphorylated at both positions threonine 231 and serine 235 turned out to be increased in patients with mild cognitive impairment (MCI) who developed AD during follow-up (Arai et al., 2000). In this study, a simultaneous evaluation of total tau and phosphorylated tau distinguished the group of patients at risk of developing AD from those who complained of having memory impairment but did not have objective memory loss.

In a recently published study, three different phospho-tau forms have been compared regarding their ability to distinguish among patients with different forms of dementia, as well as nondementia controls, showing overall equal performance by pTau181 and pTau231, with a somewhat worse performance by pTau199. Interestingly, discrimination between AD and LBD was maximized using pTau181 at a sensitivity of 94% and a specificity of 64%, and pTau231 maximized group separation between AD and FTD with a sensitivity of 88% and a specificity of 92%. There therefore seems to be a nonsignificant tendency for phospho-tau proteins to perform differently in the discrimination of particular types of dementias (Hampel et al., 2004)

14.2.3
Apolipoprotein E (ApoE) Genotype

ApoE is a protein involved in the transport of cholesterol. Apart from its presence in plasma, it is also produced by astrocytes in the CNS to support growth and repair of neurons. The ApoE gene is localized on chromosome 19 and three alleles are described, ε 2, ε 3, and ε 4. Recently, a growing body of evidence has been reported on an association between ApoE ε 4 and late-onset familial AD (for review, see Mulder et al., 2000). As many as 40–50% of AD patients posses the ε 4 allele compared to 15–25% of controls (Strittmatter et al., 1993a). Subjects homozygous for the ε 4 allele are reported to have a six- to eightfold increased risk of developing AD compared to the risk for heterozygotic subjects whose risk increased three- to fourfold (reviewed by Mulder et al., 2000). So far, the ε 4 allele is identified as a major risk factor, independently of gender, age, and ethnic origin of individuals (Farrer et al., 1997). In a large American study with 2188 patients, analysis of the ε 4 allele showed mild sensitivity and specificity of 65% and 68% respectively (Mayeux et al., 1998). Thus, it is suggested that genotyping for ApoE allele should be reserved for dementia patients, and that the presence of one or two ApoE ε 4 alleles can improve specificity of the diagnosis in patients who fulfil clinical criteria for AD. The mechanisms regulating increased risk of developing AD in cases carrying the ε 4 allele are still unclear.

14.2.4
Other Possible Factors

Apart from the classic biomarkers, i.e., tau protein(s) and their phosphorylated forms as well as Aβ peptides, several other candidate biomarkers have been tested, as recently and extensively reviewed by Frank et al. (2003). Neuronal thread proteins (NTPs) are a family of molecules expressed in the CNS. In a post mortem study, brains of AD patients expressed significant increases of NTP immunoreactivity (de la Monte and Wands, 1992). Following this finding, CSF examination for NTP revealed increased concentration of NTP which correlated with progression of dementia and neuronal degeneration (de la Monte et al., 1992). Sensitivity and specificity of this protein as a possible marker of AD, however, have not been determined in a large enough number of patients.

The soluble form of the iron-binding protein p97 has been suggested to be increased in AD patients in blood. In one report, all AD patients had elevated serum concentration of this factor, and no overlap with control subjects was observed (Kennard et al., 1996). Recently, this interesting finding has been confirmed by others, although using a different analytical method (Ujiie et al., 2002). Further studies are necessary to confirm p97 as a reliable marker of AD. Another interesting factor reported in plasma is that the homocysteine level was presented to correlate with the risk of developing AD. A value greater than 14 µmol L^{-1} was associated with a twofold increased risk of developing AD (Seshadri et al., 2002).

Interestingly, significantly increased CSF concentrations of transforming growth factor β1 has been found recently in AD by two independent groups of researchers (Tarkowski et al., 2002; Zetterberg et al., 2004).

Since there is growing evidence of the involvement of oxidative and/or nitrative cell damage in the pathogenesis of AD, factors related to lipid peroxidation have been tested as candidate AD biomarkers. Recently, Pratico et al. (Pratico et al., 2002) reported increased levels of isoprostanes (e.g., 8,12-iso-iPF$_{2\alpha}$-VI) in body fluids of patients with AD. This finding has been further supported by the elevated levels of this factor in the body fluids of transgenic mice expressing AD amyloidosis (Pratico et al., 2001).

14.2.5
Combined Analysis of CSF Parameters

There are many examples of CNS diseases where a combination of more than one CSF parameter significantly improves the accuracy of the diagnosis. Neuro-inflammatory diseases, like neuroborreliosis (Tumani et al., 1995) or multiple sclerosis (Reiber et al., 1998) are representative examples. Similarly, studies have been reported showing increased sensitivity and specificity of a combination of CSF parameters in early and differential diagnoses of AD.

Correspondingly, in an international multicenter project, combined analysis of Aβ1-42 and tau protein showed 85% diagnostic sensitivity and 58% specificity in distinguishing AD from non-Alzheimer types of dementia (Hulstaert et al., 1999). In this study, the mean sensitivity and specificity levels of the individual markers were significantly improved from 74–79% to 86% if both markers were considered simultaneously. In our study (Lewczuk et al., 2004d), we have found slightly better discrimination of patients with AD, non-Alzheimer's dementia and controls when Aβ42 was combined with Aβ40 (i.e., a concentration quotient of Aβ42/Aβ40). This discrimination was further slightly improved by a simultaneous evaluation of CSF total tau concentration, and combination of all these three parameters resulted in a correct separation of 94% of subjects in our study. Andreasen et al. reported sensitivity of 94% in the group of $n = 105$ probable AD, and 88% in the group of $n = 58$ possible AD when analysis of CSF total tau was accompanied by Aβ1-42 (Andreasen et al., 2001). Specificity in this study was high for differentiation of AD from psychiatric disorders and nondementia subjects (100%, and 89% respectively); however, low concentrations of Aβ1-42 found in several cases of LBD resulted in lower specificity in the case of this disease. The lowest specificity (48%) was found to discriminate AD from VD, probably because latter patients had concomitant pathological features of AD. A study of Kanai et al. reported similar figures of 71% and 83% for diagnostic sensitivity and specificity respectively for simultaneous tau/Aβ1-42 analysis (Kanai et al., 1998).

By plotting concentrations of Aβ1-42 versus tau, Motter et al. found a 96% predictive value for having AD for subjects with high tau/low Aβ1-42, and 100% predictive value not to have AD for subjects with low tau/high Aβ1-42 (Motter et al., 1995). Similar figures, with positive and negative predictive values of 90% and 95%, respectively, at a prevalence of probable AD of 44% were also obtained by Andreasen et al. (Andreasen et al., 2001). To evaluate data from the simultaneous analysis of CSF Aβ1-42 and tau, Galasko et al. used the binary tree-structured classification system obtaining 85.2% of correct diagnoses with sensitivity and specificity of 90 and 80%, respectively (Galasko et al., 1998). In a recently published study, a combination of low CSF Aβ42 and high CSF pTau181 allowed early-onset AD patients to be distinguished from these with frontotemporal lobar degeneration with a sensitivity of 72% and a specificity of 93% (Schoonenboom et al., 2004).

Somewhat discrepant results have been presented when CSF tau or CSF phosphorylated tau has been related to the ApoE genotype. While Arai et al. (1995) reported no correlation of total tau to the number of ApoE ε 4 alleles, and Itoh et al. (Itoh et al., 2001) reported a similar finding regarding pTau199. Golombowski et al., and Blomberg et al. found that AD patients with the ApoE ε 4 allele had higher values of CSF tau than those without ApoE ε 4 (Blomberg et al., 1996; Golombowski et al., 1997). With regard to the combination of ApoE genotype and Aβ1-42 levels, Riemenschneider et al., Galasko et al., and Hulstaert et al. reported the peptide's highest level in AD patients with no ε 4 allele, intermediate in those with ε 4 heterozygous, and lowest in those with ε 4 homozygous (Galasko et al., 1998; Hulstaert et al., 1999; Riemenschneider et al., 2000). As a possible explanation of this correlation, a high-affinity of binding Aβ1-42 to ApoE is suggested (Strittmatter et al., 1993b). Moreover, sensitivity for the combination of CSF tau and CSF Aβ1-42 in patients possessing the ε 4 allele increased from 94% to 99%

for probable and from 88% to 100% for possible AD (Andreasen et al., 2001). On the other hand, in a study including more that 400 AD cases, no effect on CSF tau levels was found for the ApoE e4 allele (Andreasen et al., 1999). These observations lead to the conclusion that the ApoE genotype should be taken into consideration in interpretation of Aβ1-42 levels, and that combination of ApoE genotyping with other parameters may significantly improve specificity and sensitivity of diagnosis.

We have recently reported a decreased *fraction* of Aβ1-42 of all Aβ peptides in CSF of patients with AD. In this study, all the AD patients and none of the controls appeared below the discrimination line of 8.5%. Plotting the results of relative ratios of Aβ1-42 against Aβ1-38 showed all AD cases and none of the controls in the region with decreased Aβ1-42; however some AD cases had increased Aβ1-38, similarly to subjects with chronic neuroinflammatory diseases (Wiltfang et al., 2002) (Figure 14.3). This observation points to a presence of chronic (micro)-inflammatory reaction in the CNS of subjects with AD. Moreover, this report also presents evidence that severity of dementia correlates negatively with percentage but not absolute value of Aβ peptides shorted at their carboxyl ends.

One of the most demanding aspects of neurochemical analysis of dementia disorders is to find biomarkers capable of predicting development of AD in patients with mild cognitive impairment (MCI). The hope is that such an early diagnosis would allow sufficiently early treatment. According to the epidemiological data of Petersen (1999), approximately 10–15% of MCI subjects develop AD within 1 year. Recently, Andreasen et al. (Andreasen et al., 2003) reported an increased positive likelihood ratio for total tau (8.45), phospho-tau (7.49) and Aβ42 (8.2) in the 1-year follow-up of 44 MCI patients progressing to AD. These data suggest that these biomarkers are already altered in an early phase of dementia and that these factors

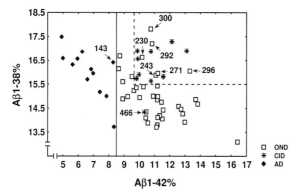

Figure 14.3 Plot of CSF Aβ1-38 relative concentration ratio (*vertical axis*, Aβ1-38%) and CSF Aβ1-42 relative concentration ratio (*horizontal axis*, Aβ1-42%). Cases of Alzheimer's diseases (*AD*), nondementive neurologic diseases (*OND*), and chronic neuroinflammatory diseases (*CID*) have been investigated. With an Aβ1-42% discriminatory value of 8.5%, all AD cases are separated from other subjects resulting in 100% sensitivity and 100% specificity. Increased Aβ1-38% above 15.5% observed in some AD cases corresponds to increased Aβ1-38% in CID and indicates signs of neuroinflammation in these cases of AD. According to Wiltfang et al. (2002), with modifications.

may help to identify MCI subjects who will progress to AD. Similarly, Andreasen et al. show elevated tau and decreased Aβ42 levels in MCI patients at baseline (Andreasen et al., 2001).

In a recently published study, a combination of three CSF biomarkers, namely tau, pTau181 and Aβ42, could detect incipient AD among patients fulfilling the criteria for MCI with a sensitivity of 68% (95% CI, 45–86%) and a specificity of 97% (95% CI, 83–100%), therefore suggesting the possibility of identifying the subgroup of patients with MCI who would eventually develop AD from those who would not, to offer early treatment for the subjects at risk (Zetterberg et al., 2003).

A limitation in the interpretation of the performance of CSF markers is the lack of standardization of assays. In addition, several preanalytical and biological confounding factors may influence the analytical outcome for CSF analyses, such as concentration gradients of the protein along the spinal cord, influence of lumbar puncture (LP) hemorrhage, presence of the protein in plasma and passage over the blood–brain and/or blood–CSF barrier, and degradation or loss of the protein *in vitro* after the CSF tap. For CSF tau and Aβ, the only preanalytical confounding factor is that all of the proteins have a tendency to stick to the walls of test tubes made of glass and hard plastic, resulting in falsely low levels (Andreasen et al., 1999; Lewczuk et al., submitted). Therefore, it is important to tap CSF into nonabsorbing test tubes made of polypropylene.

14.2.6
Perspectives: Novel Techniques to Search for AD Biomarkers –
Mass Spectrometry (MS), Differential Gel Electrophoresis (DIGE), and Multiplexing

Surface-enhanced laser desorption/ionization–time-of-flight mass spectrometry (SELDI–TOF MS) is currently applied in many fields of biological and medical sciences, as extensively reviewed by Merchant and Weinberger (2000). Briefly, crude biological samples (e.g., plasma, CSF, cell culture supernatants, etc.) are applied to the surface of ProteinChips mimicking chromatographic interactions (e.g., cation/anion exchangers, hydrophobic interactions surfaces, metal affinity surfaces, etc.), and after washing steps, the only molecules retained on the surface are bound to it. The straightforward opportunity offered by this technology is immunoaffinity chips with specific antibodies covalently bound to the chip surface to capture adequate antigens. These molecules are then desorbed and ionized with a laser shot, and the time of flight of the molecules in the vacuum of the mass spectrometer correlates with the molecular mass of the ions. Applying the mass standards, the molecular mass of the peptides/proteins analyzed can be measured with imprecision lower than 0.05%, i.e., about 2–3 Da in the range of the peptides' molecular masses. Polypeptides of interest can be further characterized by gas-phase sequencing, and peptide mapping technology using additional MS technology (e.g., liquid chromatography–tandem MS). SELDI–TOF MS has recently been successfully used to analyze Aβ peptides in supernatants of transfected HEK 293 cells (Frears et al., 1999), and we have been recently the first group to analyze and compare the patterns of Aβ peptides in human CSF and post mortem brain tissue from patients with AD and nondemented controls (Lewczuk et al., 2003; Lewczuk et al., 2004c).

Another potentially promising technique is two-dimensional fluorescence difference gel electrophoresis (DIGE), which offers advanced opportunity to study protein expression alterations in biological material (Karp et al., 2004). Briefly, on a single gel, up to three samples can be coelectrophoresed, and they are then scanned with a three-beam fluorescent laser imager (Typhoon scanner). The crucial step of the procedure is a labeling of proteins with a fluorescent probes (CyDye), which allows accurate quantification of differences in the protein expression patterns (coefficients of variations < 20%, up to 3000 individual spots) using a sophisticated software package (DeCyder). As samples from different sources (e.g., from patients and controls) are simultaneously run on a single gel, any imprecision of the method afflicts similar proteins in a similar way, i.e., methodological errors are compensated for. The detection sensitivity for proteins is in the low nanogram range. In spite of the fluorescent labeling, the protein of interest can be further characterized by gas-phase sequencing and peptide mapping. The CSF samples of patients with neurodegenerative disorders and controls will first be depleted of high-abundance proteins by multiple immunoaffinity columns. Following desalting and concentration, the samples will be studied by DIGE using immobilized ph gradients (IPG, nonlinear gradient 3–10) in the first dimension and SDS-PAGE step gradient gels in the second dimension (molecular mass range: 5000–600 000 Da).

Since it is likely that combined analysis of several CSF markers will be necessary to improve diagnostic performance, multiplex methods will be of importance. One such method is the microsphere-based xMap technology (Luminex, Austin, Texas, USA), which is a multiplex flow cytometric method based on antibodies coupled to spectrally specific fluorescent microspheres (Vignali, 2000). Another fluorescent reporter antibody binds the protein captured on the microspheres. Each microsphere is spectrally identified, and quantification is based on the intensity of the reporter signal. In comparison to conventional ELISA methods, this multiplex technology allows simultaneous quantification of up to 100 proteins in a small sample volume, and provides higher reproducibility than multiple ELISA methods. We have used the xMap technology to design multiplex assays for simultaneous quantification of several Aß and tau isoforms. Preliminary data show equal, or better, diagnostic performance than with ELISA (Lewczuk et al., submitted).

14.3
Conclusions

The increasing number of patients with dementia demands improved standards of *intra vitam* diagnosis. The growing body of evidence summarized briefly in this paper may support the use of CSF analysis as part of a diagnostic workup for dementia. When clinical suspicion for an infectious or inflammatory disorder is strong, or in subacute dementia, routine CSF analysis combined with tests for causes of dementia such as infection may yield a specific and often treatable diagnosis. This is further supported by a finding of relatively low incidences of postpuncture complications, especially in the relevant group of patients (Andreasen et al., 2001; Blennow et al., 1993). CSF/serum analysis should be performed according to the generally accepted, theoretically based and practically-confirmed concept of CSF-flow rate-related protein diffusion (Reiber and Peter, 2001). CSF/serum analysis may provide useful information in patients with conditions currently considered untreatable. An early and definitive diagnosis may provide a reasonable time window for starting treatment during a period when the patient has relatively mild impairment and is best able to benefit. As more definitive treatment is tested, very early diagnosis of AD will become imperative. Simultaneous analysis of two or more factors in CSF can significantly improve the accuracy of the AD diagnosis. Further studies are required to confirm sensitivity and specificity of the markers, including studies in patients who have had post mortem confirmation of diagnosis. With the introduction of novel, very sensitive techniques, e.g., fluorescence correlation spectroscopy (FCS), further advances in early diagnosis of AD are expected.

Acknowledgements

This study is part of the German Research Network on Dementia and was funded by the German Federal Ministry of Education and Research (grant O1 GI 0102). JW and PL are supported by BMBF-funded project NGFN2 (Sub-project No. PPO-S10T10), and are members of the Human Brain Proteome Project of the Human Proteome Organisation (HUPO).

References

ALAFUZOFF, I., IQBAL, K., FRIDEN, H., ADOLFSSON, R., WINBLAD, B. (1987). Histopathological criteria for progressive dementia disorders: clinical–pathological correlation and classification by multivariate data analysis. *Acta Neuropathol. (Berl.)* 74, 209–225.

ANDREASEN, N., MINTHON, L., CLARBERG, A., DAVIDSSON, P., GOTTFRIES, J., VANMECHELEN, E., VANDERSTICHELE, H., WINBLAD, B., BLENNOW, K. (1999). Sensitivity, specificity, and stability of CSF-tau in AD in a community-based patient sample. *Neurology* 53, 1488–1494.

ANDREASEN, N., MINTHON, L., DAVIDSSON, P., VANMECHELEN, E., VANDERSTICHELE, H., WINBLAD, B.,

BLENNOW, K. (2001). Evaluation of CSF-tau and CSF-Aβ42 as diagnostic markers for Alzheimer disease in clinical practice. *Arch. Neurol.* 58, 373–379.

ANDREASEN, N., VANMECHELEN, E., VANDERSTICHELE, H., DAVIDSSON, P., BLENNOW, K. (2003). Cerebrospinal fluid levels of total-tau, phospho-tau and A beta 42 predicts development of Alzheimer's disease in patients with mild cognitive impairment. *Acta Neurol. Scand. Suppl.* 179, 47–51.

ARAI, H., ISHIGURO, K., OHNO, H., MORIYAMA, M., ITOH, N., OKAMURA, N., MATSUI, T., MORIKAWA, Y., HORIKAWA, E., KOHNO, H., SASAKI, H., IMAHORI, K. (2000). CSF phosphorylated tau protein and mild cognitive impairment: a prospective study. *Exp. Neurol.* **166**, 201–203.

ARAI, H., TERAJIMA, M., MIURA, M., HIGUCHI, S., MURAMATSU, T., MACHIDA, N., SEIKI, H., TAKASE, S., CLARK, C. M., LEE, V. M., et al. (**1995**). Tau in cerebrospinal fluid: a potential diagnostic marker in Alzheimer's disease. *Ann. Neurol.* **38**, 649–652.

ARRIAGADA, P. V., GROWDON, J. H., HEDLEY-WHYTE, E. T., HYMAN, B. T. (**1992**). Neurofibrillary tangles but not senile plaques parallel duration and severity of Alzheimer's disease. *Neurology* **42**, 631–639.

AUGUSTINACK, J. C., SCHNEIDER, A., MANDELKOW, E. M., HYMAN, B. T. (**2002**). Specific tau phosphorylation sites correlate with severity of neuronal cytopathology in Alzheimer's disease. *Acta Neuropathol. (Berl.)* **103**, 26–35.

BAAS, P. W., PIENKOWSKI, T. P., KOSIK, K. S. (**1991**). Processes induced by tau expression in Sf9 cells have an axon-like microtubule organization. *J. Cell Biol.* **115**, 1333–1344.

BLENNOW, K., VANMECHELEN, E., HAMPEL, H. (**2001**). CSF total tau, Aβ42 and phosphorylated tau protein as biomarkers for Alzheimer's disease. *Mol. Neurobiol.* **24**, 87–97.

BLENNOW, K., WALLIN, A., HAGER, O. (**1993**). Low frequency of post-lumbar puncture headache in demented patients. *Acta Neurol. Scand.* **88**, 221–223.

BLOMBERG, M., JENSEN, M., BASUN, H., LANNFELT, L., WAHLUND, L. O. (**1996**). Increasing cerebrospinal fluid tau levels in a subgroup of Alzheimer patients with apolipoprotein E allele epsilon 4 during 14 months follow-up. *Neurosci. Lett.* **214**, 163–166.

BRAAK, H., BRAAK, E. (**1991**). Neuropathological stageing of Alzheimer-related changes. *Acta Neuropathol. (Berl.)* **82**, 239–259.

BUEE, L., BUSSIERE, T., BUEE-SCHERRER, V., DELACOURTE, A., HOF, P. R. (**2000**). Tau protein isoforms, phosphorylation and role in neurodegenerative disorders. *Brain Res. Brain Res. Rev.* **33**, 95–130.

BUERGER, K., ZINKOWSKI, R., TEIPEL, S. J., ARAI, H., DEBERNARDIS, J., KERKMAN, D., MCCULLOCH, C., PADBERG, F., FALTRACO, F., GOERNITZ, A., TAPIOLA, T., RAPOPORT, S. I., PIRTTILA, T., MOLLER, H. J., HAMPEL, H. (**2003**). Differentiation of geriatric major depression from Alzheimer's disease with CSF tau protein phosphorylated at threonine 231. *Am. J. Psychiatry* **160**, 376–379.

BUERGER, K., ZINKOWSKI, R., TEIPEL, S. J., TAPIOLA, T., ARAI, H., BLENNOW, K., ANDREASEN, N., HOFMANN-KIEFER, K., DEBERNARDIS, J., KERKMAN, D., MCCULLOCH, C., KOHNKEN, R., PADBERG, F., PIRTTILÄ, T., SCHAPIRO, M. B., RAPOPORT, S. I., MÖLLER, H. J., DAVIES, P., HAMPEL, H. (**2002**). Differential diagnosis of Alzheimer disease with cerebrospinal fluid levels of tau protein phosphorylated at threonine 231. *Arch. Neurol.* **59**, 1267–1272.

BULLOCK, R. (**2002**). New drugs for Alzheimer's disease and other dementias. *Br. J. Psychiatry* **180**, 135–139.

DAWSON, H. N., FERREIRA, A., EYSTER, M. V., GHOSHAL, N., BINDER, L. I., VITEK, M. P. (**2001**). Inhibition of neuronal maturation in primary hippocampal neurons from tau deficient mice. *J. Cell Sci.* **114**, 1179–1187.

DE LA MONTE, S. M., VOLICER, L., HAUSER, S. L., WANDS, J. R. (**1992**). Increased levels of neuronal thread protein in cerebrospinal fluid of patients with Alzheimer's disease. *Ann. Neurol.* **32**, 733–742.

DE LA MONTE, S. M., WANDS, J. R. (**1992**). Neuronal thread protein over-expression in brains with Alzheimer's disease lesions. *J. Neurol. Sci.* **113**, 152–164.

EIKELENBOOM, P., ROZEMULLER, A. J., HOOZEMANS, J. J., VEERHUIS, R., VAN GOOL, W. A. (**2000**). Neuroinflammation and Alzheimer disease: clinical and therapeutic implications. *Alzheimer Dis. Assoc. Disord.* **14**, S54–61.

FARRER, L. A., CUPPLES, L. A., HAINES, J. L., HYMAN, B., KUKULL, W. A., MAYEUX, R., MYERS, R. H., PERICAK-VANCE, M. A., RISCH, N., VAN DUIJN, C. M. (**1997**). Effects of age, sex, and ethnicity on the association between apolipoprotein E genotype and Alzheimer disease A meta-analysis APOE and Alzheimer Disease Meta Analysis Consortium. *JAMA* **278**, 1349–1356.

FRANK, R. A., GALASKO, D., HAMPEL, H., HARDY, J., DE LEON, M. J., MEHTA, P. D., ROGERS, J., SIEMERS, E., TROJANOWSKI, J. Q. (**2003**). Biological markers for therapeutic trials in Alzheimer's disease Proceedings of the biological markers working group; NIA initiative on neuroimaging in Alzheimer's disease. *Neurobiol. Aging* **24**, 521–536.

FREARS, E. R., STEPHENS, D. J., WALTERS, C. E., DAVIES, H., AUSTEN, B. M. (**1999**). The role of cholesterol in the biosynthesis of β-amyloid. *Neuroreport* **10**, 1699–1705.

GALASKO, D., CHANG, L., MOTTER, R., CLARK, C. M., KAYE, J., KNOPMAN, D., THOMAS, R., KHOLODENKO, D., SCHENK, D., LIEBERBURG, I., MILLER, B., GREEN, R., BASHERAD, R., KERTILES, L., BOSS, M. A., SEUBERT, P. (**1998**). High cerebrospinal fluid tau and low amyloid β42 levels in the clinical diagnosis of Alzheimer disease and relation to apolipoprotein E genotype. *Arch. Neurol.* **55**, 937–945.

GOLOMBOWSKI, S., MULLER-SPAHN, F., ROMIG, H., MENDLA, K., HOCK, C. (**1997**). Dependence of cerebrospinal fluid Tau protein levels on apolipoprotein E4 allele frequency in patients with Alzheimer's disease. *Neurosci. Lett.* **225**, 213–215.

GRUNDKE-IQBAL, I., IQBAL, K., TUNG, Y. C., QUINLAN, M., WISNIEWSKI, H. M., BINDER, L. I. (**1986**). Abnormal phosphorylation of the microtubule-associated protein tau (tau) in Alzheimer cytoskeletal pathology. *Proc. Natl. Acad. Sci. USA* **83**, 4913–4917.

HAMPEL, H., BUERGER, K., KOHNKEN, R., TEIPEL, S. J., ZINKOWSKI, R., MOELLER, H. J., RAPOPORT, S. I., DAVIES, P. (**2001**). Tracking of Alzheimer's disease progression with cerebrospinal fluid tau protein phosphorylated at threonine 231. *Ann. Neurol.* **49**, 545–546.

HAMPEL, H., BUERGER, K., ZINKOWSKI, R., TEIPEL, S. J., GOERNITZ, A., ANDREASEN, N., SJOEGREN, M., DEBERNARDIS, J., KERKMAN, D.,

ISHIGURO, K., OHNO, H., VANMECHELEN, E., VANDERSTICHELE, H., McCULLOCH, C., MOLLER, H. J., DAVIES, P., BLENNOW, K. (**2004**). Measurement of phosphorylated tau epitopes in the differential diagnosis of Alzheimer disease: a comparative cerebrospinal fluid study. *Arch. Gen. Psychiatry* **61**, 95–102.

HARADA, A., OGUCHI, K., OKABE, S., KUNO, J., TERADA, S., OHSHIMA, T., SATO-YOSHITAKE, R., TAKEI, Y., NODA, T., HIROKAWA, N. (**1994**). Altered microtubule organization in small-calibre axons of mice lacking tau protein. *Nature* **369**, 488–491.

HESSE, C., ROSENGREN, L., ANDREASEN, N., DAVIDSSON, P., VANDERSTICHELE, H., VANMECHELEN, E., BLENNOW, K. (**2001**). Transient increase in total tau but not phospho-tau in human cerebrospinal fluid after acute stroke. *Neurosci. Lett.* **297**, 187–190.

HU, Y. Y., HE, S. S., WANG, X., DUAN, Q. H., GRUNDKE-IQBAL, I., IQBAL, K., WANG, J. (**2002**). Levels of nonphosphorylated and phosphorylated tau in cerebrospinal fluid of Alzheimer's disease patients: an ultrasensitive bienzyme-substrate-recycle enzyme-linked immunosorbent assay. *Am J. Pathol.* **160**, 1269–1278.

HULSTAERT, F., BLENNOW, K., IVANOIU, A., SCHOONDERWALDT, H. C., RIEMENSCHNEIDER, M., DE DEYN, P. P., BANCHER, C., CRAS, P., WILTFANG, J., MEHTA, P. D., IQBAL, K., POTTEL, H., VANMECHELEN, E., VANDERSTICHELE, H. (**1999**). Improved discrimination of AD patients using β-amyloid$_{(1-42)}$ and tau levels in CSF. *Neurology* **52**, 1555–1562.

IQBAL, K., ALONSO ADEL, C., EL-AKKAD, E., GONG, C. X., HAQUE, N., KHATOON, S., PEI, J. J., TSUJIO, I., WANG, J. Z., GRUNDKE-IQBAL, I. (**2002**). Significance and mechanism of Alzheimer neurofibrillary degeneration and therapeutic targets to inhibit this lesion. *J. Mol. Neurosci.* **19**, 95–99.

IQBAL, K., GRUNDKE-IQBAL, I., ZAIDI, T., MERZ, P. A., WEN, G. Y., SHAIKH, S. S., WISNIEWSKI, H. M., ALAFUZOFF, I., WINBLAD, B. (**1986**). Defective brain microtubule assembly in Alzheimer's disease. *Lancet* **2**, 421–426.

ITOH, N., ARAI, H., URAKAMI, K., ISHIGURO, K., OHNO, H., HAMPEL, H., BUERGER, K., WILTFANG, J., OTTO, M., KRETZSCHMAR, H., MOELLER, H. J., IMAGAWA, M., KOHNO, H., NAKASHIMA, K., KUZUHARA, S., SASAKI, H., IMAHORI, K. (2001). Large-scale, multicenter study of cerebrospinal fluid tau protein phosphorylated at serine 199 for the antemortem diagnosis of Alzheimer's disease. *Ann. Neurol.* **50**, 150–156.

KANAI, M., MATSUBARA, E., ISOE, K., URAKAMI, K., NAKASHIMA, K., ARAI, H., SASAKI, H., ABE, K., IWATSUBO, T., KOSAKA, T., WATANABE, M., TOMIDOKORO, Y., SHIZUKA, M., MIZUSHIMA, K., NAKAMURA, T., IGETA, Y., IKEDA, Y., AMARI, M., KAWARABAYASHI, T., ISHIGURO, K., HARIGAYA, Y., WAKABAYASHI, K., OKAMOTO, K., HIRAI, S., SHOJI, M. (1998). Longitudinal study of cerebro-spinal fluid levels of tau, A β1–40, and A β1–42(43) in Alzheimer's disease: a study in Japan. *Ann. Neurol.* **44**, 17–26.

KANG, J., LEMAIRE, H. G., UNTERBECK, A., SALBAUM, J. M., MASTERS, C. L., GRZESCHIK, K. H., MULTHAUP, G., BEYREUTHER, K., MULLER-HILL, B. (1987). The precursor of Alzheimer's disease amyloid A4 protein resembles a cell-surface receptor. *Nature* **325**, 733–736.

KARP, N. A., KREIL, D. P., LILLEY, K. S. (2004). Determining a significant change in protein expression with DeCyder during a pair-wise comparison using two-dimensional difference gel electrophoresis. *Proteomics* **4**, 1421–1432.

KENNARD, M. L., FELDMAN, H., YAMADA, T., JEFFERIES, W. A. (1996). Serum levels of the iron binding protein p97 are elevated in Alzheimer's disease. *Nat. Med.* **2**, 1230–1235.

KNOPMAN, D. (2001). Pharmacotheraphy for Alzheimer's disease. *Curr. Neurol. Neurosci. Rep.* **1**, 428–434.

KNOPS, J., KOSIK, K. S., LEE, G., PARDEE, J. D., COHEN-GOULD, L., MCCONLOGUE, L. (1991). Overexpression of tau in a nonneuronal cell induces long cellular processes. *J. Cell Biol.* **114**, 725–733.

LEWCZUK, P., ESSELMANN, H., BIBL, M., BECK, G., MALER, J. M., OTTO, M., KORNHUBER, J., WILTFANG, J. (2004a). Tau protein phosphorylated at threonine 181 in csf as a neurochemical biomarker in Alzheimer's disease: original data and review of the literature. *J. Mol. Neurosci.* **23**, 115–122.

LEWCZUK, P., ESSELMANN, H., BIBL, M., PAUL, S., SVITEK, J., MIERTSCHISCHK, J., MEYER, R., SMIRNOV, A., MALER, J. M., KLEIN, C., OTTO, M., BLEICH, S., SPERLING, W., KORNHUBER, J., RUTHER, E., WILTFANG, J. (2004b). Electrophoretic separation of amyloid β peptides in plasma. *Electrophoresis* **25**, 3336–3343.

LEWCZUK, P., ESSELMANN, H., GROEMER, T. W., BIBL, M., MALER, J. M., STEINACKER, P., OTTO, M., KORNHUBER, J., WILTFANG, J. (2004c). Amyloid β peptides in cerebrospinal fluid as profiled with surface enhanced laser desorption/ionization time-of-flight mass spectrometry: evidence of novel biomarkers in Alzheimer's disease. *Biol. Psychiatry* **55**, 524–530.

LEWCZUK, P., ESSELMANN, H., MEYER, M., WOLLSCHEID, V., NEUMANN, M., OTTO, M., MALER, J. M., RÜTHER, E., KORNHUBER, J., WILTFANG, J. (2003). The amyloid-β (Aβ) peptide pattern in cerebrospinal fluid in Alzheimer's disease: evidence of a novel carboxy-terminally elongated Aβ peptide. *Rapid Commun. Mass Spectrom.* **17**, 1291–1296.

LEWCZUK, P., ESSELMANN, H., OTTO, M., MALER, J. M., HENKEL, A. W., HENKEL, M. K., EIKENBERG, O., ANTZ, C., KRAUSE, W. R., REULBACH, U., KORNHUBER, J., WILTFANG, J. (2004d). Neurochemical diagnosis of Alzheimer's dementia by CSF Aβ42, Aβ42/Aβ40 ratio and total tau. *Neurobiol. Aging* **25**, 273–281.

MAWAL-DEWAN, M., HENLEY, J., VAN DE VOORDE, A., TROJANOWSKI, J. Q., LEE, V. M. (1994). The phosphorylation state of tau in the developing rat brain is regulated by phosphoprotein phospha-tases. *J. Biol. Chem.* **269**, 30981–30987.

MAYEUX, R. (1998). Evaluation and use of diagnostic tests in Alzheimer's disease. *Neurobiol. Aging* **19**, 139–143.

MAYEUX, R., SAUNDERS, A. M., SHEA, S., MIRRA, S., EVANS, D., ROSES, A. D., HYMAN, B. T., CRAIN, B., TANG, M. X., PHELPS, C. H. (**1998**). Utility of the apolipoprotein E genotype in the diagnosis of Alzheimer's disease Alzheimer's Disease Centers Consortium on Apolipoprotein E and Alzheimer's Disease. *N. Engl. J. Med.* **338**, 506–511.

McGEER, P. L., McGEER, E. G. (**2001**). Inflammation, autotoxicity and Alzheimer disease. *Neurobiol. Aging* **22**, 799–809.

MERCHANT, M., WEINBERGER, S. R. (**2000**). Recent advancements in surface-enhanced laser desorption/ionization-time of flight-mass spectrometry. *Electrophoresis* **21**, 1164–1177.

MOTTER, R., VIGO-PELFREY, C., KHOLODENKO, D., BARBOUR, R., JOHNSON-WOOD, K., GALASKO, D., CHANG, L., MILLER, B., CLARK, C., GREEN, R., et al. (**1995**). Reduction of beta-amyloid peptide42 in the cerebrospinal fluid of patients with Alzheimer's disease. *Ann. Neurol.* **38**, 643–648.

MULDER, C., SCHELTENS, P., VISSER, J. J., VAN KAMP, G. J., SCHUTGENS, R. B. (**2000**). Genetic and biochemical markers for Alzheimer's disease: recent developments. *Ann. Clin. Biochem.* **37**, 593–607.

NÄGGA, K., GOTTFRIES, J., BLENNOW, K., MARCUSSON, J. (**2002**). Cerebrospinal fluid phospho-tau, total tau and β-amyloid$_{1-42}$ in the differentiation between Alzheimer's disease and vascular dementia. *Dement. Geriatr. Cogn. Disord.* **14**, 183–190.

OTTO, M., ESSELMANN, H., SCHULZ-SHAEFFER, W., NEUMANN, M., SCHROTER, A., RATZKA, P., CEPEK, L., ZERR, I., STEINACKER, P., WINDL, O., KORNHUBER, J., KRETZSCHMAR, H. A., POSER, S., WILTFANG, J. (**2000**). Decreased β-amyloid1–42 in cerebrospinal fluid of patients with Creutzfeldt-Jakob disease. *Neurology* **54**, 1099–1102.

OTTO, M., WILTFANG, J., TUMANI, H., ZERR, I., LANTSCH, M., KORNHUBER, J., WEBER, T., KRETZSCHMAR, H. A., POSER, S. (**1997**). Elevated levels of tau-protein in cerebrospinal fluid of patients with Creutzfeldt-Jakob disease. *Neurosci. Lett.* **225**, 210–212.

PANEGYRES, P. K. (**1997**). The amyloid precursor protein gene: a neuropeptide gene with diverse functions in the central nervous system. *Neuropeptides* **31**, 523–535.

PAPASSOTIROPOULOS, A., STREFFER, J. R., TSOLAKI, M., SCHMID, S., THAL, D., NICOSIA, F., IAKOVIDOU, V., MADDALENA, A., LÜTJOHANN, D., GHEBREMEDHIN, E., HEGI, T., PASCH, T., TRÄXLER, M., BRÜHL, A., BENUSSI, L., BINETTI, G., BRAAK, H., NITSCH, R. M., HOCK, C. (**2003**). Increased brain β-amyloid load, phosphorylated tau, and risk of Alzheimer disease associated with an intronic CYP46 polymorphism. *Arch. Neurol.* **60**, 29–35.

PARNETTI, L., LANARI, A., AMICI, S., GALLAI, V., VANMECHELEN, E., HULSTAERT, F. (**2001**). CSF phosphorylated tau is a possible marker for discriminating Alzheimer's disease from dementia with Lewy bodies Phospho-Tau International Study Group. *Neurol. Sci.* **22**, 77–78.

PETERSEN, R. C., SMITH, G. E., WARING, S. C., IVNIK, R. J., TANGALOS, E. G., KOKMEN, E. (**1999**). Mild cognitive impairment: clinical characterization and outcome. *Arch. Neurol.* **56**, 303–308.

PITSCHKE, M., PRIOR, R., HAUPT, M., RIESNER, D. (**1998**). Detection of single amyloid beta-protein aggregates in the cerebrospinal fluid of Alzheimer's patients by fluorescence correlation spectroscopy. *Nat. Med.* **4**, 832–834.

PRATICO, D., CLARK, C. M., LIUN, F., ROKACH, J., LEE, V. Y., TROJANOWSKI, J. Q. (**2002**). Increase of brain oxidative stress in mild cognitive impairment: a possible predictor of Alzheimer disease. *Arch. Neurol.* **59**, 972–976.

PRATICO, D., URYU, K., LEIGHT, S., TROJANOSWKI, J. Q., LEE, V. M. (**2001**). Increased lipid peroxidation precedes amyloid plaque formation in an animal model of Alzheimer amyloidosis. *J. Neurosci.* **21**, 4183–4187.

REIBER, H. (**2001**). Dynamics of brain-derived proteins in cerebrospinal fluid. *Clin. Chim. Acta* **310**, 173–186.

REIBER, H., PETER, J. B. (**2001**). Cerebrospinal fluid analysis: disease related data patterns and evaluation programs. *J. Neurol. Sci.* **184**, 101–122.

REIBER, H., UNGEFEHR, S., JACOBI, C. (1998). The intrathecal, polyspecific and oligoclonal immune response in multiple sclerosis. *Mult. Scler.* **4**, 111–117.

RIEMENSCHNEIDER, M., SCHMOLKE, M., LAUTENSCHLAGER, N., GUDER, W. G., VANDERSTICHELE, H., VANMECHELEN, E., KURZ, A. (2000). Cerebrospinal beta-amyloid $_{(1–42)}$ in early Alzheimer's disease: association with apolipoprotein E genotype and cognitive decline. *Neurosci. Lett.* **284**, 85–88.

ROSNER, H., REBHAN, M., VACUN, G., VANMECHELEN, E. (1995). Developmental expression of tau proteins in the chicken and rat brain: rapid down-regulation of a paired helical filament epitope in the rat cerebral cortex coincides with the transition from immature to adult tau isoforms. *Int. J. Dev. Neurosci.* **13**, 607–617.

SCHOONENBOOM, N. S., PIJNENBURG, Y. A., MULDER, C., ROSSO, S. M., VAN ELK, E. J., VAN KAMP, G. J., VAN SWIETEN, J. C., SCHELTENS, P. (2004). Amyloid beta(1–42) and phosphorylated tau in CSF as markers for early-onset Alzheimer disease. *Neurology* **62**, 1580–1584.

SENGUPTA, A., KABAT, J., NOVAK, M., WU, Q., GRUNDKE-IQBAL, I., IQBAL, K. (1998). Phosphorylation of tau at both Thr 231 and Ser 262 is required for maximal inhibition of its binding to microtubules. *Arch. Biochem. Biophys.* **357**, 299–309.

SESHADRI, S., BEISER, A., SELHUB, J., JACQUES, P. F., ROSENBERG, I. H., D'AGOSTINO, R. B., WILSON, P. W., WOLF, P. A. (2002). Plasma homocysteine as a risk factor for dementia and Alzheimer's disease. *N. Engl. J. Med.* **346**, 476–483.

SHAHANI, N., BRANDT, R. (2002). Functions and malfunctions of the tau proteins. *Cell Mol. Life Sci.* **59**, 1668–1680.

SINGH, T. J., WANG, J. Z., NOVAK, M., KONTZEKOVA, E., GRUNDKE-IQBAL, I., IQBAL, K. (1996). Calcium/calmodulin-dependent protein kinase II phosphorylates tau at Ser-262 but only partially inhibits its binding to microtubules. *FEBS Lett.* **387**, 145–148.

SJÖGREN, M., VANDERSTICHELE, H., AGREN, H., ZACHRISSON, O., EDSBAGGE, M., WIKKELSO, C., SKOOG, I., WALLIN, A., WAHLUND, L. O., MARCUSSON, J., NAGGA, K., ANDREASEN, N., DAVIDSSON, P., VANMECHELEN, E., BLENNOW, K. (2001). Tau and Abeta42 in cerebrospinal fluid from healthy adults 21–93 years of age: establishment of reference values. *Clin. Chem.* **47**, 1776–1781.

SJÖGREN, M., GISSLEN, M., VANMECHELEN, E., BLENNOW, K. (2001). Low cerebrospinal fluid β-amyloid 42 in patients with acute bacterial meningitis and normalization after treatment. *Neurosci. Lett.* **314**, 33–36.

STAMER, K., VOGEL, R., THIES, E., MANDELKOW, E., MANDELKOW, E. M. (2002). Tau blocks traffic of organelles, neurofilaments, and APP vesicles in neurons and enhances oxidative stress. *J. Cell Biol.* **156**, 1051–1063.

STRITTMATTER, W. J., SAUNDERS, A. M., SCHMECHEL, D., PERICAK-VANCE, M., ENGHILD, J., SALVESEN, G. S., ROSES, A. D. (1993a). Apolipoprotein E: high-avidity binding to beta-amyloid and increased frequency of type 4 allele in late-onset familial Alzheimer disease. *Proc. Natl. Acad. Sci. USA* **90**, 1977–1981.

STRITTMATTER, W. J., WEISGRABER, K. H., HUANG, D. Y., DONG, L. M., SALVESEN, G. S., PERICAK-VANCE, M., SCHMECHEL, D., SAUNDERS, A. M., GOLDGABER, D., ROSES, A. D. (1993b). Binding of human apolipoprotein E to synthetic amyloid beta peptide: isoform-specific effects and implications for late-onset Alzheimer disease. *Proc. Natl. Acad. Sci. USA* **90**, 8098–8102.

SUNDERLAND, T., LINKER, G., MIRZA, N., PUTNAM, K. T., FRIEDMAN, D. L., KIMMEL, L. H., BERGESON, J., MANETTI, G. J., ZIMMERMANN, M., TANG, B., BARTKO, J. J., COHEN, R. M. (2003). Decreased beta-amyloid 1-42 and increased tau levels in cerebrospinal fluid of patients with Alzheimer disease. *JAMA* **289**, 2094–2103.

TAKEI, Y., TENG, J., HARADA, A., HIROKAWA, N. (2000). Defects in axonal elongation and neuronal migration in mice with disrupted tau and map1b genes. *J. Cell Biol.* **150**, 989–1000.

TARKOWSKI, E., ISSA, R., SJOGREN, M., WALLIN, A., BLENNOW, K., TARKOWSKI, A., KUMAR, P. (2002). Increased intrathecal levels of the angiogenic factors VEGF and TGF-beta in Alzheimer's disease and vascular dementia. *Neurobiol. Aging* 23, 237–243.

TATEBAYASHI, Y., HAQUE, N., TUNG, Y. C., IQBAL, K., GRUNDKE-IQBAL, I. (2004). Role of tau phosphorylation by glycogen synthase kinase-3beta in the regulation of organelle transport. *J. Cell Sci.* 117, 1653–1663.

THE RONALD AND NANCY REAGAN RESEARCH INSTITUTE OF THE ALZHEIMER'S ASSOCIATION AND THE NATIONAL INSTITUTE ON AGING WORKING GROUP (1998). Consensus report of the Working Group on: "Molecular and Biochemical Markers of Alzheimer's Disease". *Neurobiol. Aging* 19, 109–116.

TUMANI, H., NOLKER, G., REIBER, H. (1995). Relevance of cerebrospinal fluid variables for early diagnosis of neuro-borreliosis. *Neurology* 45, 1663–1670.

UJIIE, M., DICKSTEIN, D. L., JEFFERIES, W. A. (2002). p97 as a biomarker for Alzheimer disease. *Front Biosci.* 7, E42–47.

VANMECHELEN, E., VANDERSTICHELE, H., DAVIDSSON, P., VAN KERSCHAVER, E., VAN DER PERRE, B., SJÖGREN, M., ANDREASEN, N., BLENNOW, K. (2000). Quantification of tau phosphorylated at threonine 181 in human cerebrospinal fluid: a sandwich ELISA with a synthetic phosphopeptide for standardization. *Neurosci. Lett.* 285, 49–52.

VANMECHELEN, E., VAN KERSCHAVER, E., BLENNOW, K., DE DEYN, P. P., GARTNER, F. H., PARNETTI, L., SINDIC, C. J. M., ARAI, H., RIEMENSCHNEIDER, M., HAMPEL, H., POTTEL, H., VALGAEREN, A., HULSTAERT, F., VANDERSTICHELE, H. (2001). CSF-phospho-tau (181P) as a promising marker for discriminating Alzheimer's disease from dementia with Lewy bodies. In IQBAL, K., SISODIA, S. S., WINBLAD, B. (Eds.), *Alzheimer's Disease: Advances in Etiology, Pathogenesis and*

Therapeutics. Wiley, Chichester, pp. 285–291.

VIGNALI, D. A. (2000). Multiplexed particle-based flow cytometric assays. *J. Immunol. Methods* 243, 243–255.

WILTFANG, J., AROLD, N., NEUHOFF, V. (1991). A new multiphasic buffer system for sodium dodecyl sulfate–polyacrylamide gel electrophoresis of proteins and peptides with molecular masses 100 000–1000, and their detection with picomolar sensitivity. *Electrophoresis* 12, 352–366.

WILTFANG, J., ESSELMANN, H., BIBL, M., SMIRNOV, A., OTTO, M., PAUL, S., SCHMIDT, B., KLAFKI, H. W., MALER, M., DYRKS, T., BIENERT, M., BEYERMANN, M., RÜTHER, E., KORNHUBER, J. (2002). Highly conserved and disease-specific patterns of carboxyterminally truncated Aβ peptides 1–37/38/39 in addition to 1–40/42 in Alzheimer's disease and in patients with chronic neuroinflammation. *J. Neurochem.* 81, 481–496.

WILTFANG, J., ESSELMANN, H., CUPERS, P., NEUMANN, M., KRETZSCHMAR, H., BEYERMANN, M., SCHLEUDER, D., JAHN, H., RÜTHER, E., KORNHUBER, J., ANNAERT, W., DE STROOPER, B., SAFTIG, P. (2001a). Elevation of β-amyloid peptide 2-42 in sporadic and familial Alzheimer's disease and its generation in PS1 knockout cells. *J. Biol. Chem.* 276, 42645–42657.

WILTFANG, J., ESSELMANN, H., MALER, J. M., BLEICH, S., HÜTHER, G., KORNHUBER, J. (2001b). Molecular biology of Alzheimer's dementia and its clinical relevance to early diagnosis and new therapeutic strategies. *Gerontology* 47, 65–71.

ZETTERBERG, H., ANDREASEN, N., BLENNOW, K. (2004). Increased cerebro-spinal fluid levels of transforming growth factor-beta1 in Alzheimer's disease. *Neurosci. Lett.* 367, 194–196.

ZETTERBERG, H., WAHLUND, L. O., BLENNOW, K. (2003). Cerebrospinal fluid markers for prediction of Alzheimer's disease. *Neurosci. Lett.* 352, 67–69.

15
Proteomics in Alzheimer's Disease

Michael Fountoulakis, Sophia Kossida and Gert Lubec

Abstract

Proteomics has been widely used in the investigation of neurodegeneration, in particular in the quantification of differences in the postmortem brains of controls and patients with Alzheimer's disease. In brain proteomics, total protein extract from several brain regions has mainly been analyzed by two-dimensional electrophoresis followed by matrix-assisted laser desorption ionization time-of-flight mass spectrometry since this approach allows protein level quantification. The proteins with deranged levels and modifications in the Alzheimer's disease brain are mainly signal transduction, guidance, antioxidant, and heat-shock proteins. In this article, we provide a protocol for the proteomic analysis of the brain and a summary of the major changes of protein levels and modifications related to Alzheimer's disease found by proteomics technologies.

15.1
Introduction

Alzheimer's disease (AD) is a well-studied dementia that mostly affects elderly people and for which no reliable premortem diagnostic marker exists today (Pasinetti and Ho, 2001; Jacobsen, 2002; Dodel et al., 2003). Loss of hippocampal and cortical neurons leads to impairment of memory and cognitive functions and is characterized by amyloid deposits with neurofibrillary tangles. Despite extensive studies, however, no casual relationship has been established between amyloid deposits in the brain and the neuronal and cognitive deficits seen in AD (Selkoe, 2000, 2002). It is likely that AD is more than one disease, due to multiple etiologies with similar clinical and pathological pathways. Age appears to be the most important risk factor for AD and an array of age-related changes are most likely the triggers that initiate a cascade of events resulting in neuronal and synaptic loss factors, dementia, and the hallmark pathologic changes.

Proteomics in Drug Research
Edited by M. Hamacher, K. Marcus, K. Stühler, A. van Hall, B. Warscheid, H. E. Meyer
Copyright © 2006 Wiley-VCH Verlag GmbH & Co. KGaA, Weinheim
ISBN: 3-527-31226-9

With the recent genomics methods, it has become possible to map and identify genes that are mutated in a wide variety of neurodegenerative disorders, like presenilins in AD. Next to genomics, which only provides a partial view of the biological problem, proteomics is widely used in clinical diagnosis today and many changes in protein levels resulting from specific disorders have been identified. To study AD, proteomics has mainly been used in the analysis of the human brain, the brains of animals serving as models for AD, and of cerebrospinal fluid from patients with AD with the goal to collect information about gene products involved in these diseases, i.e., alterations in protein levels and posttranslational modifications (Fountoulakis, 2001). In addition to AD, brains from patients with Down syndrome (DS) and of the corresponding animal models have been studied. DS is the most frequent genetic cause of dementia, the pathomechanisms of which are not well understood. Almost all subjects with DS over 40 years show neuropathological and neurochemical abnormalities on postmortem brain examinations that are indistinguishable from those seen in Alzheimer's disease. Thus, results from the study of the DS brain may be useful in AD studies (Antonarakis, 1998; Cairns, 1999; Engidawork and Lubec, 2001, 2003).

In the proteomic analysis of the brain, two-dimensional gels for protein separation, followed by matrix-assisted laser desorption ionization time-of-flight mass spectrometry for protein identification, have mainly been employed. This classical proteomics approach allows for the quantification of changes in protein levels and modifications. Simultaneously, it is a robust, well-established method that finds wide application in the study of biological systems. In this article, we provide a description of the protocols of the proteomic analysis used in our laboratory and a summary of the major findings from our group and other neuroproteomics groups.

15.2
Proteomic Analysis

15.2.1
Sample Preparation

Careful sample selection and preparation are the prerequisites for a successful analysis (Boguski and McIntosh, 2003). Postmortem brain samples for our studies were obtained from the MRC London Brain Bank for Neurodegenerative Diseases, Institute of Psychiatry. The AD patients fulfilled the National Institute of Neurological and Communicative Disorders and Stroke and Alzheimer's Disease and Related Disorders Association (NINCDS/ADRDA) criteria for probable AD (Mirra et al., 1991). A "definite" diagnosis of AD was performed by historical analysis of the brain samples, which was consistent with the CERAD criteria (Tierney et al., 1988). The brain regions temporal, frontal, occipital, parietal cortex, and cerebellum of patients with AD (72.3 ± 7.6 years old) and controls (72.6 ± 9.6 years old) (Seidl et al., 1997) were used for the studies at the protein level. The controls were brains from individuals with no history of neurological or psychiatric illness. The major cause of death was bronchopneumonia in the AD patients and heart disease in the control patients. The postmortem interval of brain dissection in AD and controls was 34.1 ± 13.7 and 34.8 ± 15.0 h, respectively. Tissue samples were stored at –80°C and the freezing chain was never interrupted.

We prepare total brain protein extract according to a general protocol (Fountou-lakis et al., 2002). Brain tissue (ca. 0.25 g) is placed in Bio101 tubes (fast RNA tubes). To each tube, 0.8 mL of 20 mM Tris, which contains 7 M urea, 2 M thiourea, 4% CHAPS, 10 mM 1,4-dithioerythritol, 1 mM EDTA, and protease inhibitors (1 mM PMSF and 1 tablet Complete per 50 mL of suspension buffer), and phosphatase inhibitors (0.2 mM Na_2VO_3 and 1 mM NaF) are added. Homogeni-zation is performed twice, for 20 s each time, in a FastPrep FP120 lab shaker, at speed 4. The suspension is centrifuged at 8000 × g for 10 min. The supernatant is separated from the beads, centrifuged further at 100 000 × g for 30 min, and the supernatant of this step is applied onto IPG strips.

For the preparation of cytosolic, membrane, and mitochondrial fractions, brain tissue (0.5 g) is suspended in 2.5 mL of 20 mM Hepes-OH, pH 7.5, that contains 320 mM sucrose, 1 mM EDTA, 5 mM dithierythritol, and the protease inhibitors mixture. The suspension is homogenized with the use of a Teflon potter homo-genizer, and is centrifuged at 800 × g for 10 min to remove nuclei and undissolved material. The supernatant is centrifuged at 10 000 × g for 15 min to separate the mitochondrial proteins. The supernatant of that centrifugation step is centrifuged

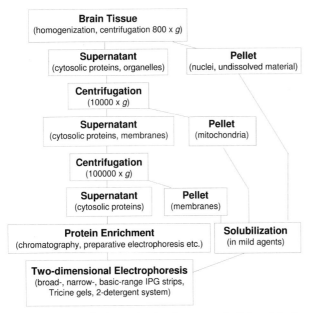

Figure 15.1 General scheme showing the steps to be followed for the preparation of cytosolic, mitochondrial, and membrane brain proteins and for the enrichment of the cytosolic proteins prior to the 2-D gel electrophoresis analysis.

at 100 000 × *g* for 1 h. The supernatant of the last step mainly contains the cytosolic proteins and is used for 2-D electrophoresis without change. Each of the two precipitates, which mainly contain the mitochondrial and microsomal proteins, is suspended in 1 mL of sample buffer that consists of 20 mM Tris, 7 M urea, 2 M thiourea, 4% CHAPS, 10 mM 1,4-dithioerythritol, 1 mM EDTA, and the protease and phosphatase inhibitors. The suspension is centrifuged at 100 000 × *g* for 1 h. Figure 15.1 shows the general scheme followed for the preparation of cytosolic, membrane, and mitochondrial fractions.

15.2.2
Two-Dimensional Electrophoresis

We perform the two-dimensional electrophoresis essentially as reported (Langen et al., 1997), using immobilized pH-gradient (IPG) strips of broad (3–10) and narrow (3–7) pH range. For isoelectric focusing, samples are applied in sample cups at their basic and acidic ends. Focusing starts at 200 V. The voltage is then gradually increased to 5000 V at 3 V/min and is kept constant for a further 24 h. The second-dimensional separation is performed either on 9–16% gradient or on 12% constant SDS polyacrylamide gels run at 40 mA per gel. Large-format gels (ca. 20 × 20 cm) allow an efficient spot resolution and such gels are preferred when the goal is the detection of posttranslational modifications. Because of the large gel size, the spots are relatively large, protein losses may occur, and a large protein amount should be applied on such gels (ca. 0.3–1.5 mg total protein). In the small-format gels (ca. 10 × 10 cm), the spot resolution is often compromised; however, a low protein amount (ca. 0.1 mg) can be applied. After protein fixation for 12 h in 40% methanol containing 5% phosphoric acid, the gels are stained with colloidal Coomassie blue for 24 h. Molecular masses are determined by analyzing standard protein markers that cover the range 10–200 kDa. pI values are used as given by the supplier of the IPG strips. Excess dye is washed from the gels with H_2O, and the gels are scanned in an Agfa DUOSCAN densitometer (resolution 400). Electronic images of the gels are recorded with Photoshop (Adobe) software. The images were stored in TIFF (about 5 Mbytes/file) and JPEG (about 50 Kbytes/file) format. The gels can be immediately used for MS analysis or sealed in plastic foil and stored at 4°C. On a 2-D gel stained with Coomassie blue, on which 1 mg of total protein has been applied, about 3000 spots can be detected with the ImageMaster 2D Elite software. The use of narrow pH range IPG strips results in a higher resolution and the detection of additional spots that represent posttranslational modifications of already detected proteins. Less often it results in the detection of new gene products not found with the broad range IPG strips.

15.2.3
Protein Quantification

Protein quantification is usually performed with commercially available software for spot detection and quantification. Spots are first outlined automatically and further manually outlined and quantified using the ImageMaster 2D Elite software (Amersham Biosciences). The percentage of the volume of the spot(s) that represent a certain protein is determined in comparison with the total proteins present in the 2-D gel after background subtraction. Several gels (at least three, usually five) that carry the same sample are evaluated to determine average values and reduce the effect of gel artifacts. Only changes in the protein levels that are statistically significant are considered ($P < 0.05$). That approach has the advantage that it is relatively simple and relies exclusively on 2-D gels. A disadvantage is the artifacts common to 2-D gels (the whole amount of a protein does not always enter the IPG strip, in particular, for large components of a protein mixture) and the inaccuracy of level determination because of the large volume difference between weak and strong spots.

15.2.4
Matrix-Assisted Laser Desorption/Ionization Time-of-Flight Mass Spectroscopy

The most efficient and most widely used protein identification method in proteomics is matrix-assisted laser desorption/ionization time-of-flight mass spectroscopy (MALDI-TOF MS) (Lahm and Langen, 2000). We perform MALDI-TOF MS analyses essentially as described elsewhere with certain modifications due to improved technologies and software (Jiang et al., 2003). The spots representing the proteins of interest are excised with a spot picker and are placed into 96-well microtiter plates. The spots are destained with 100 µL of 30% acetonitrile in 50 mM ammonium bicarbonate in a CyBi-Well apparatus (Cybio AG, Jena, Germany). After destaining, each gel piece is washed with 100 µL of H_2O for 5 min, and is dried in a speedvac evaporator without heating for 45 min. Each dried gel piece is rehydrated with 5 µL of 1 mM ammonium bicarbonate containing 50 ng trypsin. After 16 h at room temperature, 20 µL of 50% acetonitrile containing 0.3% trifluoroacetic acid is added to each gel piece. The gel pieces are incubated for 15 min with constant shaking. A peptide mixture (1.5 µL) is simultaneously applied with 1 µL of matrix solution consisting of 0.025% α-cyano-4-hydroxycinnamic acid and containing the standard peptides des-Arg-bradykinin (20 nM, 904.4681 Da) and adrenocorticotropic hormone fragment 18–39 (20 nM, 2465.1989 Da) in 65% ethanol, 32% acetonitrile, and 0.03% trifluoroacetic acid, to the AnchorChip. The sample application is performed with the CyBi-Well apparatus. The use of the hydrophobic surface target with subsequent peptide concentration allows for a higher identification rate, in particular, from weak spots. Samples are reflected in a time-of-flight mass spectrometer (Ultraflex TOF-TOF, Bruker Daltonics) in the reflectron mode. An accelerating voltage of 20 kV is used. Proteins are identified on the basis of peptide-mass matching. Peptides of masses between 850 and 3500 kDa are acquired, and the masses are corrected with the standard peptides added to each sample. The corrected masses are compared to the theoretical peptide masses of all available proteins from all species using software developed in-house (Berndt et al., 1999) and searching in the major, public-domain databases, which are managed and updated in-house.

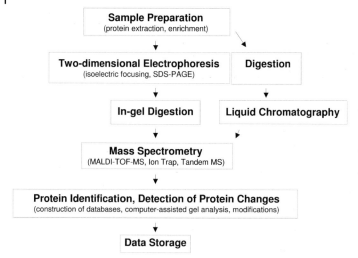

Figure 15.2 Workflow in proteomics.

Monoisotopic masses are considered and a mass tolerance of 0.0025% or lower is allowed. Four matching peptides is the minimal requirement for an identity assignment. The probability of a false positive match with a given mass spectrum is determined for each analysis. Unmatched peptides or miscleavage are usually not considered for further, automated protein searches. The most abundant brain proteins detected in 2-D gels are structural proteins, like tubulin chains and actin, heat shock proteins, dihydropyrimidinase-related proteins-2 and -3, and house-keeping enzymes. The less abundant proteins are mainly hypothetical proteins, enzymes, as well as structural proteins. The relative levels of the low-abundance brain proteins vary in the range of 0.005–0.01% and of the high-abundance proteins about 3%. Figure 15.2 shows the general workflow scheme usually followed in the proteomic analysis of the brain.

15.3
Proteins with Deranged Levels and Modifications in AD

Most proteomics analyses involve 2-D electrophoresis and MALDI-TOF MS steps, as this approach allows for a reliable quantification of changes in protein levels. In one study, the LC-MS/MS approach was used for the detection of proteins enriched in amyloid plaques in the AD brain (Liao et al., 2004). Employing the former proteomic approach, levels of about 100 proteins were found to be changed in the AD brain and cerebrospinal fluid (CSF) in comparison with the control brain and CSF, respectively (Table 15.1). These proteins have various functions, mainly involved in neurotransmission, guidance, signal transduction, metabolism, detoxification, and conformational changes. The CSF proteins are plasma proteins.

Table 15.1 Brain proteins with deranged levels in AD. Human brain proteins reported to show changed levels or to be modified in Alzheimer's disease (AD) and Down syndrome (DS) brain are listed. The differences were found by application of proteomics technologies. In the column "Change", increase (I) or decrease (D) at the protein levels are indicated together with the modification observed. In the column "Reference", the corresponding literature is listed.

Protein	Change	Reference
14-3-3 Beta/alpha	Enriched in amyloid plaques	Liao et al. 2004
14-3-3 Gamma protein	I	Fountoulakis et al. 1999; Fountoulakis et al. 2004
14-3-3 Epsilon protein	I	Fountoulakis et al. 1999; Fountoulakis et al. 2004
14-3-3 Zeta	Enriched in amyloid plaques	Liao et al. 2004
2'-3'-Cyclic nucleotide 3-phosphodiesterase	D	Vlkolinsky et al. 2001
Alcohol dehydrogenase (ADH)	I	Balcz et al. 2001
Alpha-1 antitrypsin	CSF, I	Puchades et al. 2003
Alpha-1 beta glycoprotein	CSF, D	Puchades et al. 2003
Alpha crystallin B chain	I	Yoo et al. 2001d
Alpha-endosulfine	D	Kim et al. 2001c
Alpha-enolase	Oxidation in AD/Nitration	Castegna et al. 2002b
Amyloid beta-peptide	Enriched in amyloid plaques	Liao et al. 2004
Antitrypsin	Enriched in amyloid plaques	Liao et al. 2004
Apoptosis repressor with caspase recruitment domain (ARC)	I	Engidawork et al. 2001c
ARPP-19	D	Kim et al. 2001d
ATP synthase alpha chain, mitochondrial	D	Tsuji et al. 2002
ATP synthase beta chain, mitochondrial	I	Tsuji et al. 2002
Vacuolar ATP synthase subunit B, brain isoform	D in DS	Kim et al. 2000b
Apolipoprotein A1	CSF, D	Puchades et al. 2003
Apolipoprotein E	CSF, D	Puchades et al. 2003
Apolipoprotein J	CSF, D	Puchades et al. 2003
ATPase, Ca^{++} transporting	Enriched in amyloid plaques	Liao et al. 2004
ATPase, H^{+} transporting, lysosomal V0 subunit A	Enriched in amyloid plaques	Liao et al. 2004

Table 15.1 (continued)

Protein	Change	Reference
ATPase, H⁺ transporting, lysosomal V1 subunit B	Enriched in amyloid plaques	Liao et al. 2004
ATPase, H⁺ transporting, lysosomal V1 subunit D	Enriched in amyloid plaques	Liao et al. 2004
ATPase, H⁺ transporting, lysosomal V1 subunit E	Enriched in amyloid plaques	Liao et al. 2004
Beta-2 microglobulin	CSF, I	Davidsson et al. 2002
Beta-actin	Oxidation in AD/D/Nitration	Butterfield et al. 2002; Butterfield et al. 2004; Castegna et al. 2003
Beta-enolase	D	Tsuji et al. 2002
Beta-soluble N-ethylmaleimide-sensitive factor attachment protein (Beta SNAP)	D	Yoo et al. 2001a
Bim/BOD (Bcl-2 interacting mediator of cell death/Bcl-2 related ovarian death gene)	I	Engidawork et al. 2001b
cAMP-dependent protein kinase (PKA)	D	Kim et al. 2001d
c, large [catalytic] subunit [precursor]	I	Tsuji et al. 2002
Carbonyl reductase (CBR)	I	Balcz et al. 2001
Cathepsin D	Enriched in amyloid plaques	Liao et al. 2004
Cell cycle progression 8 protein	CSF, D	Puchades et al. 2003
Creatine kinase BB	Oxidation in AD/D	Castegna et al. 2002a; Aksenov et al. 2000
Collagen I, alpha-1, polypeptide	Enriched in amyloid plaques	Liao et al. 2004
Coronin, actin binding protein	Enriched in amyloid plaques	Liao et al. 2004
Cystatin B	Enriched in amyloid plaques	Liao et al. 2004
Cystatin C	Enriched in amyloid plaques	Liao et al. 2004
Ubiquinol-cytochrome-c reductase complex III core protein I	D	Kim et al. 2000b
Drebrin	D	Shim et al. 2002
Didyropyrimidinase-related protein 2	Oxidation in AD/D	Castegna et al. 2002b; Tsuji et al. 2002
Didyropyrimidinase-related protein 2 (DRP-2)	D	Lubec et al. 1999

Table 15.1 (continued)

Protein	Change	Reference
DNA fragmentation factor 45 (DFF45)	D	Engidawork et al. 2001c
Dynamin 1	Enriched in amyloid plaques	Liao et al. 2004
Dynein, heavy polypeptide 1	Enriched in amyloid plaques	Liao et al. 2004
Fatty acid-binding protein, heart	D	Tsuji et al. 2002
Fibrinogen, gamma	Enriched in amyloid plaques	Liao et al. 2004
Fas associated death domain (FADD)-like interleukin-1beta-converting enzyme inhibitory proteins (FLIP)		Engidawork et al. 2001c
Gamma-Enolase	Oxidation in AD/Nitration	Castegna et al. 2003
Glial fibrillary acidic protein (GFAP)	I	Greber et al. 1999
Glucose regulated protein 75 kDa = GRP75	D	Yoo et al. 2001d
Glutamine synthase	Oxidation in AD/D	Castegna et al. 2002a; Bajo et al. 2001
Glutamate transporter, EAAT2	Oxidation in AD/D	Butterfield et al. 2004
Guanine nucleotide-binding protein beta subunit	D	Tsuji et al. 2002
GRP 94	I	Yoo et al. 2001c
Heat shock cognate 71	Oxidation in AD	Castegna et al. 2002b
Heat shock 60 kDa, mitochondrial [precursor]	D	Tsuji et al. 2002
Heat shock protein 70 RY	I	Yoo et al. 2001c
Heat shock 90kDa protein 1, beta	Enriched in amyloid plaques	Liao et al. 2004
Histamin-releasing factor	D	Kim et al. 2001a
Kininogen	CSF, D	Puchades et al. 2003
L-Lactate dehydrogenase	Oxidation in AD/Nitration	Castegna et al. 2003
NADH-ubiquinone oxidoreductase 75 kDa subunit, mitochondrial [Precursor]	D	Kim et al. 2001e
NADH-ubiquinone oxidoreductase 24 kDa subunit, mitochondrial [Precursor]	D	Kim et al. 2001e
Neuronal acetylcholine receptor protein, alpha-3 chain [Precursor] 26-kDa 3 subunit nAChR	D	Engidawork et el. 2001a

Table 15.1 (continued)

Protein	Change	Reference
Neuronal acetylcholine receptor protein, alpha-3 chain [Precursor] 45-kDa 3 subunit nAChR	I	Engidawork et el. 2001a
Neuronal acetylcholine receptor protein, alpha-7 chain [Precursor]	D	Engidawork et el. 2001a
Neuropolypeptide h3	Oxidation in AD/Nitration	Butterfield et al. 2004; Castegna et al. 2003
NF-L	D	Bajo et al. 2001
Nucleoside diphosphate kinase	D	Kim et al. 2002
p21	I	Engidawork et al. 2001b
Peroxiredoxin I (Prx-I)	I	Kim et al. 2001b
Peroxiredoxin II(Prx-II)	I	Kim et al. 2001b; Krapfenbauer et al. 2003a
Peroxiredoxin III (Prx-III)	D	Kim et al. 2001b
Peroxiredoxin VI	I	Krapfenbauer et al. 2003a
Antioxidant protein 2 (1-Cys peroxiredoxin) (1-Cys Prx)	I	Krapfenbauer et al. 2002
Phosphofructokinase, muscle ttype	Enriched in amyloid plaques	Liao et al. 2004
Phosphofructokinase, platelet type	Enriched in amyloid plaques	Liao et al. 2004
Proapolipoprotein	CSF, D	Davidsson et al. 2002
Procaspase-3	D	Engidawork et al. 2001c
Procaspase-8	D	Engidawork et al. 2001c
Procaspase-9	D	Engidawork et al. 2001c
Receptor interacting protein (RIP)- like interacting CLARP kinase (RICK)	I	Engidawork et al. 2001c
Retinol-binding protein	CSF, D	Puchades et al. 2003
Stathmin	D	Cheon et al. 2001
Synaptosomal associated protein 25 kDa (Snap-25)	D	Greber et al. 1999
Synaptotagmin I	D	Yoo et al. 2001a
Tau	Enriched in amyloid plaques	Liao et al. 2004
TCP-1	D	Schuller et al. 2001
Transthyretin	CSF, I	Davidsson et al. 2002
Triosephosphate isomerase	Oxidation in AD/Nitration	Bajo et al. 2001; Butterfield et al. 2002

Table 15.1 (continued)

Protein	Change	Reference
Ubiquitin-activating enzyme E1	Enriched in amyloid plaques	Liao et al. 2004
Ubiquitin carboxy-terminal hydrolase L-1	Oxidation in AD	Castegna et al. 2002a
Vacuolar ATPase subunit H	Enriched in amyloid plaques	Liao et al. 2004
VDAC1 pI 10.0	D	Yoo et al. 2001b
VDAC2	I	Yoo et al. 2001b
Vimentin	Enriched in amyloid plaques	Liao et al. 2004
Zipper interacting protein kinase	I	Engidawork et al. 2001b

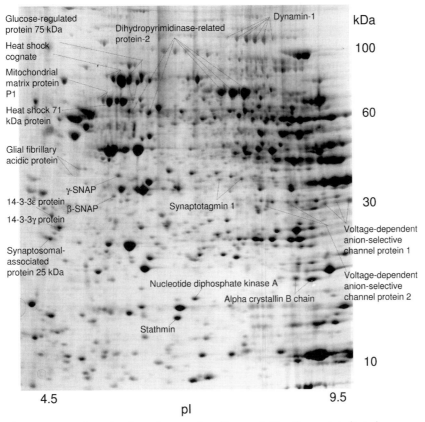

Figure 15.3 Two-dimensional protein map of the human brain. Proteins from the control brain were analyzed on a pH 3–10 nonlinear IPG strip, followed by a 9–16% SDS-poly-acrylamide gel. The gel was stained with Coomassie blue. The spots indicated represent proteins for which we found deranged levels in AD and DS. The proteins are listed in Table 15.1 (not all proteins are indicated in the figure).

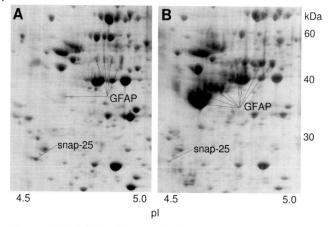

Figure 15.4 Partial 2D-gel images showing examples of altered protein levels in the AD brain. Brain samples were from the parietal cortex of a control (A) and a patient with AD (B). The spots that represent glial fibrillary acidic protein (GFAP) and synaptosomal protein 25 kDa (snap25) are indicated. Higher GFAP levels and lower snap25 are seen in the AD brain (B).

Figure 15.3 shows the spots representing proteins with changed levels in the AD brain, which were found in our laboratory using proteomics technologies.

The observed changes are usually on the order of 20–100% in comparison with the average levels of the control group. The largest changes were observed for the glial fibrillary acidic protein (GFAP), which distinguishes astrocytes from other glial cells during the development of the brain. For this protein, ten-fold or even higher levels were observed in individual AD brains, compared to the average of the control group (Greber et al., 1999). In Figure 15.4, a comparison of the GFAP signals in control and AD brain is shown. More spots are seen in the AD brain, in particular for the shorter GFAP forms.

The proteins with deranged levels or modifications, which were found in the AD and DS brain and in CSF by applying proteomics technologies (Table 15.1) are potential drug targets and diagnostic markers. The changes found by proteomics need to be investigated by application of other approaches, for example, to confirm the observed changes by immunoblots and specific assays and to design experiments to study how specific the alterations observed are for AD.

15.3.1
Synaptosomal Proteins

Synaptosomal proteins are involved in synaptic transmission, which is deranged in AD. The machinery consists of synaptosomal proteins, including soluble N-ethylmaleimide-sensitive factor attachment proteins (SNAPs) present as three isoforms, α, β and γ, synaptosomal-associated protein 25 (SNAP 25), synapto-tagmin, and vesicular proteins. Proteomics studies showed reduced expression for SNAP 25 (Figure 15.4) (Greber et al., 1999), β-SNAP and synaptotagmin I (Yoo et al., 2001a) in the AD brain. These findings suggest that impaired neurotransmission is apparent in AD, secondary to deterioration of the exocytotic machinery.

15.3.2
Guidance Proteins

Guidance is facilitated by a variety of diffusible proteins, signaling the correct path to the growing axon. Proteomic studies of these molecular cues revealed deranged expression (Lubec et al., 1999) and oxidative modification (Castegna et al., 2002b) of dihydropyrimidinase-related protein (DRP-2) in the AD brain. DRP-2 is also known as collapsin response mediator protein-2. It is a cytosolic protein that shows a high homology to the rat, turned on after division 64 kDa protein and to the chicken collapsin response mediator protein-62, both of which are involved in pathfinding and migration of neurons. Two important functions are ascribed to DRP-2 in neurons. It interacts with and modulates the activity of collapsin, a protein that elongates dendrites and directs them to adjacent neurons, and neuronal repair. The reduced levels of DRP-2 in the AD brain may be responsible for synaptic loss in AD as it can not respond to the formation of new synapses. Moreover, DRP-2 is immediately modified by oxidation and loses activity, which results in shortened dendrites that would cause decreased interneuronal communication and memory impairment (Butterfield et al., 2003).

15.3.3
Signal Transduction Proteins

Altered expression in AD of a series of proteins related to synaptic function, neurite outgrowth, and signaling processes has been reported. Thus, reduced levels were found for 2', 3'-cyclic nucleotide-3'-phosphodiesterase (CNPase), nucleoside diphosphate kinase (NDK)-A, and stathmin (Cheon et al., 2001; Vlkolinsky et al., 2001; Kim et al., 2002) as well as increased expression for 14-3-3 gamma and epsilon proteins (Fountoulakis et al., 1999; Greber et al., 1999). CNPase is associated with oligodendroglia and hence myelination. NDKs are essential oligomeric enzymes that catalyze the transfer of a phosphate from a nucleoside triphosphate to a diphosphate, thereby playing a key role in the maintenance of the intracellular pool of all deoxynucleotide triphosphates and nucleotide triphosphates, which endows them with a regulatory role in activation of the classical signaling proteins, G-proteins (Engidawork and Lubec, 2003). Stathmin, a major substrate of cyclin-dependent kinases and protein kinase A, plays a key role in multiple signal transduction pathways. Its negative correlation with neurofibrillary tangles, a hallmark of AD neuropathology, and interaction with tubulin to cause microtuble destabilization, give stathmin a role in neurodegeneration and impaired signaling (Cheon et al., 2001). 14-3-3 proteins form homo- or heterodimers and are primarily, but not exclusively, expressed in neurons, and bind to and modulate the function of a wide variety of cellular proteins. They are involved in neuronal development, signal transduction, cell growth, and cell death. They show increased levels in AD, which may be linked to apoptosis rather than to other cellular activities, as they are highly enriched in areas of massive neuronal death caused by pathological processes (Engidawork and Lubec, 2001d). The increased levels of 14-3-3 proteins in AD may represent a protective response against apoptosis.

15.3.4
Oxidized Proteins

Oxidative stress, detected as protein oxidation, lipid peroxidation, DNA oxidation, and 3-nitrotyrosine formation, is considered to be a common mechanism for different age-related neurodegenerative pathologies. It is observed in the AD brain in regions where amyloid β-peptide 1-42 is present, but not in the cerebellum, where AD pathology is not observed (Butterfield, 2002). Battlefield and coworkers used 2-D gels to separate AD and control brain proteins. The separated proteins were treated with 2,4-dinitrophenylhydrazine to form hydrazone and the modified proteins were detected by immunoblots and identified by MALDI-TOF MS (Table 15.1). One of the oxidized proteins identified is creatine kinase (BB isoform). Amyloid β-peptide inhibits creatine kinase and causes lipid peroxidation leading to decreased energy utilization, alterations in cytoskeletal proteins, and increased excitotoxicity, all of which are reported for AD. The other oxidized proteins are glutamine synthase, which is related to increased excitotoxicity, ubiquitin C-terminal hydrolase L-1, associated with aberrant proteasomal degradation of damaged or aggregated proteins, α-enolase, linked to altered energy production, and dihydropyrimidinase-related protein 2, related to reduced growth cone elongation (Butterfield, 2002). The ubiquitin C-terminal hydrolases are suggested to hydrolyze small adducts of ubiquitin and generate free monomeric ubiquitins, thereby allowing recycling of ubiquitin, which is essential to the function of the proteolytic machinery. Thus, putative dysfunction of this oxidized deubiquitinating enzyme may be one way by which oxidative stress promotes ubiquitin-positive inclusion biogenesis, a pathological characteristic of the AD brain, which leads to neurodegeneration.

In addition to modifications, the levels of antioxidant proteins were also studied using proteomics technologies. Superoxide dismutase (CuZn, SOD1), which catalyzes the dismutation of superoxide anion into hydrogen peroxide, showed decreased levels in AD (Gulesserian et al., 2001). The decrease may reflect cell loss in AD. Peroxiredoxin I and II showed increased and peroxiredoxin III decreased levels in the AD brain (Kim et al., 2001b; Krapfenbauer et al., 2003a). Peroxiredoxins are a family of peroxidases that protect biomolecules by reducing hydrogen peroxide and alkyl hydroperoxides. There is evidence that overexpression of peroxiredoxins, in particular of peroxiredoxins I and II, in various cell lines abrogates the effects of agents that positively modulate apoptosis, including hydrogen peroxide (Kim et al., 2000a). Hence, up-regulation of these enzymes could represent a cellular response initiated against apoptosis, as enhanced apoptosis is demonstrated in AD (Engidawork et al., 2001b). Other neurodegenerative disorders also show a similar pattern of expression with subtle differences. Thus, expression of peroxiredoxin I and II are increased, while that of III is decreased in DS and Pick's disease (Kim et al., 2001b; Krapfenbauer et al., 2003a). Other supporting evidence for the role of oxidative stress in AD comes from the observation of increased carbonyl reductase and alcohol dehydrogenase in AD (Balcz et al., 2001). These two cytosolic enzymes catalyze reduction of carbonyls, which are toxic metabolic intermediates that serve as markers for oxidative stress. Increase in carbonyl content in the AD brain has been reported (Smith et al., 1998; Castegna et al., 2002a,b), suggesting a possible role of oxidation-related decrease in protein function in the process of neurodegeneration. Up-regulation of these enzymes may represent an adaptive mechanism to detoxify the toxic intermediates.

15.3.5
Heat Shock Proteins

Heat shock proteins play a critical role in mediating protein folding and assembly into oligomeric structures. The major chaperone families are heat shock protein 70 (HSP70), HSP40, HSP90, the small HSPs and the chaperonins (Agashe and Hartl, 2000; Ellis, 2000). Heat shock proteins showed deranged levels in the AD brain, with increased expression of alpha crystallin B, glucose regulated protein 94 and HSP70 RY, and decreased expression of heat shock cognate (HSC71), GRP 75, HSP60, and T-complex protein-1 (Yoo et al., 2001c). Perez et al. (1991) have also reported increased expression of HSP72 and HSP 73, an HSP70 family of proteins, in the AD brain. Differential response of individual HSPs to stress conditions may explain why there is inconsistency in the direction of change between different HSPs. Increased expression of HSPs in AD has been proposed as a defensive mechanism of response to amyloid fibril formation, because these chaperone proteins are capable of preventing aggregation of other proteins and in particular inhibiting self-assembly of polyglutamine proteins into amyloid-like fibrils. Loss of the HSP 70 proteins, particularly HSC 71, might be linked to their protective role in neuronal death associated with AD. The protein might be involved in the structural maintenance of the proteasome and conformational recognition of misfolded proteins by proteases (Castegna et al., 2002a,b). Thus, decreased activity of HSC-71, caused by either reduced levels (Yoo et al., 2001c,d) or increased oxidative modification of the protein (Castegna et al., 2002a,b), may be responsible for impaired protein clearance and deposition of amyloid-β peptide in the AD brain.

15.3.6
Proteins Enriched in Amyloid Plaques

Using LC-MS/MS, Liao et al. (2004) identified 26 brain proteins enriched in amyloid plaques excised with laser capture microdissection from AD brains.

The proteins include cytoskeletal proteins (coronin, tau), membrane trafficking proteins (clathrin, dynamin 1, dynein), proteases (ATPases, cathepsin), 14-3-3 proteins, chaperons, and others.

15.4
Limitations

The proteomic analysis of the brain has certain limitations that are related either to the sample and/or analytical approach. In the analysis of the brain, many factors may be involved, such as differences among individuals, differences in age and sex, possible other diseases, treatment with medicines, as well as technical factors, disease-unrelated factors, such as postmortem time, improper treatment of the samples, etc., all of which can affect a clear discrimination between healthy and diseased states of interest. The technical limitations involve inefficient detection of low-abundance gene products, hydrophobic proteins (they do not enter the IPG strips), and acidic, basic, high-, and low-molecular mass proteins. All these protein classes are underrepresented in 2-D gels (Lubec et al., 2003; Fountoulakis, 2004). A combination of proteomics methods with protein separation, enriching techniques, and alternative methodologies for detection will improve the detection of additional differences between AD and control brains. Such differences may be essential in the discovery of early disease markers and therapeutic approaches.

References

AGASHE, V. R., HARTL, F. U. (2000). Roles of molecular chaperones in cytoplasmic protein folding. *Semin. Cell Dev. Biol.* **11**, 15–25.

AKSENOV, M., AKSENOVA, M., BUTTERFIELD, D. A., MARKESBERY, W. R. (2000). Oxidative modification of creatine kinase BB in Alzheimer's disease brain. *J. Neurochem.* **74**, 2520–2527.

ANTONARAKIS, S. E. (1998). 10 years of Genomics, chromosome 21, and Down syndrome. *Genomics* **51**, 1–16.

BAJO, M., YOO, B. C., CAIRNS, N., GRATZER, M., LUBEC, G. (2001). Neurofilament proteins NF-L, NF-M and NF-H in brain of patients with Down syndrome and Alzheimer's disease. *Amino Acids* **21**, 293–301.

BALCZ, B., KIRCHNER, L., CAIRNS, N., FOUNTOULAKIS, M., LUBEC, G. (2001). Increased brain protein levels of carbonyl reductase and alcohol dehydrogenase in Down syndrome and Alzheimer's disease. *J. Neural. Transm. Suppl.* 193–201.

BERNDT, P., HOBOHM, U., LANGEN, H. (1999). Reliable automatic protein identification from matrix-assisted laser desorption/ionization mass spectrometric peptide fingerprints. *Electrophoresis* **20**, 3521–3526.

BOGUSKI, M. S., MCINTOSH, M. W. (2003). Biomedical informatics for proteomics. *Nature* **422**, 233–237.

BUTTERFIELD, D. A. (2002). Amyloid beta-peptide (1-42)-induced oxidative stress and neurotoxicity: implications for neurodegeneration in Alzheimer's disease brain. A review. *Free Radic. Res.* **36**, 1307–1313.

BUTTERFIELD, D. A., BOYD-KIMBALL, D., CASTEGNA, A. (2003). Proteomics in Alzheimer's disease: insights into potential mechanisms of neurodegeneration. *J. Neurochem.* **86**, 1313–1327.

BUTTERFIELD, D. A. (2004). *Proteomics*: a new approach to investigate oxidative stress in Alzheimer's disease brain. *Brain Res.* **1000**, 1–7.

Cairns, N. J. (**1999**). Neuropathology. *J. Neural. Transm. Suppl.* **57**, 61–74.

Castegna, A., Aksenov, M., Aksenova, M., Thongboonkerd, V., Klein, J. B., Pierce, W. M., Booze, R., Markesbery, W. R., Butterfield, D. A. (**2002a**). Proteomic identification of oxidatively modified proteins in Alzheimer's disease brain. Part I: creatine kinase BB, glutamine synthase, and ubiquitin carboxy-terminal hydrolase L-1. *Free Radic. Biol. Med.* **33**, 562–571.

Castegna, A., Aksenov, M., Thongboonkerd, V., Klein, J. B., Pierce, W. M., Booze, R., Markesbery, W. R., Butterfield, D. A. (**2002b**). Proteomic identification of oxidatively modified proteins in Alzheimer's disease brain. Part II: dihydropyrimidinase-related protein 2, alpha-enolase and heat shock cognate 71. *J. Neurochem.* **82**, 1524–1532.

Castegna, A., Thongboonkerd, V., Klein, J. B., Lynn, B., Markesbery, W. R., Butterfield, D. A. (**2003**). Proteomic identification of nitrated proteins in Alzheimer's disease brain. *J. Neurochem.* **85**, 1394–1401.

Cheon, M. S., Fountoulakis, M., Cairns, N. J., Dierssen, M., Herkner, K., Lubec, G. (**2001**). Decreased protein levels of stathmin in adult brains with Down syndrome and Alzheimer's disease. *J. Neural. Transm. Suppl.* 281–288.

Davidsson, P., Westman-Brinkmalm, A., Nilsson, C. L., Lindbjer, M., Paulson, L., Andreasen, N., Sjogren, M., Blennow, K. (**2002**). Proteome analysis of cerebrospinal fluid proteins in Alzheimer patients. *Neuroreport* **16**, 611–615.

Dodel, R. C., Hampel, H., Du, Y. (**2003**). Immunotherapy for Alzheimer's disease. *Lancet Neurol.* **2**, 215–220.

Ellis, R. J. (**2000**). Moledular chaperones ten years. Introduction. *Semin. Cell Dev. Biol.* **11**, 1–5.

Engidawork, E., Gulesserian, T., Balic, N., Cairns, N., Lubec, G. (**2001a**). Changes in nicotinic acetylcholine receptor subunits expression in brain of patients with Down syndrome and Alzheimer's disease. *J. Neural. Transm. Suppl.* 211–222.

Engidawork, E., Gulesserian, T., Seidl, R., Cairns, N., Lubec, G. (**2001b**). Expression of apoptosis related proteins in brains of patients with Alzheimer's disease. *Neurosci. Lett.* **303**, 79–82.

Engidawork, E., Gulesserian, T., Yoo, B. C., Cairns, N., Lubec, G. (**2001c**). Alteration of caspases and apoptosis-related proteins in brains of patients with Alzheimer's disease. *Biochem. Biophys. Res. Commun.* **281**, 84–93.

Engidawork, E., Lubec, G. (**2001d**). Protein expression in Down syndrome brain. *Amino Acids* **21**, 331–361.

Engidawork, E., Lubec, G. (**2003**). Molecular changes in fetal Down syndrome brain. *J. Neurochem.* **84**, 895–904.

Fountoulakis, M., Cairns, N., Lubec, G. (**1999**). Increased levels of 14-3-3 gamma and epsilon proteins in brain of patients with Alzheimer's disease and Down syndrome. *J. Neural. Transm. Suppl.* **57**, 323–335.

Fountoulakis, M. (**2001**). *Proteomics*: current technologies and applications in neurological disorders and toxicology. *Amino Acids* **21**, 363–381.

Fountoulakis, M., Juranville, J. F., Dierssen, M., Lubec, G. (**2002**). Proteomic analysis of the fetal brain. *Proteomics* **2**, 1547–1576.

Fountoulakis, M. (**2004**). Application of proteomics technologies in the investigation of the brain. *Mass Spectrom. Rev.* **23**, 231–258.

Greber, S., Lubec, G., Cairns, N., Fountoulakis, M. (**1999**). Decreased levels of synaptosomal associated protein 25 in the brain of patients with Down syndrome and Alzheimer's disease. *Electrophoresis* **20**, 928–934.

Gulesserian, T., Seidl, R., Hardmeier, R., Cairns, N., Lubec, G. (**2001**). Superoxide dismutase SOD1, encoded on chromosome **21**, but not SOD2 is overexpressed in brains of patients with Down syndrome. *J. Investig. Med.* **49**, 41–46.

Jacobsen, J. S. (**2002**). Alzheimer's disease: an overview of current and emerging therapeutic strategies. *Curr. Top. Med. Chem.* **2**, 343–352.

JIANG, L., LINDPAINTNER, K., LI, H. F., GU, N. F., LANGEN, H., HE, L., FOUNTOULAKIS, M. (**2003**). Proteomic analysis of the cerebrospinal fluid of patients with schizophrenia. *Amino Acids* **25**, 49–57.

KIM, H., LEE, T. H., PARK, E. S., SUH, J. M., PARK, S. J., CHUNG, H. K., KWON, O. Y., KIM, Y. K., RO, H. K., SHONG, M. (**2000a**). Role of peroxiredoxins in regulating intracellular hydrogen peroxide and hydrogen peroxide-induced apoptosis in thyroid cells. *J. Biol. Chem.* **275**, 18266–18270.

KIM, S. H., VLKOLINSKY, R., CAIRNS, N., LUBEC, G. (**2000b**). Decreased levels of complex III core protein 1 and complex V beta chain in brains from patients with Alzheimer's disease and Down syndrome. *Cell Mol. Life Sci.* **57**, 1810–1816.

KIM, S. H., CAIRNS, N., FOUNTOULAKISC, M., LUBEC, G. (**2001a**). Decreased brain histamine-releasing factor protein in patients with Down syndrome and Alzheimer's disease. *Neurosci. Lett.* **300**, 41–44.

KIM, S. H., FOUNTOULAKIS, M., CAIRNS, N., LUBEC, G. (**2001b**). Protein levels of human peroxiredoxin subtypes in brains of patients with Alzheimer's disease and Down syndrome. *J. Neural. Transm. Suppl.* 223–235.

KIM, S. H., LUBEC, G. (**2001c**). Brain alpha-endosulfine is manifold decreased in brains from patients with Alzheimer's disease: a tentative marker and drug target? *Neurosci. Lett.* **310**, 77–80.

KIM, S. H., NAIRN, A. C., CAIRNS, N., LUBEC, G. (**2001d**). Decreased levels of ARPP-19 and PKA in brains of Down syndrome and Alzheimer's disease. *J. Neural. Transm. Suppl.* 263–272.

KIM, S. H., VLKOLINSKY, R., CAIRNS, N., FOUNTOULAKIS, M., LUBEC, G. (**2001e**). The reduction of NADH ubiquinone oxidoreductase 24- and 75-kDa subunits in brains of patients with Down syndrome and Alzheimer's disease. *Life Sci.* **68**, 2741–2750.

KIM, S. H., FOUNTOULAKIS, M., CAIRNS, N. J., LUBEC, G. (**2002**).

Human brain nucleoside diphosphate kinase activity is decreased in Alzheimer's disease and Down syndrome. *Biochem. Biophys. Res. Commun.* **296**, 970–975.

KRAPFENBAUER, K., YOO, B. C., FOUNTOULAKIS, M., MITROVA, E., LUBEC, G. (**2002**). Expression patterns of antioxidant proteins in brains of patients with sporadic Creutzfeldt-Jacob disease. *Electrophoresis* **23**, 2541–2547.

KRAPFENBAUER, K., ENGIDAWORK, E., CAIRNS, N., FOUNTOULAKIS, M., LUBEC, G. (**2003**). Aberrant expression of peroxiredoxin subtypes in neuro-degenerative disorders. *Brain Res.* **967**, 152–160.

LAHM, H. W., LANGEN, H. (**2000**). Mass spectrometry: a tool for the identification of proteins separated by gels. *Electrophoresis* **21**, 2105–2114.

LANGEN, H., RODER, D., JURANVILLE, J. F., FOUNTOULAKIS, M. (**1997**). Effect of protein application mode and acryl-amide concentration on the resolution of protein spots separated by two-dimensional gel electrophoresis. *Electrophoresis* **18**, 2085–2090.

LIAO, L., CHENG, D., WANG, J., DUONG, D. M., LOSIK, T. G., GEARING, M., REES, H. D., LAH, J. J., LEVEY, A. I., PENG, J. (**2004**). Proteomic characterization of postmortem amyloid plaques isolated by laser capture microdissection. *J. Biol. Chem.* **279**, 37061–37068.

LUBEC, G., NONAKA, M., KRAPFENBAUER, K., GRATZER, M., CAIRNS, N., FOUNTOULAKIS, M. (**1999**). Expression of the dihydropyrimidinase related protein 2 (DRP-2) in Down syndrome and Alzheimer's disease brain is down-regulated at the mRNA and dysregulated at the protein level. *J. Neural. Transm. Suppl.* **57**, 161–177.

LUBEC, G., KRAPFENBAUER, K., FOUNTOULAKIS, M. (**2003**). Proteomics in brain research: potentials and limi-tations. *Prog. Neurobiol.* **69**, 193–211.

MIRRA, S. S., HEYMAN, A., McKEEL, D., SUMI, S., CRAIN, B. J. (**1991**). The consortium to establish a registry for Alzheimer disease (CERAD). II. Standardisation of the neuropathological assessment of Alzheimer's disease. *Neurology* **41**, 479–486.

PASINETTI, G. M., HO, L. (**2001**). From cDNA microarrays to high-throughput proteomics. Implications in the search for preventive initiatives to slow the clinical progression of Alzheimer's disease dementia. *Restor. Neurol. Neurosci.* **18**, 137–142.

PEREZ, N., SUGAR, J., CHARYA, S., JOHNSON, G., MERRIL, C., BIERER, L., PERL, D., HAROUTUNIAN, V., WALLACE, W. (**1991**). Increased synthesis and accumulation of heat shock 70 proteins in Alzheimer's disease. *Brain Res. Mol. Brain Res.* **11**, 249–254.

PUCHADES, M., HANSSON, S. F., NILSSON, C. L., ANDREASEN, N., BLENNOW, K., DAVIDSSON, P. (**2003**). Proteomic studies of potential cerebrospinal fluid protein markers for Alzheimer's disease. *Brain Res. Mol. Brain Res.* **118**, 140–146.

SCHULLER, E., GULESSERIAN, T., SEIDL, R., CAIRNS, N., LUBE, G. (**2001**). Brain t-complex polypeptide 1 (TCP-1) related to its natural substrate beta1 tubulin is decreased in Alzheimer's disease. *Life Sci.* **69**, 263–270.

SEIDL, R., GREBER, S., SCHULLER, E., BERNERT, G., CAIRNS, N., LUBEC, G. (**1997**). Evidence against increased oxidative DNA-damage in Down Syndrome. *Neurosci. Lett.* **235**, 137–140.

SELKOE, D. J. (**2000**). Toward a comprehensive theory for Alzheimer's disease. Hypothesis: Alzheimer's disease is caused by the cerebral accumulation and cytotoxicity of amyloid beta-protein. *Ann. NY Acad. Sci.* **924**, 17–25.

SELKOE, D. J. (**2002**). Alzheimer's disease is a synaptic failure. *Science* **298**, 789–791.

SHIM, K. S., LUBEC, G. (**2002**). Drebrin, a dendritic spine protein, is manifold decreased in brains of patients with Alzheimer's disease and Down syndrome. *Neurosci. Lett.* **324**, 209–212.

SMITH, M. A., SAYRE, L. M., ANDERSON, V. E., HARRIS, P. L., BEAL, M. F., KOWALL, N., PERRY, G. (**1998**). Cytochemical demonstration of oxidative damage in Alzheimer disease by immunochemical enhancement of the carbonyl reaction with 2,4-dinitrophenylhydrazine. *J. Histochem. Cytochem.* **46**, 731–735.

TIERNEY, M. C., FISHER, R. H., LEWIS, A. J., TORZITTO, M. L., SNOW, W. G., REID, D. W. NIEUWSTRAATEN, P., VAN ROOIJEN, L. A. A., DERKS, H. J. G. M., VAN WIJK, R., BISCHOP, A. (**1988**). The NINCDA-ADRDA work group criteria for the clinical diagnosis of probable Alzheimer's disease. *Neurology* **38**, 359–364.

TSUJI, T., SHIOZAKI, A., KOHNO, R., YOSHIZATO, K., SHIMOHAMA, S. (**2002**). Proteomic profiling and neurodegeneration in Alzheimer's disease. *Neurochem. Res.* **27**, 1245–1253.

VLKOLINSKY, R., CAIRNS, N., FOUNTOULAKIS, M., LUBEC, G. (**2001**). Decreased brain levels of 2′,3′-cyclic nucleotide-3′-phosphodiesterase in Down syndrome and Alzheimer's disease. *Neurobiol. Aging* **22**, 547–553.

YOO, B. C., CAIRNS, N., FOUNTOULAKIS, M., LUBEC, G. (**2001a**). Synaptosomal proteins, beta-soluble N-ethylmaleimide-sensitive factor attachment protein (beta-SNAP), gamma-SNAP and synaptotagmin I in brain of patients with Down syndrome and Alzheimer's disease. *Dement. Geriatr. Cogn. Disord.* **12**, 219–225.

YOO, B. C., FOUNTOULAKIS, M., CAIRNS, N., LUBEC, G. (**2001b**). Changes of voltage-dependent anion-selective channel proteins VDAC1 and VDAC2 brain levels in patients with Alzheimer's disease and Down syndrome. *Electrophoresis* **22**, 172–179.

YOO, B. C., KIM, S. H., CAIRNS, N., FOUNTOULAKIS, M., LUBEC, G. (**2001c**). Deranged expression of molecular chaperones in brains of patients with Alzheimer's disease. *Biochem. Biophys. Res. Commun.* **280**, 249–258.

YOO, B. C., VLKOLINSKY, R., ENGIDAWORK, E., CAIRNS, N., FOUNTOULAKIS, M., LUBEC, G. (**2001d**). Differential expression of molecular chaperones in brain of patients with Down syndrome. *Electrophoresis* **22**, 1233–1241.

16
Cardiac Proteomics

Emma McGregor and Michael J. Dunn

Abstract

The majority of proteomic investigations employ two-dimensional gel electrophoresis (2-D) with immobilized pH gradients (IPGs) to separate the proteins in a sample and combine this with mass spectrometry (MS) technologies to identify proteins.

In spite of the development of novel gel-free technologies, 2-D remains the only technique that can be routinely applied to parallel quantitative expression profiling of large sets of complex protein mixtures, such as whole cell and tissue lysates. 2-D involves solubilized proteins being separated in the first dimension according to their charge properties (isoelectric point, pI) by isoelectric focusing (IEF) under denaturing conditions. This is then followed by sodium dodecyl sulphate polyacrylamide gel electrophoresis (SDS-PAGE), where proteins are separated according to their relative molecular mass (M_r), the second dimension.

As the charge and mass properties of proteins are essentially independent parameters, this orthogonal combination of charge (pI) and size (M_r) separations results in the sample proteins being distributed across the two-dimensional gel profile. In excess of 5000 proteins can readily be resolved simultaneously (~2000 proteins routinely), with 2-D allowing detection of < 1 ng of protein per spot. Furthermore, a map of intact proteins is provided, reflecting changes in protein expression level, individual isoforms, or posttranslational modifications. In this chapter we will review proteomic investigations of cardiac and blood vessel proteins focusing on their application to the study of heart and cardiovascular disease in the human and in animal models of cardiac dysfunction. The majority of these cardiac proteome studies have involved protein separation, visualization and quantification using the traditional 2-D approach, as described above, combined with mass spectrometry technologies used for the identification of proteins.

Proteomics in Drug Research
Edited by M. Hamacher, K. Marcus, K. Stühler, A. van Hall, B. Warscheid, H. E. Meyer
Copyright © 2006 Wiley-VCH Verlag GmbH & Co. KGaA, Weinheim
ISBN: 3-527-31226-9

16.1
Heart Proteomics

16.1.1
Heart 2-D Protein Databases

There are four gel protein databases of cardiac proteins, established by three independent groups, that can be accessed via the World Wide Web (Table 16.1). These databases facilitate proteomic research into heart diseases containing information on several hundred cardiac proteins that have been identified by protein chemical methods. They all conform to the rules for federated 2-D protein databases (Appel et al., 1996). In addition, 2-D protein databases for other mammals, such as the mouse, rat (Li et al., 1999), dog (Dunn et al., 1997), pig and cow, are also under construction to support work on animal models of heart disease and heart failure.

Table 16.1 2-D heart protein databases accessible via the World Wide Web.

Database name	Institute	Organ	Web site	Reference
HSC-2DPAGE	Heart Science Centre, Harefield Hospital	Human heart (ventricle), Rat heart (ventricle), Dog heart (ventricle)	http://www.doc.ic.ac.uk/ vip/hsc-2dpage/	Evans et al., 1997
HP-2DPAGE	Heart 2-D Database, MDC, Berlin	Human heart (ventricle)	http://www.mdc-berlin.de/ ~emu/heart/	Muller et al., 1996
HEART-2DPAGE	German Heart Institute, Berlin	Human heart (ventricle) Human heart (atrium)	http://userpage.chemie.fu-berlin.de/~pleiss/dhzb.html	Pleissner et al., 1996
RAT HEART-2DPAGE	German Heart Institute, Berlin	Rat heart	http://www.mpiib-berlin.mpg.de/2D-PAGE/ RAT-HEART/2d/	Li et al., 1999

16.1.2
Dilated Cardiomyopathy

To date, proteomic investigations into human heart disease have centered on dilated cardiomyopathy (DCM). DCM is a disease of unknown etiology, characterized by impaired systolic function resulting in heart failure. Known contributory factors of DCM are viral infections, cardiac-specific autoantibodies, toxic agents, genetic factors, and sustained alcohol abuse. As many as 100 cardiac proteins have been observed to exhibit differential expression in DCM, with the majority of these proteins decreasing in abundance in the diseased heart. This has been presented in several studies (Corbett et al., 1998; Dunn et al., 1997; Knecht et al., 1994a,b; Li et al., 1999; Pleissner et al., 1995; Scheler et al., 1999). Mass spectrometry techniques have been used to identify many of these proteins (Corbett et al., 1998; Otto et al., 1996; Pleissner et al., 1997; Thiede et al., 1996), which have included cytoskeletal and myofibrillar proteins, proteins associated with mitochondria and energy production, and proteins associated with stress responses.

Investigating the consequences of these changes with respect to altered cellular function underlying cardiac dysfunction now poses a major challenge to researchers. For example, 59 isoelectric isoforms of HSP27 have been observed in human myocardium using traditional 2DE large-format gels. Twelve of these protein spots in the pI range of 4.9–6.2 and mass range of 27 000–28 000 Da were significantly altered in intensity in myocardium taken from patients with DCM. Ten of these protein spots were significantly changed in myocardium taken from patients with ischemic heart failure (Scheler et al., 1999).

16.1.3
Animal Models of Heart Disease

Appropriate animal models of human heart disease are an attractive alternative for proteomic investigations as human diseased tissue samples can often be compromised by factors such as the disease stage, tissue heterogeneity, genetic variability, and the patient's medical history/therapy. Avoiding any of the above complications when working with human samples can prove to be extremely difficult.

Several models of cardiac hypertrophy, heart disease, and heart failure in small animals, particularly the rat, exist and have been used for proteomics analysis. These investigations have focused on changes in cardiac proteins in response to alcohol (Patel et al., 1997; Patel et al., 2000) and lead toxicity (Toraason et al., 1997). Unfortunately, the cardiac physiology of small animal models and their normal pattern of gene expression (e.g., isoforms of the major cardiac contractile proteins) differ from that in larger mammals such as humans. Therefore, investigations have moved into higher mammals with two relatively recent proteomic studies of heart failure in large animals. One study of the dog investigated pacing-induced heart failure (Heinke et al., 1998; Heinke et al., 1999), while the second, based in our laboratory, investigated bovine DCM (Weekes et al., 1999). Both studies shared similarities with the proteome analysis of human DCM, with the majority of protein expression changes being a reduction in abundance in the diseased heart.

Identifying altered canine and bovine proteins has proved to be particularly challenging since these species are poorly represented in current genomic databases. As a result of this, new bioinformatic tools (MultiIdent, http://expasy.org) have had to be developed to facilitate cross-species protein identification (Wilkins et al., 1998). The most significant change observed for bovine DCM was a seven-fold increase in the enzyme ubiquitin carboxyl-terminal hydrolase (UCH) (Weekes et al., 1999). This could potentially facilitate increased protein ubiquitination in the diseased state, leading to proteolysis via the 26S proteosome pathway. Interestingly, there is evidence to suggest that inappropriate ubiquitination of proteins could contribute to the development of heart failure (Field and Clark, 1997).

More recently we have investigated whether the ubiquitin-proteosome system is perturbed in the heart of human DCM patients (Weekes et al., 2003). As in bovine DCM, expression of the enzyme UCH was elevated more than 8-fold at the protein level and elevated more than 5-fold at the mRNA level in human DCM. Moreover, this increased expression of UCH was shown by immunocytochemistry to be associated with the myocytes, which do not exhibit detectable staining in control hearts. Overall protein ubiquitination was increased 5-fold in DCM relative to control hearts. Using a selective affinity purification method we were able to demonstrate enhanced ubiquitination of a number of distinct proteins in DCM hearts. We have identified a number of these proteins by mass spectrometry. Interestingly many of these proteins were the same proteins previously found to be present at reduced abundance in DCM hearts (Corbett et al., 1998). This new evidence strengthens our hypothesis that inappropriate ubiquitin conjugation leads to proteolysis and depletion of certain proteins in the DCM heart and may contribute to loss of normal cellular function in the diseased heart.

16.1.4
Subproteomics of the Heart

16.1.4.1 Mitochondria
Mitochondria are involved in a variety of cellular processes. Their primary role is the production of ATP, but they are also involved in ionic homeostasis, apoptosis, oxidation of carbohydrates and fatty acids, and a variety of other catabolic and anabolic pathways. Mitochondrial dysfunction, therefore, can cause a variety of diseases including heart disease (Hirano et al., 2001). The characterization of the mitochondrial proteome could provide interesting new insights into cardiac dysfunction in heart disease (Lopez and Melov, 2002).

The prediction is that the human mitochondrial proteome comprises around 1500 distinct proteins (Taylor et al., 2003). Such complexity should be capable of being addressed using the currently available panel of proteomic technologies. Providing that sufficient tissue is available, mitochondria can be purified relatively easily by differential centrifugation (Lopez and Melov, 2002), and the mitochondrial proteins separated by 2-D and identified by peptide mass fingerprinting (PMF). Fountoulakis et al. (2002) used broad and narrow range IPG 2-D gels to study the rat liver mitochondria and were able to identify 192 different gene products from a total of approximately 1800 protein spots. Seventy percent of the identified proteins were found to be enzymes with a broad spectrum of catalytic activities (Fountoulakis et al., 2002). A similar study identified 185 different gene products from around 600 protein spots in the mitochondrial proteome of the neuroblastoma cell line IMR-32 (Fountoulakis and Schlaeger, 2003). This approach has not been systematically applied to the study of the mitochondrial proteome of the heart, although the 2-D/PMF approach has been applied in differential expression studies of hearts from knockout mouse strains deficient in creatine kinase (Kernec et al., 2001) and mitochondrial superoxide dismutase (Lopez and Melov, 2002). In a more recent study, Liu and colleagues have applied 2-D and matrix-assisted laser desorption/ionization mass spectrometry (MALDI-TOF MS) to study mitochondrial proteins in cardiomyocytes from chronic stressed rats (Liu et al., 2004). After comparing the protein profiles of myocardial mitochondria between a chronic restraint stress group and a control group, 11 protein spots were found to change, of which seven were identified. Five of these proteins, carnitine palmitoyltransferase 2, mitochondrial acyl-CoA thioesterase 1, isocitrate dehydrogenase 3 (NAD$^+$) alpha, fumarate hydratase 1, and pyruvate dehydrogenase beta, were found to decrease in abundance following chronic restraint stress with functional roles in the Krebs cycle and lipid metabolism in mitochondria. The other two proteins, creatine kinase and prohibitin, increased after chronic restraint stress (Liu et al., 2004).

The 2-D approach does have its limitations, one of the main ones being that it cannot be used to separate and analyze membrane proteins efficiently, so any such mitochondrial proteins will be poorly represented on 2-D profiles. In addition, many of the mitochondrial proteins are more basic than cytosolic proteins, mitochondria are rich in low molecular weight (< 10 kDa) proteins, and mitochondrial proteins are poorly described in databases (Lescuyer et al., 2003). In an attempt to overcome some of these hurdles, proteins from isolated yeast mitochondria were separated by one-dimensional SDS-PAGE. This approach avoided the problems associated with the IEF dimension of 2-D (Pflieger et al., 2002). The SDS gel was then cut into 27 slices of around 2 mm and tryptic digests of the proteins contained in these bands were then analyzed by LC-MS/MS. This approach resulted in 179 gene products being identified, i.e., similar to the number identified from isolated rat liver mitochondria by 2-D/MS (Fountoulakis et al., 2002). However, these proteins represented a broader range of proteins than covered by the 2-D approach, with their physicochemical properties spanning a wide range of pI, M_r, and hydrophobicity.

An alternative approach to increasing proteomic coverage is based on the analysis of isolated intact mitochondrial protein complexes. One strategy is to use sucrose density gradients to separate intact mitochondrial complexes solubilized with *n*-dodecyl-β-D-maltoside (Hanson et al., 2001). Initially the proteins from the individual fractions were analyzed using the 2-D method. However, subsequently this approach has been coupled with 1-D SDS-PAGE and MALDI-PMF analysis of tryptic digests of excised protein bands (Taylor et al., 2003). Applying this approach to human heart mitochondria resulted in the identification of 615 *bona fide* or potential mitochondrial proteins, many of which had not been previously reported using 2-D (Taylor et al., 2003). Proteins with a wide range of pI, M_r and hydrophobicities were reported with a high coverage of the known subunits of the oxidative machinery of the inner mitochondrial membrane. A considerable proportion of the identified proteins are associated with signaling, RNA, DNA, protein synthesis, lipid metabolism, and ion transport (Taylor et al., 2003). In a recent study of complex I purified from bovine heart mitochondria, three independent separation methods (1-D SDS-PAGE, 2-D, and reverse phase HPLC) combined with MALDI-PMF and ESI-MS/MS were employed and the intact enzyme was shown to be an assembly of 46 different proteins (Carroll et al., 2003).

Alternative types of 2-D separations are available that can help overcome the limitations inherent in the isoelectric focusing of 2-D. Two-dimensional blue native (BN) electrophoresis (Schagger and von Jagow, 1991) can be used to separate membrane and other functional protein complexes as intact, enzymatically active complexes in the first dimension. Coomassie dyes are introduced to induce a charge shift on the proteins and aminocaproic acid serves to improve solubilization of membrane proteins. Protein complexes are then resolved into their component subunits using second-dimension separation by Tricine-SDS-PAGE. Combining this method with protein identification by MALDI-PMF, has been applied to several studies of the mitochondrial proteome (Brookes et al., 2002; Kruft et al., 2001). In a study of human heart mitochondria using BN/SDS-PAGE, the individual subunits of all five complexes of the oxidative phosphorylation system were represented and a novel variant of cytochrome *c* oxidase subunit Vic was reported (Devreese et al., 2002).

16.1.4.2 PKC Signal Transduction Pathways

Protein kinase signal transduction pathways have been extensively studied and characterized in the myocardium. Ischemic preconditioning (IP) and the role of individual kinases involved is where much of the research efforts have been focused. IP is the reduction in susceptibility to myocardial infarction that follows brief periods of sublethal ischemia (Murry et al., 1986). This reduction can manifest itself as a 4-fold reduction in infarct size, this being secondary to a delay in the onset and rate of cell necrosis during the subsequent lethal ischemia (Marber et al., 1994).

The involvement of protein kinase C (PKC) in preconditioning was first suggested by Downey and colleagues (Ytrehus et al., 1994). They were able to clearly demonstrate that pharmacological inhibition of PKC blocked ischemic preconditioning and that the infarct-sparing effects of ischemic preconditioning could be mimicked by activating PKC using phorbol esters. However, it is still uncertain which PKC isoform(s) is/are responsible even though many groups have repeated this experiment (Goto et al., 1995; Kitakaze et al., 1996; Liu et al., 1995; Mitchell et al., 1995).

Isoform-specific inhibitory peptides, which are able to abolish protection in response to IP, have shown that PKC-ε activation plays a significant role in protection (Gray et al., 1997; Liu et al., 1999). Here, the binding of a specific PKC isoform is prevented from localizing with its substrate(s) and there is a loss of function. However, the selectivity of this approach has been brought into question. Another approach to validating PKC-ε as having a cardio-protective role is to use animal models where the gene of interest, in this case that encoding PKC-ε, has been knocked out (Saurin et al., 2002).

Ping and colleagues have adopted a functional proteomic approach to investigate PKC-ε-mediated cardioprotection (Vondriska et al., 2001a). In this approach PKC-ε monoclonal antibodies are used to immunoprecipitate and isolate PKC-ε complexes. This subproteome is then separated out using 1-D/2-D electrophoresis and putative candidate proteins, which associate with PKC-ε, are then identified using western blotting-/mass spectrometry-based techniques. In order to validate this proteomic data, the colocalization of candidate proteins with the PKC-ε complex is established using PKC-ε-GST affinity pull-down assays. Expression of these candidate proteins in cardiac cells is then confirmed using isolated mouse cardiac myocytes (Vondriska et al., 2001a). This approach has allowed Ping and colleagues to demonstrate that within the PKC-ε subproteome, PKC-ε forms complexes with at least 93 proteins in the mouse heart (Baines et al., 2002; Edmondson et al., 2002; Li et al., 2000; Ping et al., 2001; Vondriska et al., 2001b). The identified proteins can be separated into six different classes of molecule incorporating structural proteins, signaling molecules, stress-activated proteins, metabolism-related proteins, transcription- and translation-related proteins, and PKC-ε binding domain containing proteins (Edmondson et al., 2002; Ping et al., 2001).

16.1.5
Proteomics of Cultured Cardiac Myocytes

Cell culture systems are ideal for detailed proteomic investigations of responses in protein expression to controlled stimuli. This is because they should provide defined systems with much lower inherent variability between samples, particularly if established cell lines are used. However, cells that are maintained in culture respond by alterations in their gene pattern, and consequently the protein expression, such that it can be quite different from that found *in vivo*. This process can occur quite rapidly in primary cultures of cells established from tissue samples and is even more profound in cells maintained long-term, particularly where transformation has been used to establish immortal cell lines. Cardiac myocytes can pose an even bigger challenge. While neonatal cardiac myocytes can be maintained and grown *in vitro*, adult cells are terminally differentiated and can be maintained for relatively short times *in vitro* but are not capable of cell division.

In fact there are relatively few published proteomic investigations of isolated cardiac myocytes. In a study of beating neonatal rat cardiac myocytes, 2-D was used to investigate the regulation of protein synthesis by catecholamines (He et al., 1992a,b). In response to treatment with norepinephrine, significant changes in protein expression were observed (He et al., 1992a). The use of the α-adrenoceptor blocker, prazosin, allowed a clear classification of α- and non-α (probably β) adrenoceptor-mediated catecholamine effects on protein expression (He et al., 1992b). Unfortunately none of the proteins could be identified at the time this study was carried out.

Arnott et al. (1998) have used phenylephrine treated neonatal rat cardiac myocytes as a model of cardiac hypertrophy for proteomic analysis. In this 2-D-based study, 11 protein spots displayed statistically significant changes in expression upon induction of hypertrophy. Of these, eight showed higher expression and three were decreased in abundance in hypertrophied cells. All of these proteins were successfully identified by a combination of PMF by MALDI-TOF and partial sequencing by LC-MS/MS. The atrial isoforms of myosin light chains (MLC) 1 (two spots) and 2 were increased, as was the ventricular isoforms of MLC2 (two spots). Other proteins that increased were chaperonin cofactor a, nucleoside diphosphate kinase a, and the 27 kDa heat shock protein (HSP27). The proteins that were decreased were identified as mitochondrial matrix protein p1 (two spots) and NADH ubiquinone oxidoreductase 75 kDa subunit. The changes in expression of MLC isoforms are consistent with previous studies of the expression of MLC isoforms in cardiac hypertrophy both by Northern blot analysis of mRNA and by immunofluorescence. In contrast, the changes in expression of the other proteins had not previously been shown to be associated with cardiac hypertrophy (Arnott et al., 1998).

In a similar study, endothelin (ET) was used to induce hypertrophy in neonatal rat cardiac myocytes. ET treatment was found by 2-D to result in a two-fold decrease in 21 proteins compared to the levels in untreated cells (Macri et al., 2000). The ET-induced hypertrophy was accompanied by a 30% increase in MLC1 and MLC2.

Isolated adult rabbit cardiac myocytes have been used in a proteomic study of myocardial ischemic preconditioning, induced pharmacologically with adenosine (Arrell et al., 2001). Here a subproteomic approach was used in which cytosolic and myofilament-enriched fractions were analyzed by 2-D. Various adenosine-mediated changes in protein expression were detected in the cytosolic fraction but these proteins were not subsequently identified. The most striking finding was a novel posttranslational modification of MLC1 that was shown to be due to phosphorylation at two sites (Arrell et al., 2001). The functional significance of this finding to preconditioning remains to be established. In a recent study of adult human cardiac myocytes from patients with end stage heart failure, two protein spots associated with MLC1 were observed on 2-D gels (van der Velden et al., 2003). These two spots may represent phosphorylated and nonphosphorylated MLC as described by Arrell et al. (2001), but the amount of the putatively phosphorylated form was not found to differ between failing and control (transplant donor) samples.

16.1.6
Proteomic Characterization of Cardiac Antigens in Heart Disease and Transplantation

Proteomic technologies can be used to identify cardiac-specific antigens that elicit antibody responses in heart disease and following cardiac transplantation. Western blot transfers of 2-D gel separations of cardiac proteins are used in this instance. These are probed with patient serum samples and developed using appropriately conjugated antihuman immunoglobulins. Several cardiac antigens that are reactive with autoantibodies in DCM (Latif et al., 1993; Pohlner et al., 1997) and myocarditis (Pankuweit et al., 1997) have been revealed in this way. Cardiac antigens that are associated with antibody responses following cardiac transplantation have also been characterized and may be involved in acute (Latif et al., 1995) and chronic (Wheeler et al., 1995) rejection.

16.1.7
Markers of Acute Allograft Rejection

Acute rejection is the major complication affecting patients during the first year following cardiac transplantation. Routine endomyocardial biopsy, a highly invasive and expensive procedure, remains the most reliable method of detecting rejection following cardiac transplantation. Despite numerous attempts to detect rejection using a minimally-invasive blood assay, none have proved reliable enough to replace biopsy. While proteomics can be used to search for disease biomarkers directly in serum samples, in practice this is made extremely difficult by the diversity of proteins comprising the serum proteome and the very high dynamic range of protein abundance, estimated at around 10^{10}. Thus, while albumin, the most abundant serum protein, is present at around 35 mg/mL, disease biomarkers are expected to be present in the range of ng/mL to pg/mL (Anderson, 2004). As an alternative approach, we have used proteomics to identify those proteins that are upregulated in endomyocardial biopsy specimens of the heart during acute rejection (Borozdenkova et al., 2004). We have then hypothesized that those proteins may then be detectable in the serum of patients at the time of acute rejection and could form the basis of an ELISA-based blood test for this disease. We have validated this approach in a study of 33 sequential biopsies from four patients, where 2-D analysis detected 13 proteins that were upregulated more than 5-fold in association with acute rejection in all four patients. Subsequently, two of these proteins (αB-crystallin, tropomyosin) were measured by ELISA and found to be present at significantly higher levels in serum samples associated with biopsy-proven acute rejection (Borozdenkova et al., 2004).

16.2
Vessel Proteomics

Normal blood vessels are made up of many different cell types, including smooth muscle cells, endothelial cells, and fibroblasts. The situation is even more complex in atherosclerotic vessels due to the inflammatory nature of the disease, so that cell types, such as T-lymphocytes and macrophages, are also present in large numbers. Thus, proteomic analysis of intact vessels, particularly if the study is designed to investigate differential protein expression in diseased and normal vessels, is very challenging.

16.2.1
Proteomics of Intact Vessels

In spite of these limitations, 2-D was first used nearly 20 years ago to compare the protein composition of fibro-fatty lesions with lesion-free segments of the human aortic intima (Stastny et al., 1986a,b). Although these studies were performed before important recent advances in proteomics, including sophisticated quantitative computer analysis of 2-D protein profiles and protein identification by mass spectrometry, they were nevertheless able to demonstrate that proteins originating from the plasma accumulated in diseased vessels. However, even with the advent of the sophisticated tool box that is now available, proteomic analysis of healthy and diseased vessels remains a challenge. Recently, You et al. (2003) showed a significant increase in ferritin light chain expression in diseased vessels and concluded that this is consistent with the "iron" hypothesis, which is that iron storage contributes to atherosclerosis by modulating lipid oxidation.

Accelerated graft atherosclerosis, often referred to as chronic rejection or cariac allograft vasculopathy (CAV), is one of the most serious long-term complications following heart transplantation. In our laboratory, we have applied a proteomic approach to investigate the hypothesis that the grafts of long term survivors of heart transplantation are protected from CAV through the expression of cyto-protective genes. The most significant finding of these studies was that a particular 2-D protein spot was found only in late biopsies from patients without CAV (De Souza et al., 2005). This protein has been shown by MS to be a specific di-phosphorylated form of the 27-kDa heat shock protein (HSP27) (De Souza et al., 2005). Like all heat shock proteins, HSP27 functions as a molecular chaperone and may play a cytoprotective role by protection from apoptosis, modulation of oxidative stress, and regulation of cytoskeletal organization. Little is known about expression of HSP27 in normal and diseased human blood vessels. Interestingly, Martin-Ventura et al. (2004) have recently reported that nontransplant athero-sclerotic vessels maintained in culture release very little HSP27 protein into the medium compared with control vessels; they also showed smooth muscle cells of mammary arteries to be positive for HSP27. These findings therefore suggest that vascular expression of HSP27 may be associated with protection against both transplant and nontransplant atherosclerosis.

16.2.2
Proteomics of Isolated Vessel Cells

At present, a more generally applicable approach to overcoming the problem of heterogeneity in the cellular composition of blood vessels is to investigate specifically isolated cell types. One approach that has been successfully used is to isolate primary cells from vessels, e.g., endothelial cells, smooth muscle cells and fibroblasts, and to solubilize proteins directly from these cells for proteomic analysis. The larger the blood vessel, the easier this approach becomes, e.g., there are many studies where human umbilical vein endothelial cells (HUVECs) have been successfully isolated and analyzed using a 2-D approach (Bruneel et al. 2003; Sprenger et al. 2004). Bruneel et al. (2003) have established a 2-D web database of HUVEC proteins (http://www.huvec.org) that contains more than 1000 proteins, but less than 50 of these have yet been identified by MS. In a recent study using isolated bovine aortic endothelial cells (Pellieux et al., 2003), altered expression of the protein CapG, a member of the gelsolin family, was shown to be modulated by exposure of the cells to oscillatory and laminar flow conditions.

McGregor et al. have demonstrated the relative ease of dissecting intact contractile smooth muscle from human saphenous vein (HSV), and analyzing quantitative protein expression of this tissue using 2-D (McGregor et al., 2001, 2003). These studies have established a comprehensive 2-D map of vascular smooth muscle cells with around 130 identified proteins (McGregor et al., 2001) (Figure 16.1) and have also shown that F-actin capping (CapZ) and other contractile saphenous vein smooth muscle proteins are altered by hemodynamic stress (McGregor et al., 2003). More recently, Mayr et al. (2004) identified 250 proteins in mouse aortic smooth muscle cells and this map is accessible at http://www.vascular-proteomics.com. These approaches can also be applied to microvessels, but with their very small size and the difficulty in isolating these vessels, this becomes a much more challenging task.

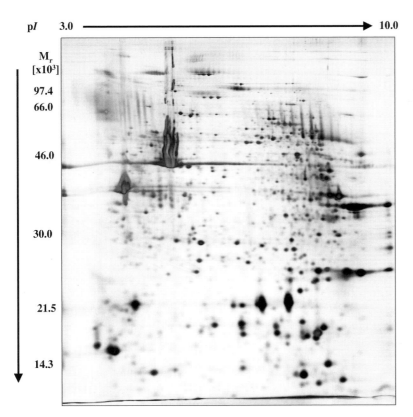

Figure 16.1 2-D separation of intact contractile human saphenous vein smooth muscle. Proteins (400 µg load) were separated by isoelectric focusing in the first dimension using a pH 3–10 nonlinear (NL), 18-cm IPG DryStrip (Amersham Biosciences). The pH gradient is illustrated across the top of the 2-D gel. The most acidic pH is to the left of the gel image and the most basic pH to the right. A 12% acrylamide gel was used in the second dimension to separate proteins by SDS-PAGE. Standard molecular weight markers (Amersham Biosciences) ranging from 14.3–97.4 kDa were run on the same gel and are annotated on the far left of the gel image. Visualization of proteins was by silver staining.

The protein yield of potential 2-D samples, following protein extraction, via standard 2-D methods, must be sufficient for running the different types of 2-D gels: analytical gels (100 µg protein) that are used for quantifying protein expression, by analyzing scanned gel images via specialized computer software packages, and preparative gels (400 µg protein), where protein spots are excised from the 2-D gel for identification by mass spectrometry techniques, respectively. Isolating primary cells and intact tissue from vessels can often only provide a limited amount of material that can be subjected to 2-D, which becomes much further reduced when this material is harvested from the microvasculature. One way to counteract this is to use small format (7 × 7 cm) mini-2-D gels, to separate limited samples, rather than large format (20 × 20 cm) 2-D gels. However, for protein identification, sample protein yield should ideally be concentrated enough to be separated on a large format 2-D preparative gel intended for subsequent analysis by mass spectrometry. One option for increasing protein yield from cell types associated with the (micro)vasculature is to grow specific cell lines in culture.

Many proteomics studies to date have used cells associated with blood vessels that have been cultured and then analyzed via 2-D. The advantage of using cultured cell lines is that there is always a ready supply of purified material, which can be concentrated by harvesting increased numbers of cells as required. Investigations of this type have used cultured HUVECs, cultures of immortalized HSV endothelial cells, as well as explanted HSV smooth muscle cells, where sections of dissected HSV medial smooth muscle are placed in culture conditions, thus stimulating the growth of smooth muscle cells outwards from the original explanted intact tissue. A recent example of this type of study, using primary human endothelial cells cultured from HUVECs and a human endothelial cell line (HUVEC-derived, immortalized EC-RF24), is that by Sprenger et al. (2004). Here, following the use of a classical 2-D approach, coupled with optimized fractionation techniques, two subdomains of similar lipid composition, caveolae and rafts, representing 0.5% of total cellular protein and less than 2% of total plasma membrane protein, were analyzed and compared (Sprenger et al., 2004). This reference study demonstrated the power of subproteomics, allowing enhanced separation and identification of membrane proteins in particular.

While using cultured cells provides abundant protein yields for many parallel experiments to be undertaken, the main disadvantage of using this approach is that one is using an *in vitro* system. Once cells are cultured through many passages they are subject to changes in their gene expression and sometimes morphology (e.g., fibroblasts), which can influence gene expression measurements/protein expression analysis by 2-D. This is a particular problem when analyzing cultured smooth muscle cells, which lose their contractile properties in culture, and is why it is preferential to use intact smooth muscle (McGregor et al., 2001, 2004) (Figure 16.1).

16.2.3
Laser Capture Microdissection

Laser capture microdissection (LCM) is a relatively new technique, which we have recently applied to the study of human cardiac tissue. The technique involves a laser beam that isolates specific regions of interest from microscope sections of tissue. While this technique generally results in the isolation of relatively small amounts of material, it has been shown to be possible to perform proteomic studies of the resulting protein samples and generate enough material for protein identification by mass spectrometry (Banks et al. 1999). Heart tissue is dominated by a particular cell type, i.e., cardiac myocytes, but also contains lower amounts of proteins from other cell types including fibroblasts, smooth muscle, and endo-thelial cells originating in the microvessels within the heart. Using LCM we have generated sufficient amounts of human cardiac tissue sections to separate on large-format 2-D gels of proteins from isolated cardiac myocytes and microvessels (Figure 16.2) (De Souza et al., 2004).

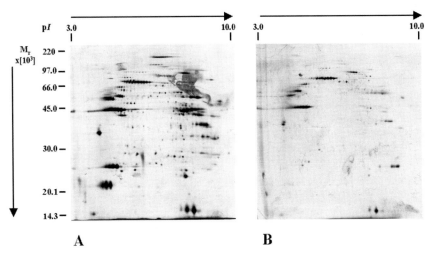

A   **B**

Figure 16.2 2-D electrophoretic separations of laser capture microdissected human cardiac tissue.
A Cardiac myocytes. **B** Blood vessels. Proteins were separated by isoelectric focusing in the first dimension using a pH 3–10 NL, 18-cm IPG DryStrip (Amersham Biosciences). The pH gradient is illustrated across the top of the 2-D gel with the most acidic pH to the left of the gel image and the most basic pH to the right. A 12% acrylamide gel was used in the second dimension to separate proteins by SDS-PAGE. Standard molecular weight markers (Amersham Biosciences), ranging from 14.3–97.4 kDa were run on the same gel and are annotated on the far left of the gel image. Visualization of proteins was by silver staining.

16.3
Concluding Remarks

Important biomedical questions can now be addressed using established proteomic technologies, together with the new and alternative strategies currently under development. Three recent reviews outline for the reader how proteomics can be applied in the clinical setting, when investigating heart disease and cardiovascular disease, providing experimental strategies for studying both (McGregor and Dunn, 2003; Stanley et al., 2004; Zerkowski et al., 2004). In this chapter we have attempted to give an overview of how proteomics technologies are being used to characterize protein expression in the human heart and associated blood vessels, and to investigate changes in protein expression associated with cardiac dysfunction in disease. The proteomics studies that have been carried out on cardiac tissue from both human patients and appropriate animal models are providing new insights into the cellular mechanisms involved in cardiac dysfunction. The proteomic analysis of blood vessels is still in its infancy and has focused mainly on the study of larger vessels. Studying blood vessels using a proteomic approach promises to yield important new information on protein expression in vessels in different biological systems. However, a major problem, if the studies are to be extended to analysis of the microvasculature, will be obtaining sufficient purified starting material and hence generating/using microvasculature reference proteomes.

The continued use of proteomics, as described in this review, should result in the discovery of new diagnostic and/or prognostic biomarkers, in addition to those already identified in the referenced studies in this chapter, facilitating the identification of potential drug targets for the development of new therapeutic approaches for combating heart and cardiovascular disease.

Acknowledgement

MJD is the recipient of a Science Foundation Ireland Research Professorship in Biomedical Proteomics and is grateful to SFI for support of his proteomic research. MJD would also like to thank the British Heart Foundation for support of his cardioproteomic research.

References

ANDERSON, N. L. (**2004**). Candidate-Based Proteomics in the search for biomarkers of cardiovascular disease. *J. Physiol.* DOI 10.1113/jphysiol.2004.080473.

APPEL, R. D., BAIROCH, A., SANCHEZ, J. C., VARGAS, J. R., GOLAZ, O., PASQUALI, C., HOCHSTRASSER, D. F. (**1996**). Federated two-dimensional electrophoresis database: a simple means of publishing two-dimensional electrophoresis data. *Electrophoresis* **17**, 540–546.

ARNOTT, D., O'CONNELL, K. L., KING, K. L., STULTS, J. T. (**1998**). An integrated approach to proteome analysis:

identification of proteins associated with cardiac hypertrophy. *Anal. Biochem.* **258**, 1–18.

ARRELL, D. K., NEVEROVA, I., FRASER, H., MARBAN, E., VAN EYK, J. E. (**2001**). Proteomic analysis of pharmacologically preconditioned cardiomyocytes reveals novel phosphorylation of myosin light chain 1. *Circ. Res.* **89**, 480–487.

BAINES, C. P., ZHANG, J., WANG, G. W., ZHENG, Y. T., XIU, J. X., CARDWELL, E. M., BOLLI, R., PING, P. (**2002**). Mitochondrial PKCepsilon and MAPK form signaling modules in the murine heart: enhanced mitochondrial PKCepsilon-MAPK interactions and differential MAPK activation in PKCepsilon-induced cardioprotection. *Circ. Res.* **90**, 390–397.

BANKS, R. E., DUNN, M. J., FORBES, M. A., STANLEY, A., PAPPIN, D., NAVEN, T., GOUGH, M., HARNDEN, P., SELBY, P. J. (**1999**). The potential use of laser capture microdissection to selectively obtain distinct populations of cells for proteomic analysis – preliminary findings. *Electrophoresis* **20**, 689–700.

BOROZDENKOVA, S., WESTBROOK, J. A., PATEL, V., WAIT, R., BOLAD, I., BURKE, M. M., BELL, A. D., BANNER, N. R., DUNN, M. J., ROSE, M. L. (**2004**). Use of proteomics to discover novel markers of cardiac allograft rejection. *J. Proteome Res.* **3**, 282–288.

BROOKES, P. S., PINNER, A., RAMACHANDRAN, A., COWARD, L., BARNES, S., KIM, H., DARLEY-USMAR, V. M. (**2002**). High throughput two-dimensional blue-native electrophoresis: a tool for functional proteomics of mitochondria and signaling complexes. *Proteomics* **2**, 969–977.

BRUNEEL, A., LABAS, V., MAILLOUX, A., SHARMA, S., VINH, J., VAUBOURDOLLE, M., BAUDIN, B. (**2003**). Proteomic study of human umbilical vein endothelial cells in culture. *Proteomics* **3**, 714–723.

CARROLL, J., FEARNLEY, I. M., SHANNON, R. J., HIRST, J., WALKER, J. E. (**2003**). Analysis of the subunit composition of complex I from bovine heart mitochondria. *Mol. Cell Proteomics* **2**, 117–126.

CORBETT, J. M., WHY, H. J., WHEELER, C. H., RICHARDSON, P. J., ARCHARD, L. C., YACOUB, M. H., DUNN, M. J. (**1998**). Cardiac protein abnormalities in dilated cardiomyopathy detected by two-dimensional polyacrylamide gel electrophoresis. *Electrophoresis* **19**, 2031–2042.

CRAVEN, R. A., JACKSON, D. H., SELBY, P. J., BANKS, R. E. (**2002**). Increased protein entry together with improved focussing using a combined IPGphor/Multiphor approach. *Proteomics* **2**, 1061–1063.

CRAVEN, R. A. AND BANKS, R. E. (**2002**). Use of laser capture microdissection to selectively obtain distinct populations of cells for proteomic analysis. *Methods Enzymol.* **356**, 33–49.

DE SOUZA, A. I., MCGREGOR, E., DUNN, M. J., ROSE, M. L. (**2004**). Preparation of human heart for laser microdissection and proteomics. *Proteomics* **4**, 578–586.

DE SOUZA, A. I., WAIT, R., MITCHELL, A. G., BANNER, N. R., DUNN, M. J., ROSE, M. L. (**2005**). Heat shcok protein 27 is associated with freedom from graft vasculopathy following human cardiac transplantation. *Circ Res.*, in press.

DEVREESE, B., VANROBAEYS, F., SMET, J., VAN BEEUMEN, J., VAN COSTER, R. (**2002**). Mass spectrometric identification of mitochondrial oxidative phosphorylation subunits separated by two-dimensional blue-native polyacrylamide gel electrophoresis. *Electrophoresis* **23**, 2525–2533.

DUNN, M. J., CORBETT, J. M., WHEELER, C. H. (**1997**). HSC-2DPAGE and the two-dimensional gel electrophoresis database of dog heart proteins. *Electrophoresis* **18**, 2795–2802.

EDMONDSON, R. D., VONDRISKA, T. M., BIEDERMAN, K. J., ZHANG, J., JONES, R. C., ZHENG, Y., ALLEN, D. L., XIU, J. X., CARDWELL, E. M., PISANO, M. R., PING, P. (**2002**). Protein kinase C epsilon signaling complexes include metabolism- and transcription/translation-related proteins: complimentary separation techniques with LC/MS/MS. *Mol. Cell Proteomics* **1**, 421–433.

EVANS, G., WHEELER, C. H., CORBETT, J. M., DUNN, M. J. (**1997**). Construction of HSC-2DPAGE: a two-dimensional gel electrophoresis database of heart proteins. *Electrophoresis* **18**, 471–479.

FIELD, M. L. AND CLARK, J. F. (**1997**). Inappropriate ubiquitin conjugation: a proposed mechanism contributing to heart failure. *Cardiovasc. Res.* **33**, 8–12.

FOUNTOULAKIS, M., BERNDT, P., LANGEN, H., SUTER, L. (**2002**). The rat liver mitochondrial proteins. *Electrophoresis* **23**, 311–328.

FOUNTOULAKIS, M. AND SCHLAEGER, E. J. (**2003**). The mitochondrial proteins of the neuroblastoma cell line IMR-32. *Electrophoresis* **24**, 260–275.

GOTO, M., LIU, Y., YANG, X. M., ARDELL, J. L., COHEN, M. V., DOWNEY, J. M. (**1995**). Role of bradykinin in protection of ischemic preconditioning in rabbit hearts. *Circ. Res.* **77**, 611–621.

GRAY, M. O., KARLINER, J. S., MOCHLY-ROSEN, D. (**1997**). A selective epsilon-protein kinase C antagonist inhibits protection of cardiac myocytes from hypoxia-induced cell death. *J. Biol. Chem.* **272**, 30945–30951.

HANSON, B. J., SCHULENBERG, B., PATTON, W. F., CAPALDI, R. A. (**2001**). A novel subfractionation approach for mitochondrial proteins: a three-dimensional mitochondrial proteome map. *Electrophoresis* **22**, 950–959.

HE, C., MULLER, U., WERDAN, K. (**1992a**). Regulation of protein biosynthesis in neonatal rat cardiomyocytes by adrenoceptor-stimulation: investigations with high resolution two-dimensional polyacrylamide gel electrophoresis. *Electrophoresis* **13**, 755–756.

HE, C., MULLER, U., OBERTHUR, W., WERDAN, K. (**1992b**). Application of high resolution two-dimensional polyacrylamide gel electrophoresis of polypeptides from cultured neonatal rat cardiomyocytes: regulation of protein synthesis by catecholamines. *Electrophoresis* **13**, 748–754.

HEINKE, M. Y., WHEELER, C. H., CHANG, D., EINSTEIN, R., DRAKE-HOLLAND, A., DUNN, M. J., DOS REMEDIOS, C. G. (**1998**). Protein changes observed in pacing-induced heart failure using two-dimensional electrophoresis. *Electrophoresis* **19**, 2021–2030.

HEINKE, M. Y., WHEELER, C. H., YAN, J. X., AMIN, V., CHANG, D., EINSTEIN, R., DUNN, M. J., DOS REMEDIOS, C. G. (**1999**). Changes in myocardial protein expression in pacing-induced canine heart failure. *Electrophoresis* **20**, 2086–2093.

HIRANO, M., DAVIDSON, M., DiMAURO, S. (**2001**). Mitochondria and the heart. *Curr. Opin. Cardiol.* **16**, 201–210.

KERNEC, F., UNLU, M., LABEIKOVSKY, W., MINDEN, J. S., KORETSKY, A. P. (**2001**). Changes in the mitochondrial proteome from mouse hearts deficient in creatine kinase. *Physiol. Genomics* **6**, 117–128.

KITAKAZE, M., NODE, K., MINAMINO, T., KOMAMURA, K., FUNAYA, H., SHINOZAKI, Y., CHUJO, M., MORI, H., INOUE, M., HORI, M., KAMADA, T. (**1996**). Role of activation of protein kinase C in the infarct size-limiting effect of ischemic preconditioning through activation of ecto-5′-nucleotidase. *Circulation* **93**, 781–791.

KNECHT, M., REGITZ-ZAGROSEK, V., PLEISSNER, K. P., EMIG, S., JUNGBLUT, P., HILDEBRANDT, A., FLECK, E. (**1994a**). Dilated cardiomyopathy: computer-assisted analysis of endomyocardial biopsy protein patterns by two-dimensional gel electrophoresis. *Eur. J. Clin. Chem. Clin. Biochem.* **32**, 615–624.

KNECHT, M., REGITZ-ZAGROSEK, V., PLEISSNER, K. P., JUNGBLUT, P., STEFFEN, C., HILDEBRANDT, A., FLECK, E. (**1994b**). Characterization of myocardial protein composition in dilated cardiomyopathy by two-dimensional gel electrophoresis. *Eur. Heart J.* **15**, Suppl. D, 37–44.

KRUFT, V., EUBEL, H., JANSCH, L., WERHAHN, W., BRAUN, H. P. (**2001**). Proteomic approach to identify novel mitochondrial proteins in Arabidopsis. *Plant Physiol.* **127**, 1694–1710.

LATIF, N., BAKER, C. S., DUNN, M. J., ROSE, M. L., BRADY, P., YACOUB, M. H. (**1993**). Frequency and specificity of antiheart antibodies in patients with dilated cardiomyopathy detected using SDS-PAGE and western blotting. *J. Am. Coll. Cardiol.* **22**, 1378–1384.

LATIF, N., ROSE, M. L., YACOUB, M. H., DUNN, M. J. (**1995**). Association of pre-transplantation antiheart antibodies with clinical course after heart transplantation. *J. Heart Lung Transplant.* **14**, 119–126.

Lescuyer, P., Strub, J. M., Luche, S., Diemer, H., Martinez, P., van Dorsselaer, A., Lunardi, J., Rabilloud, T. (2003). Progress in the definition of a reference human mitochondrial proteome. *Proteomics* **3**, 157–167.

Li, X. P., Pleissner, K. P., Scheler, C., Regitz-Zagrosek, V., Salnikow, J., Jungblut, P. R. (1999). A two-dimensional electrophoresis database of rat heart proteins. *Electrophoresis* **20**, 891–897.

Li, R. C., Ping, P., Zhang, J., Wead, W. B., Cao, X., Gao, J., Zheng, Y., Huang, S., Han, J., Bolli, R. (2000). PKCepsilon modulates NF-kappaB and AP-1 via mitogen-activated protein kinases in adult rabbit cardiomyocytes. *Am.J. Physiol. Heart Circ. Physiol.* **279**, H1679–H1689.

Liu, Y., Tsuchida, A., Cohen, M. V., Downey, J. M. (1995). Pretreatment with angiotensin II activates protein kinase C and limits myocardial infarction in isolated rabbit hearts. *J. Mol. Cell Cardiol.* **27**, 883–892.

Liu, G. S., Cohen, M. V., Mochly-Rosen, D., Downey, J. M. (1999). Protein kinase C-epsilon is responsible for the protection of preconditioning in rabbit cardiomyocytes. *J. Mol. Cell Cardiol.* **31**, 1937–1948.

Liu, X. H., Qian, L. J., Gong, J. B., Shen, J., Zhang, X. M., Qian, X. H. (2004). Proteomic analysis of mitochondrial proteins in cardiomyocytes from chronic stressed rat. *Proteomics* **4**, 3167–3176.

Lopez, M. F. and Melov, S. (2002). Applied proteomics: mitochondrial proteins and effect on function. *Circ. Res.* **90**, 380–389.

Macri, J., Dubay, T., Matteson, D. (2000). Characterization of the protein profile associated with endothelin-induced hypertrophy in neonatal rat myocytes. *J. Mol. Cell Cardiol.* **32**, A60.

Marber, M., Walker, D., Yellon, D. (1994). Ischaemic preconditioning. *BMJ* **308**, 1–2.

Martin-Ventura, J. L., Duran, M. C., Blanco-Colio, L. M., Meilhac, O., Leclercq, A., Michel, J. B., Jensen, O. N., Hermandex-Merida, S., Tunon, J., Vivanco, F., Egido, J. (2004). Identification by a Differential Proteomic Approach of Heat Shock Protein 27 as a Potential Marker of Atherosclerosis. *Circulation* **110**, 2216–2219.

Mayr, M., Mayr, U., Chung, Y. L., Yin, X., Griffiths, J. R., Xu, Q. (2004). Vascular proteomics: Linking proteomic and metabolic changes. *Proteomics* **4**, 3751–3761.

McGregor, E., Kempster, L., Wait, R., Welson, S. Y., Gosling, M., Dunn, M. J., Powel, J. T. (2001). Identification and mapping of human saphenous vein medial smooth muscle proteins by two-dimensional polyacrylamide gel electrophoresis. *Proteomics* **1**, 1405–1414.

McGregor, E. and Dunn, M. J. (2003). *Proteomics* of heart disease. *Hum. Mol. Genet.* **12**, Spec. No. 2, R135–R144.

McGregor, E., Kempster, L., Wait, R., Gosling, M., Dunn, M. J., Powell, J. T. (2004). F-actin capping (CapZ) and other contractile saphenous vein smooth muscle proteins are altered by hemodynamic stress: a proteonomic approach. *Mol. Cell Proteomics* **3**, 115–124.

Mitchell, M. B., Meng, X., Ao, L., Brown, J. M., Harken, A. H., Banerjee, A. (1995). Preconditioning of isolated rat heart is mediated by protein kinase C. *Circ. Res.* **76**, 73–81.

Muller, E. C., Thiede, B., Zimny-Arndt, U., Scheler, C., Prehm, J., Muller-Werdan, U., Wittmann-Liebold, B., Otto, A., Jungblut, P. (1996). High-performance human myocardial two-dimensional electrophoresis database: edition 1996. *Electrophoresis* **17**, 1700–1712.

Murry, C. E., Jennings, R. B., Reimer, K. A. (1986). Preconditioning with ischemia: a delay of lethal cell injury in ischemic myocardium. *Circulation* **74**, 1124–1136.

Otto, A., Thiede, B., Muller, E. C., Scheler, C., Wittmann-Liebold, B., Jungblut, P. (1996). Identification of human myocardial proteins separated by two-dimensional electrophoresis using an effective sample preparation for mass spectrometry. *Electrophoresis* **17**, 1643–1650.

Pankuweit, S., Portig, I., Lottspeich, F., Maisch, B. (1997). Autoantibodies in sera of patients with myocarditis:

characterization of the corresponding proteins by isoelectric focusing and N-terminal sequence analysis. *J. Mol. Cell Cardiol.* **29**, 77–84.

PATEL, V. B., CORBETT, J. M., DUNN, M. J., WINROW, V. R., PORTMANN, B., RICHARDSON, P. J., PREEDY, V. R. (1997). Protein profiling in cardiac tissue in response to the chronic effects of alcohol. *Electrophoresis* **18**, 2788–2794.

PATEL, V. B., SANDHU, G., CORBETT, J. M., DUNN, M. J., RODRIGUES, L. M., GRIFFITHS, J. R., WASSIF, W., SHERWOOD, R. A., RICHARDSON, P. J., PREEDY, V. R. (2000). A comparative investigation into the effect of chronic alcohol feeding on the myocardium of normotensive and hypertensive rats: an electrophoretic and biochemical study. *Electrophoresis* **21**, 2454–2462.

PELLIEUX, C., DESGEORGES, A., PIGEON, C. H., CHAMBAZ, C., YIN, H., HAYOZ, D., SILACCI, P. (2003). Cap G, a Gelsolin family protein modulating protective effects of unidirectional shear stress. *J. Biol. Chem.* **278**, 29136–29144.

PFLIEGER, D., LE CAER, J. P., LEMAIRE, C., BERNARD, B. A., DUJARDIN, G., ROSSIER, J. (2002). Systematic identification of mitochondrial proteins by LC-MS/MS. *Anal. Chem.* **74**, 2400–2406.

PING, P., ZHANG, J., PIERCE, W. M. JR., BOLLI, R. (2001). Functional proteomic analysis of protein kinase C epsilon signaling complexes in the normal heart and during cardioprotection. *Circ. Res.* **88**, 59–62.

PLEISSNER, K. P., REGITZ-ZAGROSEK, V., WEISE, C., NEUSS, M., KRUDEWAGEN, B., SODING, P., BUCHNER, K., HUCHO, F., HILDEBRANDT, A., FLECK, E. (1995). Chamber-specific expression of human myocardial proteins detected by two-dimensional gel electrophoresis. *Electrophoresis* **16**, 841–850.

PLEISSNER, K. P., SANDER, S., OSWALD, H., REGITZ-ZAGROSEK, V., FLECK, E. (1996). The construction of the World Wide Web-accessible myocardial two-dimensional gel electrophoresis protein database "HEART-2DPAGE": a practical approach. *Electrophoresis* **17**, 1386–1392.

PLEISSNER, K. P., SODING, P., SANDER, S., OSWALD, H., NEUSS, M.,

REGITZ-ZAGROSEK, V., FLECK, E. (1997). Dilated cardiomyopathy-associated proteins and their presentation in a WWW-accessible two-dimensional gel protein database. *Electrophoresis* **18**, 802–808.

POHLNER, K., PORTIG, I., PANKUWEIT, S., LOTTSPEICH, F., MAISCH, B. (1997). Identification of mitochondrial antigens recognized by antibodies in sera of patients with idiopathic dilated cardiomyopathy by two-dimensional gel electrophoresis and protein sequencing. *Am. J. Cardiol.* **80**, 1040–1045.

SAURIN, A. T., PENNINGTON, D. J., RAAT, N. J., LATCHMAN, D. S., OWEN, M. J., MARBER, M. S. (2002). Targeted disruption of the protein kinase C epsilon gene abolishes the infarct size reduction that follows ischaemic pre-conditioning of isolated buffer-perfused mouse hearts. *Cardiovasc. Res.* **55**, 672–680.

SCHAGGER, H. AND VON JAGOW, G. (1991). Blue native electrophoresis for isolation of membrane protein complexes in enzymatically active form. *Anal. Biochem.* **199**, 223–231.

SCHELER, C., LI, X. P., SALNIKOW, J., DUNN, M. J., JUNGBLUT, P. R. (1999). Comparison of two-dimensional electrophoresis patterns of heat shock protein Hsp27 species in normal and cardiomyopathic hearts. *Electrophoresis* **20**, 3623–3628.

SPRENGER, R. R., SPEIJER, D., BACK, J. W., DE KOSTER, C. G., PANNEKOEK, H., HORREVOETS, A. J. (2004). Comparative proteomics of human endothelial cell caveolae and rafts using two-dimensional gel electrophoresis and mass spectrometry. *Electrophoresis* **25**, 156–172.

STANLEY, B. A., GUNDRY, R. L., COTTER, R. J., VAN EYK, J. E. (2004). Heart disease, clinical proteomics and mass spectrometry. *Dis. Markers* **20**, 167–178.

STASTNY, J., FOSSLIEN, E., ROBERTSON, A. L. (1986). Human aortic intima protein composition during initial stages of atherogenesis. *Atherosclerosis* **62**, 131–139.

STASTNY, J., ROBERTSON, A. L., FOSSLIEN, E. (1986). Basic proteins in the human aortic intima: nonequilibrium two-dimensional electrophoretic analysis of tissue extracts. *Exp. Mol. Pathol.* **45**, 279–286.

TAYLOR, S. W., FAHY, E., ZHANG, B., GLENN, G. M., WARNOCK, D. E., WILEY, S., MURPHY, A. N., GAUCHER, S. P., CAPALDI, R. A., GIBSON, B. W., GHOSH, S. S. (2003). Characterization of the human heart mitochondrial proteome. *Nat. Biotechnol.* **21**, 281–286.

THIEDE, B., OTTO, A., ZIMNY-ARNDT, U., MULLER, E. C., JUNGBLUT, P. (1996). Identification of human myocardial proteins separated by two-dimensional electrophoresis with matrix-assisted laser desorption/ionization mass spectrometry. *Electrophoresis* **17**, 588–599.

TORAASON, M., MOORMAN, W., MATHIAS, P. I., FULTZ, C., WITZMANN, F. (1997). Two-dimensional electrophoretic analysis of myocardial proteins from lead-exposed rabbits. *Electrophoresis* **18**, 2978–2982.

VAN DER VELDEN, J., PAPP, Z., BOONTJE, N. M., ZAREMBA, R., DE JONG, J. W., JANSSEN, P. M., HASENFUSS, G., STIENEN, G. J. (2003). The effect of myosin light chain 2 dephosphorylation on Ca^{2+}-sensitivity of force is enhanced in failing human hearts. *Cardiovasc. Res.* **57**, 505–514.

VONDRISKA, T. M., KLEIN, J. B., PING, P. (2001a). Use of functional proteomics to investigate PKC epsilon-mediated cardioprotection: the signaling module hypothesis. *Am. J. Physiol. Heart Circ. Physiol.* **280**, H1434-H1441.

VONDRISKA, T. M., ZHANG, J., SONG, C., TANG, X. L., CAO, X., BAINES, C. P., PASS, J. M., WANG, S., BOLLI, R., PING, P. (2001b). Protein kinase C epsilon-Src modules direct signal transduction in nitric oxide-induced cardioprotection: complex formation as a means for cardioprotective signaling. *Circ. Res.* **88**, 1306–1313.

WEEKES, J., WHEELER, C. H., YAN, J. X., WEIL, J., ESCHENHAGEN, T., SCHOLTYSIK, G., DUNN, M. J. (1999). Bovine dilated cardiomyopathy: proteomic analysis of an animal model of human dilated cardiomyopathy. *Electrophoresis* **20**, 898–906.

WEEKES, J., MORRISON, K., MULLEN, A., WAIT, R., BARTON, P., DUNN, M. J. (2003). Hyperubiquitination of proteins in dilated cardiomyopathy. *Proteomics* **3**, 208–216.

WHEELER, C. H., COLLINS, A., DUNN, M. J., CRISP, S. J., YACOUB, M. H., ROSE, M. L. (1995). Characterization of endothelial antigens associated with transplant-associated coronary artery disease. *J. Heart Lung Transplant.* **14**, S188-S197.

WILKINS, M. R., GASTEIGER, E., WHEELER, C. H., LINDSKOG, I., SANCHEZ, J. C., BAIROCH, A., APPEL, R. D., DUNN, M. J., HOCHSTRASSER, D. F. (1998). Multiple parameter cross-species protein identification using MultiIdent – a world-wide web accessible tool. *Electrophoresis* **19**, 3199–3206.

YOU, S. A., ARCHACKI, S. R., ANGHELOIU, G., MORAVEC, C. S., RAO, S., KINTER, M., TOPOL, E. J., WANG, Q. (2003). *Physiol. Genomics* **13**, 25–30.

YTREHUS, K., LIU, Y., DOWNEY, J. M. (1994). Preconditioning protects ischemic rabbit heart by protein kinase C activation. *Am. J. Physiol.* **266**, H1145–H1152.

ZERKOWSKI, H. R., GRUSSENMEYER, T., MATT, P., GRAPOW, M., ENGELHARDT, S., LEFKOVITS, I. (2004). *Proteomics* strategies in cardiovascular research. *J. Proteome. Res.* **3**, 200–208.

IV
To the Market

Proteomics in Drug Research
Edited by M. Hamacher, K. Marcus, K. Stühler, A. van Hall, B. Warscheid, H. E. Meyer
Copyright © 2006 Wiley-VCH Verlag GmbH & Co. KGaA, Weinheim
ISBN: 3-527-31226-9

17
Innovation Processes

Sven Rüger

Abstract

This chapter is an extract from a handbook for an innovation process being generated and implemented in the organization by the Business Assessment and Planning team of the former Hoechst AG Corporate Research+Technology (CR&T) unit.

Generally it shows how one may deal with ideas and its implementation/ realization within a commercial or target-oriented environment. Not all ideas will generate a business opportunity, as not all opportunities guarantee market success. The innovation process helps one focus on those opportunities with the highest potential for success.

This chapter is designed to serve as a ready reference guide to assist employees in understanding and applying the process to build a strong business portfolio that contributes to the growth of an organization. Thus, the described workflow addresses both industry and academia, taking into account that these considerations will help implement an individual research group into a distinct scientific field.

17.1
Introduction

In the following, the term "innovation" is used to describe more than the generation of a new idea. Innovation describes an idea that has a product concept, a customer base, and a defined advantage. What may seem to be an exciting new concept does not qualify as innovation until it has been developed into a useful product and converted into profit for a corporation. In other words, successful innovations are not simply technical advances, but instead business opportunities.

The innovation process is a tool to help us develop sound concepts into profitable business opportunities. The process methodology offers a number of advantages:

Proteomics in Drug Research
Edited by M. Hamacher, K. Marcus, K. Stühler, A. van Hall, B. Warscheid, H. E. Meyer
Copyright © 2006 Wiley-VCH Verlag GmbH & Co. KGaA, Weinheim
ISBN: 3-527-31226-9

- It ensures that the concept is a good fit with portfolio targets by requiring early and ongoing participation of all stakeholders in a visible decision-making process.
- It is driven by commercial issues in the earliest stages, both for acquiring opportunities and for measuring up against the competition.
- It focuses on rewards that are both technological and commercial.
- It focuses on quality and team accountability for project execution.
- It provides a structure for encouraging better developed, evaluated, and prioritized programs for selection.

The process can be viewed as a filter mechanism, which sorts through all the ideas generated within a group. These ideas will undergo a screening phase to identify business opportunities. These opportunities are then developed and capitalized on.

17.2
Innovation Process Criteria

To move from one stage to the next, all research projects are evaluated against a set of universal criteria. The criteria consist of seven key points against which all concepts or projects will be judged and scored. The examples presented in this criteria section have not been assigned a stage since projects will often encounter similar problems at different stages of development.

Project A's vision is to "produce antibodies". This is not a vision statement because it lacks direction and focus. The statement also refers only to the technological advance and does not mention the anticipated business impact.

Project Z proposes process improvements for array technology that would increase the yield by 10%. The benefit to the customer would be lower production costs for the business, which could be passed on to its customers as lower prices.

17.3
The Concept

Since a project's success is driven by an identifiable business opportunity and not only by a technical advance, the concept should describe the targeted business opportunity. A project description should always start with the vision to allow others to gain insight quickly into the project and its objectives. A well-stated vision is plausible, inspirational, goal-oriented, and straight to the point.

The statement describing the concept should also include the targeted product the organization will ultimately produce, the technology required, and the market your business will supply. This focus is necessary to concentrate commercialization efforts and R&D resources. This focus also provides clear guidelines for the project team and steers all efforts in the same direction.

One should not be interested in being a "me too" producer, but instead be on the cutting edge of technological innovation. The expected benefits to customers through the technological advance must, therefore, be outlined in the concept. The statement of concept should be motivating and clearly state the value the project will add to your business.

17.4
Market Attractiveness

Market attractiveness covers all aspects related to the identified market need, including size, growth, needs, trends, and competitors. The purpose of these criteria is to analyze the general market conditions into which the product will be introduced and determine how well the organization will perform in that market.

Examples of what could make a market attractive include:
- low or no barriers to entry (if one is not already present in this market);
- an established market position;
- high growth rate.

Identifying the target market and quantifying the correct market data is critical. Innovations are classified into three categories: new market with existing technology, existing market with new technology, and new market with new technology. Of the three cases, the latter is the most difficult to quantify because estimates on market size and growth would be speculative. However, these cases are uncommon since most innovations will replace an existing product or technology.

The market attractiveness criterion assumes that it is possible for one to become one of the top three companies serving the chosen market. This relates directly to the fact that the profit impact of an innovation is tied closely to its market share. Empirical studies show that this normally constitutes an attainable market share of between 10 and 40%.

Once the current or future market need for the proposed product has been identified, it is important to determine whether the economic impact of the innovation is large enough to warrant R&D resource allocation. The project team leader must show that growth in the market for the targeted product is attractive. In trying to assess the future business impact, the project leader should try to visualize the future development of the target market and the effect the new product would have at the time it is to be commercialized.

17.5
Competitive Market Position

While market attractiveness focuses on identifying a market need, competitive market position is aimed at proving that the product actually meets this need.

To assess the competitive market position, it is important to receive feedback from the eventual end-user of the product, and not to rely solely on theoretical market studies or statements made by a representative of one's business.

The project team must show that one's competitive position is sustainable. One way to sustain the position would be to attain a competitive advantage within the market. A competitive advantage can be achieved, for example, by being the first to market, or by having existing customer relations.

A critical aspect of competitive market position is time to market. The faster an innovation can be brought to market, the higher the profit impact and potential for market leadership. If a marketing infrastructure already exists within one's organization to commercialize a product, time to market will be faster than if competencies need to be established prior to entry.

It is generally less risky to commercialize a product in an existing market or in one where the organization has established competencies rather than in a completely new market. This established position need not be attained by oneself alone, but could be achieved through alliances with other companies.

17.6
Competitive Technology Position

A good competitive technology position calls for the proposed product to offer performance, cost, or quality advantages over competing products.

The product must differentiate itself from the competition through technological innovation.

To sustain this competitive advantage, it is important to protect the organization's position in the long term. This is done by erecting barriers to entry, or hurdles, to prevent new competitors from entering the market. These barriers to the competition can be technological or in manufacturing (i.e., strong intellectual property position, know-how, assets in place, lower cost processes).

Conversely, it is extremely difficult to commercialize an innovation in a market where the competition has erected barriers to entry. Suppose that one is trying to commercialize technology that has already been patented by a competitor. If one is not able to access this technology, there is no use in continuing the project. The intellectual property strategy is a critical element of an innovation process and must be planned as early as possible. Intellectual property encompasses licensing, cross-licensing, cooperative agreements, technology transfer, and patentable technology.

Just as important as the sustainability of competitive advantage is the ability to close the gap between current and required technology. If the customer's needs can never be met, there is no reason to continue with the project. All projects must show that the required technology is feasible and how the team plans to reach that goal for each stage of the process in a detailed plan outlining specific milestones.

17.7
Strengthen the Fit

In order to show a strong fit, a research project should be consistent with the objectives of the whole organization. That is, the project offers sustainable development or concentrates on selected core businesses.

In addition to the compliance with overall goals, the relative position of the project within the portfolio of one's organization affects its fit. Based on the current portfolio status, management decides whether or not a particular project supplements the current spectrum of projects and thereby satisfies the desired portfolio targets.

Finally, projects that leverage or supplement know-how that is already available within the organization have a better fit. This reflects the fact that projects tend to be more successful if they are not entirely new to the organization.

17.8
Reward

The reward criterion covers those areas related to the expected financial performance of the project. An important method for determining the financial success of a project is the net present value (NPV) calculation. This method calculates the present value of future cash flows. If the NPV is positive, the investment is beneficial from a financial standpoint. Additionally, a financially successful project is expected to have an average return on capital employed (ROCE) above a selected level, which makes commercial sense with regards to industry/branch standards.

17.9
Risk

This criterion deals with the key risks of the project and how well the project plan addresses these issues.

Receiving a low score on one or two criteria will not reflect poorly upon the entire project. It simply points out the project's weak points, which must be resolved or dealt with in the present or following stage of development.

Proper evaluation of the project's risks must take into account the exposure to one's business. Exposure outlines what is at stake for the organization in terms of committed resources, time, image, etc. If the risk is high, the exposure should be relatively low to be acceptable.

17.10
Innovation Process Deliverables for each Stage

As a project progresses through the various stages of the innovation process, the required level of detail of the business plan or deliverables becomes greater and more focused. This ensures that the allocation of resources to a project is in line with the degree of certainty of success.

The deliverables are based on the fact that the economic impact of technology can be assessed based on revenues and costs, and are quantified with varying degrees of certainty, regardless of the stage of development. The process takes into account that the economic impact of a fully commercialized technology is more easily quantified than for one that is in the evaluation stage where there is a greater degree of uncertainty both in terms of size and timing.

The following innovation process deliverables apply to all projects within the research. Although all project leaders should adhere to the general guidelines of these deliverables, projects should not be forced into any type of rigid framework. The project team may decide not to include certain subcategories or add others where they see fit.

17.11
Stage Gate-Like Process

The following section explains how a selected set of deliverables may be used from stage to stage (Figure 17.1).

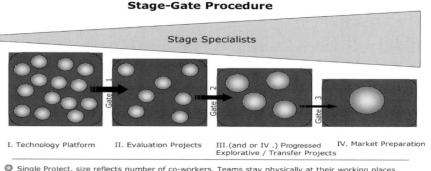

Figure 17.1 Stage gate-like process.

17.11.1
Designation as an Evaluation Project (EvP)

The objective of the EvP stage is to develop a relatively fast and inexpensive assessment of the market prospects and technical merits of the concept. The following deliverables are required for a project to move into the EvP stage.

Concept

The concept should be presented in the form of an executive summary. As the first element of evaluation, it is a GO/NO GO criterion. This means that this section of the deliverables must be clearly communicated for the project to even be considered. The concept is made up of five parts: the vision, targeted product(s), targeted technology, targeted market(s), and the intended benefits to the customer. In contrast to the other deliverables, there are no increasing hurdles for this criterion. The concept should, however, be reassessed constantly and may change as the project moves toward commercialization.

Market Attractiveness

The project team must indicate that there is a current or future market need for the product. Furthermore, the size and growth of the market should be attractive. Although attractive markets usually grow very rapidly, they also tend to attract many competitors. Therefore, the project team must show that the organization will be able to achieve a leading position in this market. A leading position means that the organization will be among the top three firms serving the respective market. Although it depends on the market structure, a leading market position usually constitutes a market share of more than 10%.

Competitive Market Position

The project team must indicate that our product meets the market need as opposed to competitive products (see market attractiveness). This indication should be based on feedback from potential customers. If the project's target customer is a customer of one's business, feedback should come from the end user.

The product must not only meet the needs of the market today, but must also be able to maintain its position in the future. The team must also indicate that there is a sustainable competitive advantage, with respect to the market. Such advantages could include being first to market, existing infrastructure, synergies, existing sales force, etc.

The team must also give a qualitative description of the value chain, and indicate where the product will be positioned.

Competitive Technology Position

To enter the EvP stage, the team must have performed a scouting experiment, or such an experiment must have been demonstrated by a third party to indicate the feasibility of the concept. Furthermore, the team must also indicate a sustainable competitive advantage with respect to the technology. Such an advantage could be intellectual property, know-how, forward/backward integration, etc.

Fit with the Organization

Fit with the organization means that the project fits the current or future guidelines under which the organization is operating. Because corporate strategy, vision and mission statements, as well as goals and objectives are constantly changing, fit with the organization means more than just adhering to today's game plan. For long-term projects, possible future changes in the organization should be taken into consideration.

Evidence of fit with the organization is a somewhat ambiguous deliverable and can be interpreted in several ways. The key to this deliverable is that the project team must make a case for why the organization should pursue the project as compared to others and where the project fits within the organization. For the EvP deliverables, the team must indicate whether they intend to establish a new business or integrate the project into an existing business.

Reward

At this early stage, because data are usually not available, any detailed financial analysis has little value. However, the team must show that the cost of the proposed product is below the intended selling price. If the market for the product does not yet exist, the price for the application it intends to displace is a good approximation. Estimates for costs will be rough as well.

The project team must also address the future revenue potential. A team may use the market potential presented under market attractiveness and use 40% of that amount as the maximum revenue potential (40% is the typical share of a market leader, according to empirical studies).

Risk

The project team must spell out the key risks of the project and explain how they will be addressed. Although risks are involved in every section of the deliverables, they should be summarized here and stated concisely.

There are three relevant risk categories: technical, commercial, and environmental, health and safety administration (EHSA).

Technical risk encompasses everything related to the technology itself, e.g., pending patents, product performance, unsolved manufacturing issues, etc.

Commercial risk refers to the market, i.e., facts that could endanger the intended market position (lack of sales infrastructure, key customer acceptance, etc.). EHSA risk covers all issues related to product safety, environmental concerns, etc.

At this stage, risk descriptions will usually be defined in relatively general statements. The project plan for the next phase (in this case, the EvP stage) must include detailed milestones including the expected completion of the EvP stage, and the exact resource requirements (manpower, cost, budget, etc.). The plan should address the key issues outlined above. This implies that milestones and resource requirements reflect the effort to manage the risks.

17.11.2
Advancement to Exploratory Project (EP)

The goal of the exploratory project stage is to provide initial confirmation of marketplace interest and technical feasibility. Business extensions and process improvement projects should provide transition details on how the project will move from the idea to the intended business. The following deliverables are required for a project to move into the EP stage.

Concept

The concept should be presented in the form of an executive summary. As the first element of evaluation, it is a GO/NO GO criterion. This means that this section of the deliverables must be clearly communicated for the project to even be considered. The concept is made up of five parts: the vision, targeted product(s), targeted technology, targeted market(s), and the intended benefits to the customer. In contrast to the other deliverables, there are no increasing hurdles for this criterion. The concept should, however, be reassessed constantly and may change as the project moves toward commercialization.

Note that the nature of the project may change extensively throughout its life creating the need to reformulate the concept. If the concept changes dramatically, the project will have to be reestablished as an EvP proposal by the project team.

Market Attractiveness

The project team must identify the individual markets the product plans to target. To avoid ambiguity, it is important to be as specific as possible. For example, if the targeted product is a new catalyst for a specific application, the respective market should be confined to the application only and not catalysts as a whole. Information pertaining to the target market should include size (volume and value), current growth and/or projected growth, as well as the general competition.

Competitive Market Position

At this stage, the team is required to get target customer feedback and/or testing of the product. If the project's target customer is a customer of a business, feedback should come from the actual end-user of the product.

The team must also do a complete value chain analysis showing the suppliers, distributors, customers, end-users of the product, and where the organization will position itself in the industry. Furthermore, our position relative to competitive products must be established and analyzed. The team must also establish a basis of pricing their product, as well as the price itself.

Competitive Technology Position

The team must establish the uniqueness of the product and how they plan to differentiate the product from the competition. A good competitive technology position calls for the proposed product to offer performance, cost, or quality advantages. In addition, a product development plan must be completed, outlining the technical milestones to commercialization. In the EP stage, the team must also have initial feasibility confirmation. Although implementation of the intellectual property strategy (issues dealing with patents, licensing, know-how, technology transfer, etc.) is not called for until the market preparation phase, it is important to start planning the implementation as soon as possible.

Reward

Fit with the organization means that the project fits the current or future guidelines under which the organization is operating. Because corporate strategy, vision, and mission statements, as well as goals and objectives are constantly changing, fit with the organization means more than just adhering to today's game plan. For long-term projects, possible future changes in the organization should be taken into consideration.

At this stage, the team should focus on sound sales assumptions and may base the cost data on simple percentages. For example, the team assumes raw material costs to be 25% of the selling price.

Usually, the costs of the plant (capital expenditure) are still a very rough estimate, but should still be included in the calculation.

The ten-year projection should be provided for three scenarios: base, optimistic, and pessimistic. These cases are not meant to be simple percentage changes of the sales projections. Instead, the team should try to identify the drivers of the project's success and construct alternatives for the future that lead to different results for the project. The base case should be the most likely case. The optimistic scenario should be based on the positive development of some (not all) key success factors. The pessimistic scenario is usually the minimum feasible case, meaning a situation where the organization would still pursue the project, but some factors do not develop in a positive way.

Risk

Under this section, the project team must spell out the key risks of the project and how they will be addressed. Although risks are involved in every section of the deliverables, they should be summarized here and stated concisely.

There are three relevant risk categories: technical, commercial, and EHSA.

Technical risk encompasses everything related to the technology itself, e.g., pending patents, product performance, unsolved manufacturing issues, etc.

Commercial risk refers to the market, i.e., facts that could endanger the intended market position (lack of sales infrastructure, key customer acceptance, etc.).

EHSA risk covers all issues related to product safety, environmental concerns, etc.

In addition, the teams must also provide an opportunity assessment for the purpose of outlining the potential impact of uncertainties on the project's success. This includes positive as well as negative factors.

The gap analysis serves to identify the gap between current project status and what is required to achieve the project's objectives. The analysis should summarize both technical and commercial requirements for the project to succeed, including timing (e.g., a statement on when the project needs to establish a sales force to achieve market penetration).

Based on the risks and the scenario analyses (see reward section), project teams should elaborate on the upside and downside potential of the project. This includes answers to questions such as, "what would happen if the technology were available sooner than expected?" or "what would be the impact if important factors did not develop as expected?" This also includes outlining an exit strategy (for example, to sell the patents to the competitor or license the technology).

The project plan for the next phase must include detailed milestones, including the expected completion of the EP phase, and exact resource requirements (manpower, cost, budget, etc.). The plan should address the key issues outlined above. This implies that milestones and resource requirements are in line with the effort to manage the risks.

17.11.3
For Advancement to Progressed Project (PP)

The following deliverables are required for a project to move into the PP stage.

Concept

Again the concept should be presented in the form of an executive summary considering the same points as in the earlier stage, but in more detail.

Market Attractiveness

The project team must identify the individual markets the product plans to target and must also study the general customer characteristics in more detail. In addition to size and growth, individual market share and specific competencies should be analyzed.

Competitive Market Position

The team is required to get target customer feedback and/or testing of the product. If the project's target customer is a customer of a business, feedback should come from the actual end-user of the product. Furthermore, it is important to establish whether there are existing supplier relationships. If a business is already supplying the team's target customers, there may be a competitive advantage. The team must also analyze the importance of the product to the target customer and establish how dependent the customer is on the innovation. This will provide valuable information on how much bargaining power the team has.

The team must also do a complete value chain analysis showing the suppliers, distributors, customers, end-users of the product, and where the organization will position itself in the industry. Furthermore, our position relative to competitive products must be established and analyzed. The team must also establish a basis of pricing the product, as well as the price itself. The market plan should also include when the product introduction will occur, if there are plans to commercialize with a partner, what the sales force requirements will be, and a distribution plan for the product.

Competitive Technology Position

The team must establish the uniqueness of the product and how they plan to differentiate the product from the competition. A good competitive technology position calls for the proposed product to offer performance, cost, or quality advantages. In addition, a product development plan must be completed, outlining the technical milestones to commercialization. The team should have developed the product to the pre-piloting optimization stage before the start of production in the pilot plant.

The team should have developed an intellectual property strategy. Intellectual property encompasses licensing, cross-licensing, co-operations, and technology transfer, in addition to patentable technology. Although implementing the intellectual property strategy is not called for until the market preparation phase, it is important to start planning the implementation as soon as possible. If the intellectual property strategy can be implemented before the market preparation stage, the team should proceed to do so.

Regarding manufacturing, the team must have a detailed manufacturing plan, outlining the quantities of product that will be produced at various stages of growth. In addition, capital investments and facility requirements must be outlined. Quality assurance of the product is a key area before entering the TP stage.

Fit with the Organization

Fit with the organization means that the project fits the current or future guidelines. Because corporate strategy, vision, and mission statements, as well as goals and objectives are constantly changing, fit with the organization means more than just adhering to today's game plan. For long-term projects, possible future changes in the organization should be taken into consideration.

As the project progresses to the EP, TP, and market preparation stages, the team must identify the organization's competitive advantage. This advantage can be in marketing, technical know-how, manufacturing, or in our present infrastructure. The more advantages the organization currently enjoys, the better the chances will be for successful commercialization of the product.

Furthermore, the project team must validate the intent to establish a new business or integrate the project into an existing business. If the team plans to establish a new business, validation means proving that this will be the best way to commercialize the product and that support from a business would not positively influence the development of the project. If the team plans to integrate the project into an existing business unit, validation means getting the business to show interest or support the project. Support does not have to be financial, but can be shown in terms of exchanging information or assisting the project team in any way.

Reward

The project team must detail all past costs that the project has incurred since its inception (start of EvP) on an annual basis. In addition, an annual project financial information table (ProFIT) data sheet should be presented. This sheet contains the revenue and cost forecasts for the upcoming ten-year period. It computes net present value (NPV) of future cash flows and return on capital employed (ROCE) automatically. At this stage, the team is expected to include detailed production costs data as well as estimates of plant costs (based on an engineering estimate, for example). The ten-year projection should be provided for three scenarios: base, optimistic, and pessimistic. These cases are not meant to be simple percentage changes of the sales projections. Instead, the team should try to identify the drivers of the project's success and construct alternatives for the future that lead to different results for the project. The base case should be the most likely case. The optimistic scenario should be based on the positive development of some (not all) key success factors. The pessimistic scenario is usually the minimum feasible case, meaning a situation where the organization would still pursue the project, but some factors do not develop in a positive way.

Risk

Under this section, the project team again must consider the points mentioned in the EP stage. In addition, the teams must also provide an opportunity assessment for the purpose of outlining the potential impact of uncertainties on the project's success. This includes positive as well as negative factors. The gap analysis serves to identify the gap between current project status and what is required to achieve the project's objectives. The analysis should summarize both technical and commercial requirements for the project to succeed, including timing (e.g., a statement on when the project needs to establish a sales force to achieve market penetration).

Based on the risks and the scenario analyses (see reward section), project teams should elaborate on the upside and downside potential of the project. This includes answers to questions such as, "what would happen if the technology were available sooner than expected?" or "what would be the impact if important factors did not develop as expected?" This also includes outlining an exit strategy (for example, selling the patents to the competitor or licensing the technology).

The project plan for the next phase (in this case the TP stage, must include detailed milestones, including the expected completion of the TP phase, and exact resource requirements (manpower, cost, budget, etc.). The plan should address the key issues outlined above. This implies that milestones and resource requirements are in line with the effort to manage the risks.

17.11.4
Advancement to Market Preparation

In the market preparation stage, the commercial viability of the business opportunity is confirmed.

The following deliverables are required for a project to move into the market preparation stage.

Concept

Again the project team must consider the points mentioned in the previous stages.

Market Attractiveness

Besides the above-described considerations the team must have confirmation of a growing market interest for their product. The team must also study the general customer characteristics in more detail. In addition to size and growth, individual market share and specific competencies should be analyzed.

Competitive Market Position

Again the project team must consider the points mentioned in the previous stages.

Competitive Technology Position

As in the stages, before the team establishes the uniqueness of the product and how they plan to differentiate the product from the competition. A good competitive technology position calls for the proposed product to offer performance, cost or quality advantages. In addition, a product development plan must be completed, outlining the technical milestones to commercialization.

The team should have developed the product to the pre-piloting optimization stage before the start of production in the pilot plant. In the TP stage, of the process, the team will also be responsible for a piloting manufacturing demonstration, before pilot plant operation begins. Furthermore, manufacturing design and economics of the commercial manufacturing process must be completed to enter this stage.

The team should have developed an intellectual property strategy. Intellectual property encompasses licensing, cross-licensing, cooperation, and technology transfer, in addition to patentable technology for the organization. Although implementing the intellectual property strategy is not called for until the market preparation phase, it is important to start planning the implementation as soon as possible.

Regarding manufacturing, the team must have a detailed manufacturing plan, outlining the quantities of product that will be produced at various stages of growth. Manufacturing cost improvements over time should also be included, if applicable. In addition, capital investments and facility requirements must be outlined. Quality assurance of the product is a key area before entering the market preparation stage.

17.12
Conclusion

The above extract of an innovation process indicates that parallel to the long and winding road of technology development, there are a lot of side and parallel processes going on with regards to fine-tuning all issues that are important for the focusing and targeting of resources.

Naturally it is always important to have goals and targets, but on the other hand creativity should not be totally planned. Therefore such a process lives and succeeds only by the people who are involved.

Subject Index

Proteomics in Drug Research
Edited by M. Hamacher, K. Marcus, K. Stühler, A. van Hall, B. Warscheid, H. E. Meyer
Copyright © 2006 Wiley-VCH Verlag GmbH & Co. KGaA, Weinheim
ISBN: 3-527-31226-9